Processes and mechanisms of welding residual stress and distortion

Related titles:

Fatigue strength of welded structures 3rd ed
(ISBN-13: 978-1-85573-506-4; ISBN-10: 1-85573-506-7)

Extensive research conducted over the years on the fatigue behaviour of welded structures has brought a greater understanding to the design methods that can be employed to reduce premature or progressive fatigue cracking. By discussing essential design rules, this book highlights the need for a design approach that incorporates an understanding of the fatigue problem and effects of welding. Specific recommendations for the design of welded joints in components or structures are discussed, as is an account of the application of fracture mechanics to fatigue, providing useful information to engineers involved in the design or assessment of welded joints. This latest edition has been revised to take into account recent advances in technology provides a practical approach with up-to-date design rules.

Cumulative damage of welded joints
(ISBN-13: 978-1-85573-938-3; ISBN-10: 1-85573-938-0)

Fatigue is a mechanism of failure that involves the formation of cracks under the action of different stresses. Stresses can be a wide range of magnitudes. To avoid fatigue it is essential to recognise, at the design stage, that the loading is such that fatigue may be a possibility. It is also important to design the structure with inherent fatigue strength. However, fatigue strength is not a constant material property. It may be assessed by fracture mechanics, but such characteristics are based on quantitative analysis and mere assumption. Fatigue cracks are exceedingly difficult to see, particularly in the early stages of crack growth. Cracks can progress to a significant extent before they are even discovered. This book is primarily concerned with fatigue under variable amplitude loading, unlike most tests performed under constant loads.

Engineering catastrophes: Causes and effects of major accidents 3rd ed
(ISBN-13: 978-1-84569-016-8; ISBN-10: 1-84569-016-8)

There is much to be gained from the study of catastrophes. Likewise, the records of accidents in industry and transport are of great importance, not only by indicating trends in the incidence of loss or casualties, but also as a measure of human behaviour. The third edition of this well-received book places emphasis on the human factor, with the first two chapters providing a method of analysing the records of accident and all-cause mortality rates, to show their relationship with levels of economic development and growth rates, and to make suggestions as to the way in which such processes may be linked.

Details of these and other Woodhead Publishing materials books and journals, as well as materials books from Maney Publishing, can be obtained by:

- visiting www.woodheadpublishing.com
- contacting Customer Services (e-mail: sales@woodhead-publishing.com; fax: +44 (0) 1223 893694; tel.: +44 (0) 1223 891358 ext. 30; address: Woodhead Publishing Ltd, Abington Hall, Abington, Cambridge CB1 6AH, England)

If you would like to receive information on forthcoming title, please send your address details to: Francis Dodds (address, tel. and fax as above; email: fancisd@woodhead-publishing.com). Please confirm which subject areas you are interested in.

Maney currently publishes 16 peer-reviewed materials science and engineering journals. For further information visit www.maney.co.uk/journals.

Processes and mechanisms of welding residual stress and distortion

Edited by

Zhili Feng

Woodhead Publishing and Maney Publishing
on behalf of
The Institute of Materials, Minerals & Mining

CRC Press
Boca Raton Boston New York Washington, DC

WOODHEAD PUBLISHING LIMITED
Cambridge England

Woodhead Publishing Limited and Maney Publishing Limited on behalf of
The Institute of Materials, Minerals and Mining

Published by Woodhead Publishing Limited, Abington Hall, Abington,
Cambridge CB1 6AH, England
www.woodheadpublishing.com

Published in North America by CRC Press LLC, 6000 Broken Sound
Parkway, NW, Suite 300, Boca Raton FL 33487, USA

First published 2005, Woodhead Publishing Limited and CRC Press LLC

British Library Cataloguing in Publication Data
A catalogue record for this book is available from the British Library.

Library of Congress Cataloging in Publication Data
A catalog record for this book is available from the Library of Congress.

Woodhead Publishing ISBN-13: 978-1-85573-771-6 (book)
Woodhead Publishing ISBN-10: 1-85573-771-X (book)
Woodhead Publishing ISBN-13: 978-1-84569-093-9 (e-book)
Woodhead Publishing ISBN-10: 1-84569-093-9 (e-book)
CRC Press ISBN-10: 0-8493-3467-5
CRC Press order number: WP3467

The publishers' policy is to use permanent paper from mills that operate a sustainable forestry policy, and which has been manufactured from pulp which is processed using acid-free and elementary chlorine-free practices. Furthermore, the publishers ensure that the text paper and cover board used have met acceptable environmental accreditation standards.

Typeset by SNP Best-set Typesetter Ltd., Hong Kong
Printed by TJ International Limited, Padstow, Cornwall, England

Contents

Contributor contact details

(*indicates main point of contact)

Editor
Dr Zhili Feng
Oak Ridge National Laboratory
PO Box 2008
Oak Ridge
TN 37831-6095
USA

email: fengz@ornl.gov

Chapter 1
Professor Chon L. Tsai
Department of Industrial Welding and Systems Engineering
The Ohio State University
1248 Arthur E. Adams Drive
Columbus
OH 43441-3560
USA

email: Tsai.1@osu.edu

Dr Dong S. Kim*
Shell Global Solutions (US)
Westhollow Technology Center
3333 Highway 6 South
PO Box 1380
Houston
TX 77082-3101
USA

email: dong.kim@shell.com

Chapter 2
Dr J. Zhou and Professor H. L. Tsai*
Department of Mechanical and Aerospace Engineering
University of Missouri-Rolla
Rolla
MO 65409-0500
USA
email: tsai@umr.edu

Chapter 3
Professor Tatsuo Inoue
Department of Mechanical Systems Engineering
Fukuyama University
Gakuen-cho 1-Sanzo
729-0292 Fukuyama
Hiroshima
Japan

email: inoue@fume.fukuyama-u.ac.jp

Chapter 4
Professor Pan Jiluan
Mechanical Engineering Department
Tsinghua University
Beijing 100084
People's Republic of China

email: Pjl-dme@tsinghua.edu.cn

Chapter 5

Dr Pan Michaleris
Department of Mechanical Engineering
The Pennsylvania State University
137 Reber Building
University Park
PA 16802-1412
USA

email: Pxm32@psu.edu

Chapter 6

Professor Yuh J. Chao
Department of Mechanical Engineering
University of South Carolina
Columbia
SC 29208
USA

email: chao@sc.edu

Chapter 7

Dr X. L. Chen* and Dr Z. Yang
Caterpillar Inc.
Technical Center, Building L
PO Box 1875
Peoria
IL 61656-1875
USA

email: Chen_Xiao_L@cat.com
email: Yang_Zhishang@cat.com

Dr F. W. Brust
Battelle Memorial Institute
505 King Avenue
Columbus
OH 43201
USA

email: brust@BATTELLE.ORG

Chapter 8

Dr F. W. Brust
Battelle Memorial Institute
505 King Avenue
Columbus
OH 43201
USA

email: brust@BATTELLE.ORG

Dr Dong S. Kim*
Shell Global Solutions (US)
Westhollow Technology Center
3333 Highway 6 South
PO Box 1380
Houston
TX 77082-3101
USA

email: dong.kim@shell.com

Chapter 9

Dr Q. Guan
Beijing Aeronautical Manufacturing Technology Research Institute
PO Box 863
Beijing 100024
People's Republic of China

email: guanq@cae.cn

Preface

Welding induced residual stress and distortion are among the most studied subjects for welded structures. The localized heating and non-uniform cooling during welding results in a complex distribution of the residual stress in the joint region, as well as the often undesirable deformation or distortion of the welded structure. As residual stress and distortion can significantly impair the performance and reliability of the welded structure, they must be properly dealt with during design, fabrication, and in-service use of the welded structures.

A number of factors influence the residual stress and distortion of a welded structure. They are related to the solidification shrinkage of the weld metal, non-uniform thermal expansion and contraction of the parent metal, the internal constraints of the structure being welded, and the external structural restraints of fixtures used in a welding operation. For many engineering materials, a transient welding thermal cycle also results in microstructural changes in the joint region, which can further complicate the formation of the residual stress field.

Since the early 1990s, considerable progress has been made on welding residual stress and distortion. Measurement techniques have improved significantly and, more importantly, the development and application of computational welding mechanics methods have been remarkable due to the explosive growth in computer capability and to the equally rapid development of numerical methods. The advances in the last decade or so have not only greatly expanded our fundamental understanding of the processes and mechanisms of residual stress and distortion during welding, they have also provided powerful tools to enable quantitative determination of the detailed residual stress and distortion information for a given welded structure. New techniques for effective residual stress and distortion mitigation and control have also been applied in different industry sectors.

This book provides an up-to-date comprehensive summary of the recent developments and is organized into the following major subject areas:

- the fundamentals of welding residual stress and distortion,
- computational methods and analysis methodology,
- measurement techniques,
- control and mitigation techniques,
- applications and practical solutions.

It is hoped that the text will be useful not only for advanced analysis of the subject, but that it will also provide sufficient examples and practical solutions for welding engineers.

The book is the result of collaboration between a number of leading experts. We would like to express our thanks to all the contributors for their hard and effective work, their promptness and their saintly patience with regard to the editorial pressures. Finally, our sincere thanks go to Emma Cooper at Woodhead Publishing for her major editorial and administrative contribution.

Zhili Feng, Ph.D.
Metals and Ceramics Division
Oak Ridge National Laboratory
Oak Ridge, Tennessee
USA

Part I

Principles

1
Understanding residual stress and distortion in welds: an overview

C. L. TSAI, The Ohio State University, USA and
D. S. KIM, Shell Global Solutions (US), USA

1.1 Introduction

One common problem associated with welding, which has been realized and documented for many years, is the dimensional tolerance and stability of the finished products. The welding-induced residual stresses in the weldment are sometimes also a concern with respect to their effect on the fracture and fatigue behaviors of the welded structures subject to dynamic loading or adverse service environment.

In making a weld, the heating and cooling cycle always causes shrinkage in both base metal and weld metal, and the shrinkage forces tend to cause a degree of distortion. Machining of a welded product may cause dimensional changes of the product owing to relaxation of weld residual stresses. As a result of welding, the finished product may not be able to perform its intended purpose because of poor fit-up, vibration problems, high reaction stresses, reduced buckling strength, premature cracking or unacceptable appearance. Control of weld residual stresses and distortion is a vital task in welding manufacturing.

Attempts have been made by many researchers since 1930 to understand weld residual stresses and distortion using predictive methodology, parametric experiments or empirical formulations. Attempts have also been made in the last three decades to predict weld residual stresses and distortion through computer simulations of welding process using the finite-element analysis (FEA) method. One significant conclusion from these studies is that the weld residual stresses and distortion are not influenced much by the weld heating cycle but instead occur as a result of shrinkage in the weld metal and its adjacent base metal during cooling when the yield strength and modulus of elasticity of the material are restored to their higher values at lower temperatures. Therefore, analysis of the shrinkage phenomena of welds alone may be sufficiently accurate to predict the state of weld residual stresses and distortion. This conclusion has led to the development of a modeling scheme referred to as the 'inherent shrinkage model' by some researchers. The root cause for welding-induced residual stresses and distortion may be described using a plasticity-based hypothesis[1].

This chapter will present the following.

1. A summary on the thermal and mechanical evolution process during welding that leads to weld residual stresses and distortion.
2. An explanation of the significance of the shrinkage plastic strains in the formation of weld residual stresses and distortion using simple analogies.
3. An overview of the common issues on analysis approaches predicting weld residual stresses and distortion.
4. A list of references in the analysis of weld residual stresses and distortion.

1.2 Thermal and mechanical processes during welding

In a welded joint, the expansion and contraction forces act on the weld metal and its adjacent base metal. As the weld metal solidifies and fuses with the base metal, it is in its maximum expanded state. However, at this point, the weld metal and its adjacent base metal are at high temperatures and have little strength or rigidity. The volume expansion causes local thickening in the weld area but is incapable of causing a significant amount of plastic strains in the cooler joint neighborhoods. On cooling, it attempts to contract to the volume that it would normally occupy at the lower temperature, but it is restrained from doing so by the adjacent cooler base metal. Stresses develop within the weld, finally reaching the yield strength of the weld metal. At this point the weld stretches, or yields, and thins out, thus adjusting to the volume requirements of the lower temperature, but only those stresses that do not exceed the yield strength of the weld metal, or the elastic mechanical strains, are relieved by this accommodation. In 1960, Blodgett[2] properly described this thermal and mechanical evolution process.

The plastic strains accumulated over the welding thermal cycles are primarily compressive and remain in the weldment. By the time that the weld reaches room temperature, the weld will contain locked-in tensile stresses (residual stresses) of yield magnitude and the base metal away from the weld is usually in compression with smaller magnitude. The internal tensile and compressive forces are in equilibrium with the joint deforming to comply with the strain compatibility. The residual stress distributions and the amount of weld distortion depend on the final state of the plastic strain distributions and their compatibility in the joint. Over three decades since 1930, the shrinkage distortion phenomena were studied by Spraragen and Ettinger[3], Guyot[4], Watanabe and Satoh[5] and others.

The welding-induced incompatible inelastic strains in the weldment during the heating and cooling weld cycles include transient thermal strains, cumulative plastic strains and final inherent shrinkage strains. At any instant during welding, the incompatible thermal strains resulting from the non-linear temperature distributions generate the mechanical strains, which lead to incremental plastic strains in the weldment if yielding occurs. The incremental plastic strains accumulate over the periods of heating and cooling. Upon completion of the welding cycles, the cumulative plastic strains interact with the weldment stiffness and the joint rigidity, resulting in the final state of residual stresses and distortion of the weldment. This final state of the inelastic strains, which are always compressive, is referred to as the 'inherent shrinkage strains'. Ueda *et al.*[6] first presented this inherent shrinkage concept to predict weld residual stresses and distortion in 1979, followed by many publications demonstrating the numerical procedures for weld residual stresses and distortion prediction using the inherent shrinkage concept[7–10]. In 1993–1995, Tsai and coworkers[11,12] used spring elements to describe the inherent shrinkage strains in the numerical simulation and analysis of angular weld distortion of tubular joints in automotive frames.

Welding-induced incompatible plastic strains (assuming a two-dimensional (2D) plane-strain condition for illustration purpose) at each heating or cooling time increment may be described mathematically as follows[1]:

$$\nabla^2(\sigma_x + \sigma_y) = -\frac{E}{1-\nu}\nabla^2(\alpha\theta) - [g(x,y) + \Delta g(x,y)] \qquad [1.1]$$

where ∇^2 is the Laplacian operator, σ_x and σ_y are thermal stress components in the respective x and y directions, E is Young's modulus, ν is Poisson's ratio, α is the thermal expansion coefficient, θ is a temperature function, $g(x, y)$ is a cumulative plastic strain function and $\Delta g(x, y)$ is the plastic strain increment function over each thermal loading step. The plastic strain functions may be written as follows:

$$g(x,y) = \frac{E}{1-\nu^2}\left(\frac{\partial^2\varepsilon_x^{\mathrm{p}}}{\partial y^2} + \frac{\partial^2\varepsilon_y^{\mathrm{p}}}{\partial x^2} - 2\frac{\partial^2\varepsilon_{xy}^{\mathrm{p}}}{\partial x\,\partial y}\right) - \frac{\nu E}{1-\nu^2}\nabla^2(\varepsilon_x^{\mathrm{p}} + \varepsilon_y^{\mathrm{p}}) \qquad [1.2]$$

$$\Delta g(x,y) = \frac{E}{1-\nu^2}\left(\frac{\partial^2(\Delta\varepsilon_x^{\mathrm{p}})}{\partial y^2} + \frac{\partial^2(\Delta\varepsilon_y^{\mathrm{p}})}{\partial x^2} - 2\frac{\partial^2(\Delta\varepsilon_{xy}^{\mathrm{p}})}{\partial x\,\partial y}\right) - \frac{\nu E}{1-\nu^2}\nabla^2(\Delta\varepsilon_x^{\mathrm{p}} + \Delta\varepsilon_y^{\mathrm{p}}) \qquad [1.3]$$

The Laplacian thermal strains are governed by the rate of enthalpy change in the weldment. Upon completion of the heating and cooling cycles the cumulative plastic strains in the weldment are usually compressive and become the inherent shrinkage strains $g^{I}(x, y)$ that interact with the structural rigidity to result in residual stresses σ_x^R and σ_y^R and distortion strains ε_x^D and ε_y^D. Residual stresses may be written with a possible reverse yielding $g^R(x, y)$ as follows:

$$\nabla^2(\sigma_x^R + \sigma_y^R) + g^R(x, y) = -g^I(x, y) \quad [1.4]$$

The distortion strains may be written in the form of final total strains as follows:

$$\varepsilon_x^D - \varepsilon_x^I = \frac{1}{E}[\sigma_x^R - \nu(\sigma_y^R + \sigma_z^R)] + \varepsilon_x^{PR} \quad [1.5]$$

$$\varepsilon_y^D - \varepsilon_y^I = \frac{1}{E}[\sigma_y^R - \nu(\sigma_z^R + \sigma_x^R)] + \varepsilon_y^{PR} \quad [1.6]$$

$$\varepsilon_z^D - (\varepsilon_x^I + \varepsilon_y^I) = \frac{1}{E}[\sigma_z^R - \nu(\sigma_x^R + \sigma_y^R)] - \varepsilon_x^{PR} - \varepsilon_y^{PR} = \text{constants} \times \tan t \quad [1.7]$$

$$\gamma_{xy}^D - \gamma_{xy}^I = \frac{2(1+\nu)}{E}\tau_{xy}^R + \gamma_{xy}^{PR} \quad [1.8]$$

with the following superscripts: D represents distortion strains, I represents the inherent (cumulative) shrinkage strains, R represents residual stresses and PR represents plastic strains due to reverse yielding.

With known distortion strains determined from the inherent shrinkage plastic strains, the distortion shape of a weldment can be determined by integrating these strains with respect to spatial coordinate variables. For a simple longitudinal plate bending case (e.g. welding along one edge of a long plate), the total strain ε_{cg}^D at the center of gravity of any cross-sections and its curvature C may be written as follows:

$$\varepsilon_{cg}^D = \frac{\iint_{A_I} \varepsilon_x^I \, dy \, dz}{A_{section}} \quad [1.9]$$

$$C = \frac{Z_{cg}^{section} - Z_{cg}^I}{I} \iint_{A_I} \varepsilon_x^I \, dy \, dz \quad [1.10]$$

where A_I is the shrinkage strain area that contains the inherent shrinkage plastic strains and $A_{section}$ is the geometric cross-sectional area. $Z_{cg}^{section}$ and Z_{cg}^I are the distances from weld to the centers of gravity of these two areas respectively. The average shrinkage strain ε_{cg}^D at the section centroid shows the amount of longitudinal shrinkage of the weldment. The curvature C is the curvature of the cross-section at a given location along the weld axis.

For an inherent plastic strain distribution uniform along the welding direction, the maximum bending distortion δ_B may be determined by integrating the curvature of all cross-sections along the weld length and can be written as

$$\delta_B = \frac{CL^2}{8} \quad [1.11]$$

Equations [1.1]–[1.8] demonstrate that the cumulative plastic strains govern the final state of weld residual stresses and distortion. Therefore, an engineering approach to estimating welding-induced residual stresses or distortion is to establish the relationships between these plastic strains and variables associated with welding process, joint design and structural detail. Some physical phenomena that occur during welding may not be described using these theoretical equations, and the effects of these phenomena can only be analyzed by the numerical simulations. During the heating process all the strains in the molten pool relax to the nil strain state. Upon solidification the weld metal shrinks from the melting temperature resulting in high shrinkage stresses. The inherent shrinkage strains should include these shrinkage strains. However, depending upon the relative stiffness of the shrinkage zone and the elastic-resistance zone in the base metal, only those inherent strains that are twice the material's yield magnitude are effective in causing the final state of residual stresses and distortion.

The inherent shrinkage strains are primarily caused by material softening and nonlinear thermal gradients in the cooler areas. The inherent shrinkage strains are uniform along the weld length, except in the areas of arc start and stop. They are nearly uniform within the softening area in the direction transverse to the weld. The strain magnitude decreases at a steep slope to zero within a short distance from the edges of the softening zone. The peak temperatures attained in the weldment are uniquely related to the longitudinal inherent shrinkage strains owing to the relatively large stiffness ratio between the soften zone and its cooler surroundings. The peak temperature distribution can be used to estimate the longitudinal inherent shrinkage strains with good accuracy. The longitudinal residual stresses and the longitudinal cambering can therefore be determined from the peak temperature distributions. Buckling can also be predicted in large thin-plate weldments[13].

The peak temperatures alone are insufficient to determine the transverse inherent shrinkage strains because of the smaller stiffness ratio and its sensitivity to the joint thickness and the external constraint conditions[13]. Nevertheless, with a procedure considering the material incompressibility, the transverse inherent shrinkage strains considering the effect of varying stiffness ratios can be used to estimate with good accuracy the transverse residual stresses and angular distortion of the weldment[14].

1.3 Modeling welding processes

To illustrate the basic formation process of the plastic strains in heating and cooling situations and the significance of the cumulative plastic strains on the final state of residual stresses and distortion, three simple thermal–mechanical models are presented as follows.

1.3.1 One-dimensional bar–spring model

A one-dimensional (1D) bar model[15] that consists of a bar and a restraining spring at one end of the bar is shown in Fig. 1.1. Assume that the bar is going through a uniform heating and cooling cycle resulting in the maximum (peak) temperature θ_p. The spring is insulated from the bar and remains at room temperature during the thermal cycle. The stiffness of the bar, $k_b = EA/L$, where E is the elastic modulus, A is the cross-sectional area and L is the original length of the bar. The spring stiffness is represented by the spring constant k_s. A stiffness ratio β is defined as

$$\beta = \frac{k_s}{k_s + k_b} \qquad [1.12]$$

The stiffness ratio represents a degree of restraint on the bar during the thermal cycle. The bar would stretch in accordance with the thermal strain

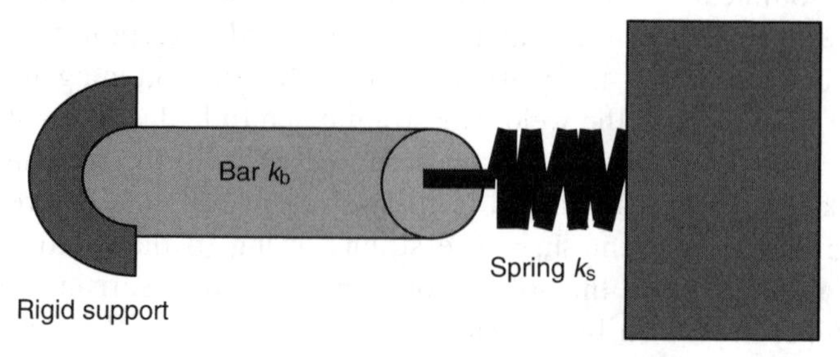

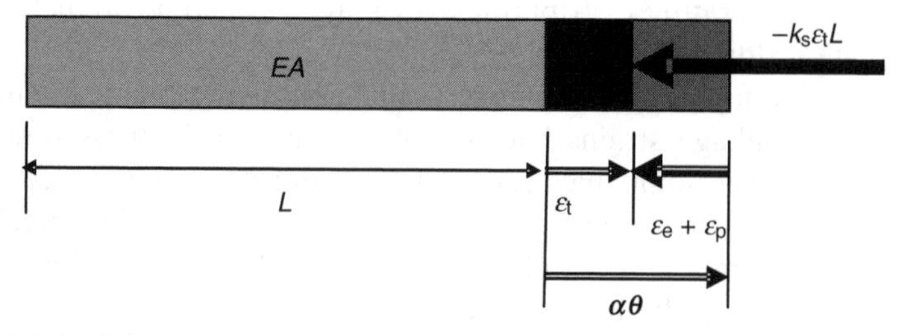

1.1 A 1D bar–spring model.

$\alpha\theta$ for a very soft spring (i.e. β approaches zero). However, owing to the spring stiffness, it compresses a portion of the free expansion, resulting in mechanical strain, which includes both elastic (ε_e) and plastic (ε_p) components if the mechanical strain surpasses the yield point of the bar. The total strain ε_t of the bar represents the actual unit elongation of the bar at a given temperature. The spring is compressed to the same amount as bar elongation $\varepsilon_t L$ because of displacement compatibility at the junction of the bar and the spring. The force exerted on the bar by the spring should be in equilibrium with the spring force.

Assuming that the bar is a perfect plastic material (i.e. no strain hardening) under monotonic loading (i.e. path independent due to one-dimensional (1D) thermal loading), by equilibrating the bar force $\varepsilon_e EA$ or $\varepsilon_e k_b L$ and the spring force $-k_s \varepsilon_t L$ or $-k_s(\alpha\theta + \varepsilon_e + \varepsilon_p)L$ the normalized (i.e. divided by the yield strain of the bar) mechanical strain components and thermal strain can be derived as

$$\varepsilon_e^* = -\beta\left(\alpha\theta^* + \varepsilon_p^*\right) \qquad [1.13]$$

Both elastic and plastic strains have negative values during heating. The elastic strain will change and may become tensile upon cooling of the bar; however, the plastic strain will remain compressive and cumulative in the bar during the entire heating and cooling cycle.

Depending upon the maximum temperature that the bar reaches during heating, the cumulative plastic strain can be derived and summarized as follows.

1. When $\alpha\theta^* \leqslant 1/\beta$, no yielding occurs. There should be no plastic strain component ($\varepsilon_p^* = 0$) and the elastic strain equals $\varepsilon_e^* = -\beta\alpha\theta^*$. The stress in the bar is $\sigma = -\beta\alpha\theta^* E$.
2. When $1/\beta \leqslant \alpha\theta^* \leqslant 2/\beta$, yielding occurs during the heating cycle. The normalized elastic strain or stress is unity when yielding occurs (i.e. $\varepsilon_e^* = 1$). There is no reverse yielding during the cooling cycle. Therefore, the normalized cumulative plastic strain will be compressive and equals

$$\varepsilon_p^* = -\alpha\theta^* + \frac{1}{\beta} \qquad [1.14]$$

3. When $2/\beta \leqslant \alpha\theta^*$, reverse tensile yielding would occur that compensates the compressive plastic strains. The normalized cumulative plastic strain will remain constant regardless of the maximum temperature. The elastic strain remains unity (i.e. $\varepsilon_e^* = 1$):

$$\varepsilon_p^* = -\frac{1}{\beta} \qquad [1.15]$$

The cumulative plastic strain is always compressive. When the spring is removed from the bar after completion of the thermal cycle, the bar will shrink by an amount $\varepsilon_p^* L$, which has the minimum possible magnitude of $1/\beta$. This cumulative plastic strain ε_p^* is often called the 'inherent shrinkage strain ε_p^i' in the analysis of weld residual stresses and distortion.

Figure 1.2 plots the relationship between the inherent shrinkage strain and the stiffness ratio, with various magnitudes of thermal strain up to ten times the yield strain of the bar material. It shows that the inherent shrinkage strain will not be generated if the stiffness ratio β is small depending upon the magnitude of the maximum temperature. The higher the maximum temperature, the less constraint would be required to start the plastic strain formation. However, on the other hand, increasing the stiffness ratio beyond the threshold value would eventually result in less cumulative plastic strain. The reverse-yielding tensile plastic strain generated from cooling compensates the compressive plastic strain developed during heating. For a highly restrained bar, the shrinkage strain has the magnitude of one material's yield strain regardless of the magnitude of the maximum temperature ever reached in the bar.

It is realized that the bar softens when the temperature reaches a threshold temperature. Under this circumstance, almost the entire thermal strain

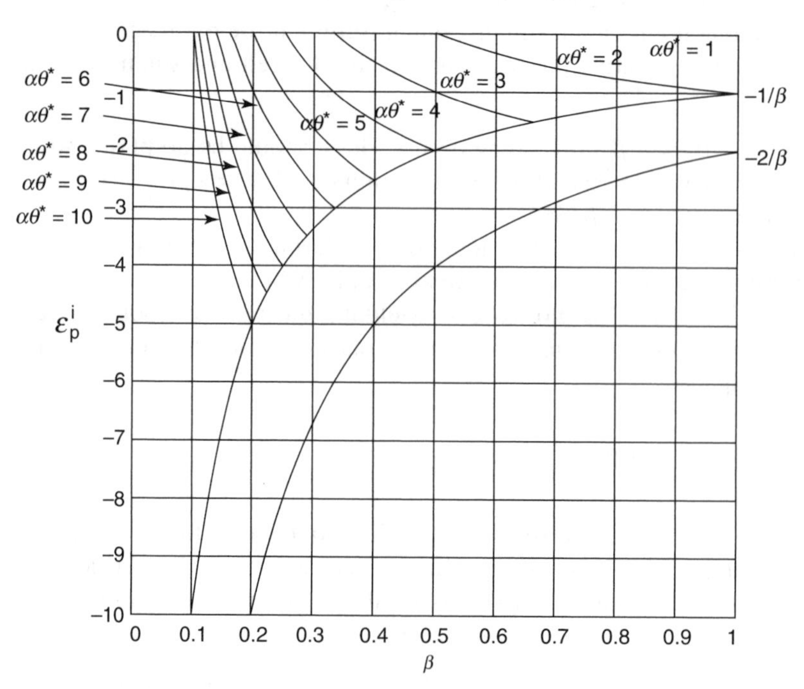

1.2 Inherent shrinkage strains as a function of the stiffness ratio and maximum temperature.

turns into compressive plastic strain with very little elastic strain in the bar. A spring of any stiffness would restrain any extension of the bar. Should the bar melt when the maximum temperature rises beyond the melting point of the material, the compressive cumulative plastic strain would be totally relaxed. Upon solidification and cooling, the bar will start to shrink upon recovering the material's stiffness at the softening temperature. The plastic strain will be in tension; however, its effect is to cause the bar to shrink. This tensile plastic strain can therefore be inverted and considered as shrinkage strain. This phenomenon would be similar to weld metal shrinkage in the actual weldment.

Figure 1.3 shows the elastic strain change with increasing thermal strain (i.e. the same as temperature change) for four different stiffness ratios: 0.25, 0.33, 0.50 and 1.00. Assuming the yield strain to be a constant value until the temperature reaches 600 °C or the normalized thermal strain $\alpha\theta^*$ equals 4.0, when the material's yield strain becomes zero, Table 1.1 summarizes the state of the elastic strain (i.e. residual stress) and the inherent shrinkage strain before and after the removal of the spring constraint.

When the bar temperature is greater than the material's softening temperature ($\alpha\theta^* > 4.0$), there will be no elastic strain or total strain in the bar until the bar cools to the softening temperature. The cumulative compressive plastic strain has the same magnitude as the thermal strain at this temperature ($\varepsilon_p^* = -4.0$). No other strain components exist. Upon further cooling, tensile elastic strain (stress) begins to develop. Reverse yielding

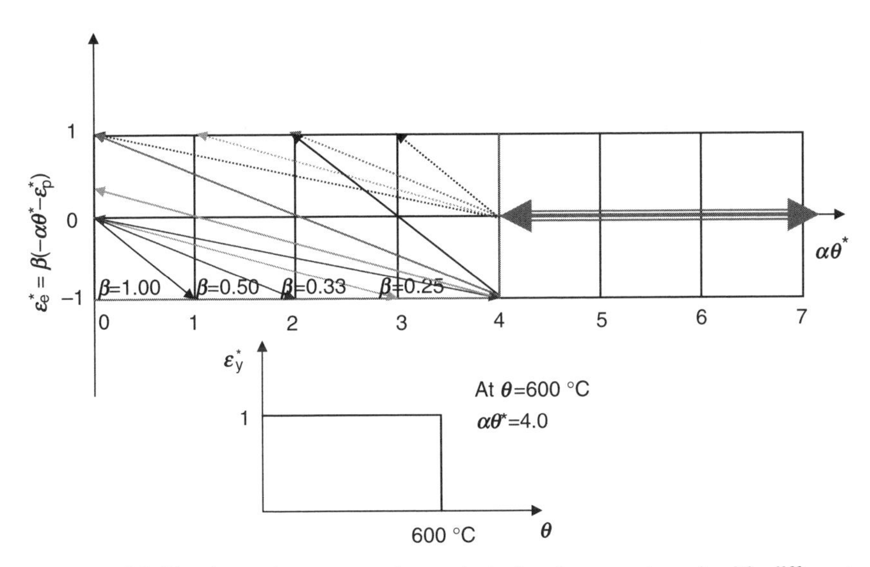

1.3 Elastic strains versus thermal strains (temperatures) with different stiffness ratios: 0.25, 0.33, 0.5 and 1.0 (on the assumption that the material's softening temperature is 600 °C).

Table 1.1 Elastic strain, inherent shrinkage strain and total strain of the bar–spring model

Maximum $\alpha\theta^*$ (with or without softening	Normalized strain type	Normalized strain by material's yield strain at room temperature for the following stiffness ratios β							
		0.25		0.33		0.50		1.00	
		Spring on	Spring removed	Spring on	Spring removed	Spring on	Spring removed	Spring on	Spring Removed
4.0 (no softening)	ε_p^1	0	0	–1.00	–1.00	–2.00	–2.00	–1.00	–1.00
	ε_e^*	0	0	0.33	0	1.00	0	1.00	0
	ε_t^*	**0**	**0**	**–0.67**	**–1.00**	**–1.00**	**–2.00**	**0**	**–1.00**
>4.0 (softening starts at 4.0)	ε_p^1	–4.00	–4.00	–3.00	–3.00	–2.00	–2.00	–1.00	–1.00
	ε_e^*	1.00	0	1.00	0	1.00	0	1.00	0
	ε_t^*	**–3.00**	**–4.00**	**–2.00**	**–3.00**	**–1.00**	**–2.00**	**0**	**–1.00**

occurs when the tensile strain reaches the material's yield point. The compressive plastic strain is reduced by the tensile plastic strain because of reverse yielding. The elastic strain (stress) remains in tension and unity ($\varepsilon_e^* = 1.0$) after reverse yielding occurs. Upon cooling to room temperature, the inherent plastic strain ε_p^i is –4.0 plus any tensile plastic strain generated during the cooling process. Table 1.1 shows that increasing the stiffness ratio will reduce the inherent shrinkage strain, as well as the final shrinkage distortion after removing the spring constraint.

A general observation from this 1D analogous model suggests that yielding would occur at a lower temperature when the degree of constraint increases. However, with increasing constraint the final shrinkage of the bar after completion of the thermal cycle and removal of the constraint would be proportionally reduced. This implies that weld joints either with no constraint (i.e. free joint) or on the other hand under strong jigging constraint would reduce the amount of inherent shrinkage strain in the weldment and, hence, result in less distortion. Weld residual stress is not much influenced by the restraint if softening of the material is considered. In such a case, the residual stress remains at yield stress level in tension regardless of the stiffness ratio except for the totally free bar.

1.3.2 Longitudinal plate shrinkage and bending model

To illustrate the theoretical basis of the inherent shrinkage behavior of a weldment, a 2D welding model that simulates a welding arc moving along one edge of a plate is presented. Figure 1.4 shows the thermal strain distribution corresponding to the peak temperature distribution (i.e. the maximum temperature trace in the cross-section during the weld thermal cycle) along a cross-section beneath the edge weld (line a–b–c–d–e–f–g–h). Line k–ι represents the softening temperature boundary (600 °C or $\alpha\theta^* = 4.0$). The total strain distribution can be approximated by a straight line (line a–b–e–f–i). The nonlinear stiffness of the softened area is insignificant in comparison with the rigid elastic cooler area in the same cross-section. The total strain line shows an average longitudinal elongation at the section centroid (middepth point n) and the bending curvature as shown by the slope C of the total strain line. When the thermal strain is subtracted from the total strain, the difference between the two is the mechanical strain, which includes both elastic and plastic strain components.

In the area where the temperature is greater than the material's softening temperature, there should not exist any elastic strain (stress) and, therefore, the effective thermal strain has the same magnitude as the total strain (line a–b). Since the plastic strain does not influence the total strain due to the assumption of a perfect plastic material, the effective thermal strain in the region represented by line ι–d is actually bounded by the material's

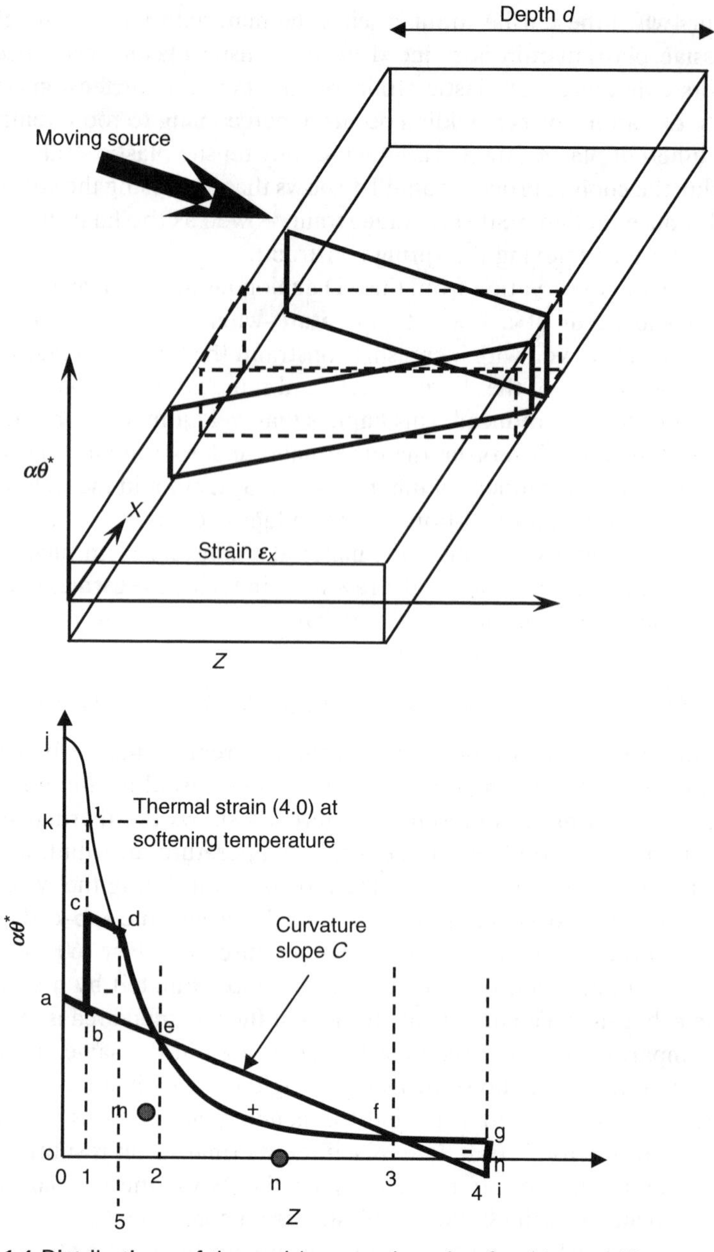

1.4 Distributions of thermal (curve a–b–c–d–e–f–g–h), total (line a–b–e–f–i) and mechanical strain components in a plate-edge weld model.

yield strain (line c–d). Assuming that no other yielding exists, the effective thermal strain distribution in the cross-section is shown by the adjusted thermal strain contour (line a–b–c–d–e–f–g). The location point m is the centroid of the effective thermal strain area. The plastic strain area is represented by area a–b–c–d–ι–j–i–k–a.

The transient thermal stress distribution just before completion of the weld thermal cycle is shown in four regions in the cross-section.

1. Zero stress between 0 and 1.
2. Compressive stress with partial yielding between 1 and 2.
3. Tensile stress between 2 and 3.
4. Compressive stress between 3 and 4.

Because of equilibrium of longitudinal forces and moments in the cross-section, the sum of the compressive areas b–c–d–e–b and f–g–h–i–f equals the tensile area bounded by the total strain line e–f and the thermal strain curve between 2 and 3.

Upon complete cooling of the weldment, the inherent shrinkage strain in the cross-section is inverted on the strain coordinate to show negative strain values. Since the thermal strain diminishes as it returns to room temperature, this inverted shrinkage strain plays a key role, similar to the thermal strain effect on the transient deformations in the plate, in determining the final state of strains of the weldment. Figure 1.5 shows the final elastic strain (residual stress) and the final total strain (distortion) components. The base width of the plastic strain (line o–e) is close to the diffusion depth of the peak temperature in the cross-section (distance 1–5 as shown in Fig. 1.4).

The inherent shrinkage strain is the cumulative plastic strain that actually affects the final total strain (distortion) and the final elastic strain (residual stress) in the cross-section. It is represented by area o–a–b–c–d–e–o as shown in Fig. 1.5. Because of the high stiffness ratio β in the longitudinal bending model, the magnitude of the inherent shrinkage strain is no more than twice the material's yield strain. The weld residual stresses shown in the three regions are as follows.

1. Tensile stress with partial yielding between 0 and 1.
2. Compressive stress between 1 and 2.
3. Tensile stress between 2 and 3.

The average longitudinal shrinkage is represented by the total strain ε_{cg} at the geometric centroid n of the cross-section, which is proportional to the inherent shrinkage strain area A_p^i but inversely proportional to the plate depth d. The average longitudinal shrinkage strain is

$$\varepsilon_{cg} = \frac{A_p^i}{d} \qquad [1.16]$$

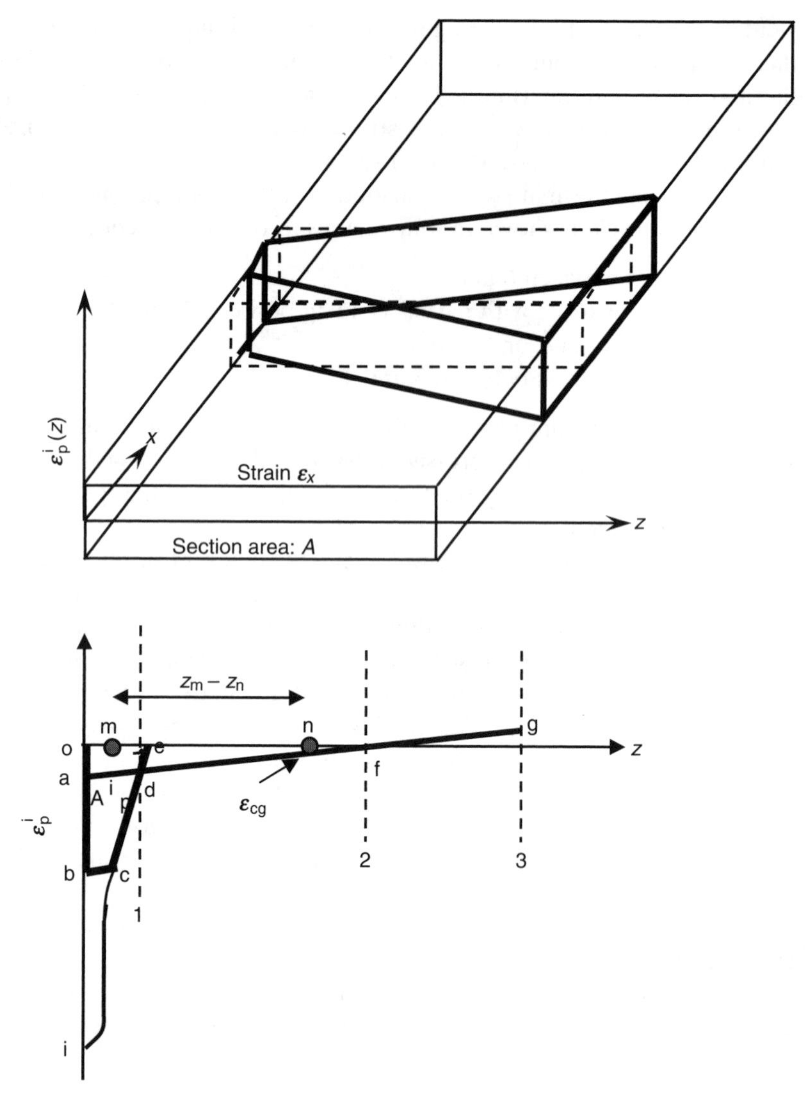

1.5 Cumulative plastic strain distribution i–c–d–e, inherent shrinkage strain area o–a–b–c–d–e–o, total strain distribution a–d–f–g and residual stresses in a plate-edge weld model. m is the centroid of the inherent shrinkage strain area on the base and n is the geometric center of the section (mid-depth location).

The curvature C is the slope of the total strain line. It is proportional to the area A_p^i of the inherent shrinkage strain multiplied by the distance $z_m - z_n$ between the two centroids, but inversely proportional to the unit-thickness moment $d^3/12$ of inertia of the cross-section:

$$C = \frac{z_m - z_n}{d^3/12} A_p^i \qquad [1.17]$$

Equations [1.16] and [1.17] hold the same relationships as Eqns [1.9] and [1.10]. The total longitudinal shrinkage (Δ_L) and bending displacements δ_B of the plate can be determined by integrating the cross-sectional shrinkage and bending curvature over the entire weld length L_w. Equation [1.11] explains the procedure for bending displacement calculation. Thus

$$\Delta_L = \varepsilon_{cg} L_w \qquad [1.18]$$

$$\delta_B = \frac{C L_w^2}{8} \qquad [1.19]$$

Assuming, firstly, that a weld bead is instantaneously deposited on one of the plate edges by a line heat source (i.e. welding completes before any place in the plate cools below the material's softening temperature), secondly, that the centroid of the plastic strain zone is close to the weld edge of the plate, thirdly, that the upper bound of the total strain at the weld edge of the plate remains the same during heating and cooling and, fourthly, that the thermal and mechanical properties of the plate material are temperature independent, the upper bound solution of the inherent shrinkage strain area may be approximately written as follows[16]:

$$A_p^i = -0.335 \frac{\alpha}{\rho C_p} \frac{0.24\eta_a E_a I_w / V}{t} = -0.335 \frac{\alpha}{\rho C_p} q_{lin} \qquad [1.20]$$

where α is the thermal expansion coefficient, ρC_p is the volumetric specific heat, η_a is the arc efficiency, E_a is the arc voltage, I_w is the welding current, V is the travel speed and t is the plate thickness. The proportional constant in Eqn [1.20] reflects the peak temperature (°C) that defines the boundary of the inherent strain depth. A_p^i (cm) may also be expressed in terms of linear heat input per unit plate thickness as q_{lin} (cal/cm^2).

For low-carbon steel, let $\alpha = 1.2 \times 10^{-5}\,°C^{-1}$ and $\rho C_p = 1.14\,cal/cm^3$; the inherent shrinkage area (cm) becomes

$$A_p^i = -3.53 \times 10^{-6} q_{lin} \qquad [1.21]$$

The final longitudinal shrinkage (cm) and bending distortion (cm) can be written as follows:

$$\Delta_L = -0.53 \times 10^{-6} q_{lin} \frac{L_w}{d} \qquad [1.22]$$

$$\delta_B = 2.64 \times 10^{-6} q_{\text{lin}} \left(\frac{L_w}{d} \right)^2 \qquad [1.23]$$

where d is the depth of the plate.

1.3.3 Transverse shrinkage and angular distortion of groove welded butt joint

The transverse shrinkage and angular distortion of groove welded butt joint without jigging constraint are practically caused by the shrinkage strains existing in the weld nugget and its vicinity bounded by the softening temperature of the material. Very little plastic strain exists outside this temperature boundary owing to the low stiffness ratio of the free joint. Therefore, it is reasonable to assume that weld shrinkage and bending curvature occur in the cross-sections beneath the width of the weld face. The inherent shrinkage strains may be assumed to be a constant value that equals the thermal expansion coefficient multiplied by the shrinkage temperature, $\alpha\theta_s$, which can be determined by the equilibrium conditions between the shrinkage strains and the total strains with adjustment of the material's yielding behavior. The shape of the inherent shrinkage strain area is influenced by the groove geometry, energy input and plate thickness. Its shape may be triangular (V-groove), rectangular (U-groove) or semicircular (bead on plate), or a more generic parabolic shape for a partial joint penetration detail. Assuming that shrinkage occurs only within the weld nugget area (i.e. this is a situation for high-speed welding), with the weld face width b and weld penetration depth s being the two geometric parameters, the shrinkage area A_s can be defined as

$$A_s = \gamma bs \qquad [1.24]$$

where γ is the geometric factor that equals unity for a rectangular shape, 0.78 for a semicircular shape, 0.67 for a parabolic shape or 0.5 for a triangular shape.

Figure 1.6 shows a schematic presentation of shrinkage deformation for a triangular nugget shape and the resulting average transverse shrinkage deformation and bending curvature of the butt joint. All the strain values can be obtained by dividing the shrinkage deformation by the weld face width b. The proportional constant k represents the portion of the shrinkage strain over the material's yield point, which does not influence the total strain or stress.

The maximum shrinkage deformation is at the weld face, which is shown as $b\alpha\theta_s$. The shrinkage deformation becomes zero at the weld root (weld penetration). The inherent shrinkage deformation area A_s^i (cm^2) is shown

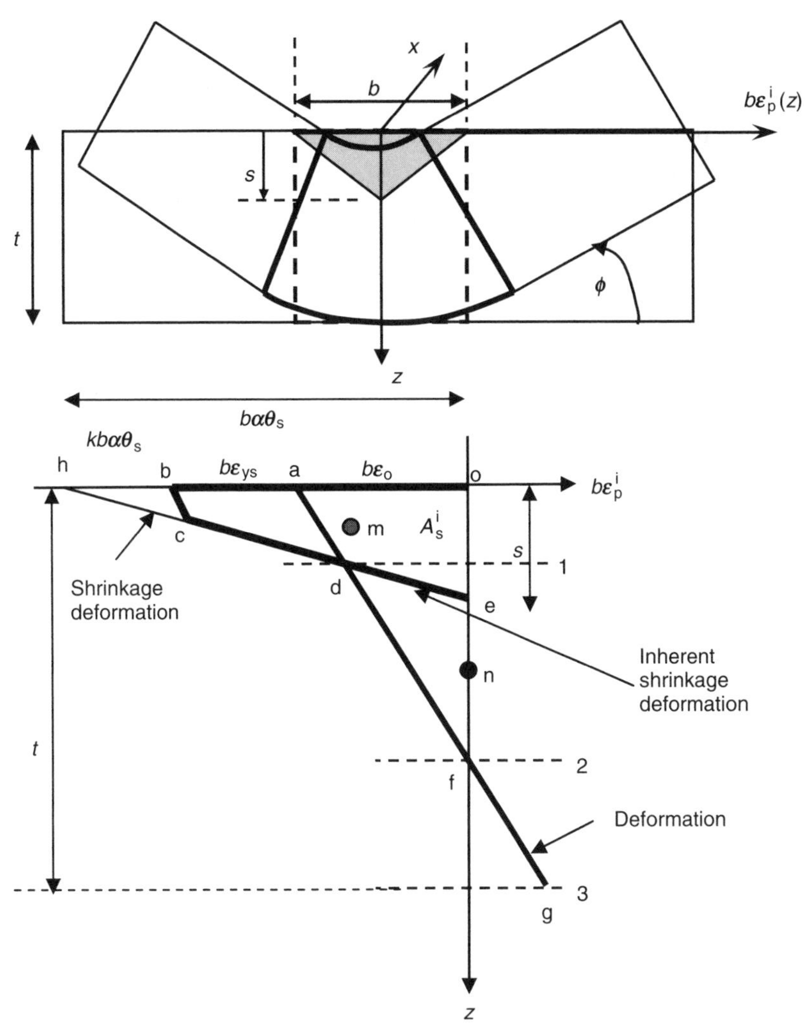

1.6 Transverse shrinkage and angular bending of butt joint; shrinkage strain, inherent shrinkage strain and total strain.

by area o–a–b–c–d–e–o and the transverse shrinkage deformation Δ_{cg} is shown by line a–d–f–g. The average transverse shrinkage is the deformation value at the midthickness n of the plate. The angular distortion ϕ is represented by the curvature, which is the slope of the deformation line. They can be written as follows:

$$\Delta_{cg} = \frac{A_s^i}{t} \quad [1.25]$$

$$\phi = \frac{t/2 - z_m}{t^3/12} A_s^i \qquad [1.26]$$

where z_m is the distance from the weld face to the centroid of the inherent shrinkage deformation area.

Equations [1.25] and [1.26] contain two unknowns, which are Δ_{cg}, and ϕ with a further two unknowns embedded in A_s^i. In addition to two mechanical conditions, namely force and moment equilibrium for Δ_{cg} and ϕ, there are also two trigonometric conditions for k and z_m within the triangular area o–a–b–h–c–d–e–o. The average transverse shrinkage and angular distortion can be written as follows:

$$\Delta_{cg} = \frac{1}{2}\frac{s}{t} b\alpha\theta_s(1-k^2) \qquad [1.27]$$

$$\phi = \frac{s}{t}\frac{b}{t}\alpha\theta_s\left(3(1-k^2) - 2\frac{s}{t}(1-k^3)\right) \qquad [1.28]$$

$$\frac{s}{t} = \frac{1-k^2}{1-k^3} \pm \left[\left(\frac{1-k^2}{1-k^3}\right)^2 - \frac{(1-k) - \varepsilon_{ys}/\alpha\theta_s}{1-k^3}\right]^{1/2} \qquad [1.29]$$

The normalized average transverse contraction r and angular distortion m for a triangular nugget shape may be written as follows:

$$r = \frac{\Delta_{cg}}{b\alpha\theta_s} = \frac{1}{2}\frac{s}{t}(1-k^2) \qquad [1.30]$$

$$m = \frac{\phi}{\alpha\theta_s b/t} = \frac{s}{t}\left(3(1-k^2) - 2\frac{s}{t}(1-k^3)\right) \qquad [1.31]$$

where k is a function of s/t as shown in Eqn [1.29] and z_m can be determined from the geometric relation of A_s^i.

Assuming that the linear heat input for a given penetration requirement s (cm) of a given joint thickness t (cm) and a given nugget shape in a given material is q_s (cal/cm), the weld nugget (shrinkage) area A_s can be expressed as $A_s = \gamma bs = q_s/L$, where L (cal/cm^3) is the melting energy density, the value of which depends on the welding process and conditions. The penetration-to-thickness ratio can therefore be written as

$$\frac{s}{t} = \frac{q_s}{L\gamma bt} \qquad [1.32]$$

Assuming that $L = 3900\,\text{cal/cm}^3$ for a carbon steel, that the nugget is triangular ($\gamma = 0.5$) and that $s/t = 5.13 \times 10^{-4} q_s/bt$, the parameter q_s/bt may be substituted for s/t in Eqns [1.30] and [1.31] for the normalized transverse contraction and normalized angular distortion respectively. Both expressions become functions of q_s/bt for a given $\alpha\theta_s/\varepsilon_{ys}$ ratio. The heat input parameter q_s/bt can be obtained for a given weld penetration s and weld face width b experimentally.

For different weld nugget shapes, the same analysis procedure as for the triangular shape can be performed to obtain relationships of s/t versus k, r versus (s/t, k) and m versus (s/t, k). A more detailed procedure has been described by Okerblom[16].

1.4 Predicting weld residual stresses and distortion

The plasticity-based analysis provides an engineering basis for study of weld residual stresses and distortion. There exist unique relationships between the inherent shrinkage strains and the weld residual stresses and distortion. This engineering analysis methodology can be extended to incorporate the effects of initial deformation, interaction between neighboring welds, welding sequence and structural stiffness[16]. To apply the plasticity-based analysis procedure for practical situations it usually requires mock-up welding tests calibrating the engineering solutions for characteristic process parameters, material nonlinearities and geometrical constraints. This engineering approach may be oversimplified to address many essential but inexplicable physical phenomena of welding. Therefore, the predicted results are often more useful towards development of heuristic knowledge than quantification of weld residual stresses and distortion in complex fabrication situations. However, heuristic knowledge, integrated with relevant practical experience, is critical to the success of any welding fabrication projects, especially when welding distortion is of a significant concern.

An alternative to the experimental calibration procedure for the plasticity-based analysis is to perform virtual simulation of the welding process and procedure using numerical modeling schemes such as the FEA method. Many of the physical phenomena associated with welding can be simulated using FEA models. The significance of this virtual simulation method is the capability to isolate essential parameters of the complex welding process and procedure to study the effects of respective parameters on the formation of welding-induced stresses and deformation. Although the simulation procedure still requires experimental calibrations on measurable parameters, the simulation procedure can be used for systematic investigations on relevant parameters, which may not be accomplished by experimental studies alone. For example, the inherent shrinkage strain can only be realistically determined using the modeling and simulation procedure. The FEA modeling and simulation procedure can be used to develop the database of inherent shrinkage strains for various welding situations of different characteristic processes and procedures.

To date, many attempts have been made by researchers to predict weld residual stresses and distortion through computer modeling and simulations of the welding process and procedure using the FEA method[7–54]. A non-linear time-dependent thermal elastic–plastic analysis of a moving source problem is usually performed to predict the thermal and mechanical responses of the connections under welding. Metallurgical evolutions due to welding have also been incorporated in the analysis for some materials. Weld residual stresses and distortion can be quantitatively determined. Through parametric studies of various welding factors by welding process and procedure simulations, appropriate weld design, welding procedure, jigging–fixturing and assembly sequence to minimize weld distortion have been realized by many researchers[17–25].

Despite recent efforts in the development of predictive weld stress and deformation methodologies using the FEA method, little success has been achieved for practical problems. The enormous temperature differential in the arc area creates a non-uniform distribution of heat in the workpiece. As the temperature increases, the yield strength decreases, the coefficient of thermal expansion increases, the thermal conductivity decreases and the specific heat increases. In addition, welding causes changes in the physical phases and metallurgical structures in the weld. To anticipate the weld stresses and deformation from a straightforward analysis of heat is difficult.

The FEA method has been used to predict weld residual stresses and distortion by many researchers in various studies. However, the weld residual stresses and distortion predicted by the thermal, elastic and plastic FEA method have been reported to be much smaller than the magnitude of the experimentally measured distortion[26]. This is because of the existence of compressive plastic strains in the weld metal created during the heating cycle of the welding process in the FEA model. In reality, these heating-induced plastic strains are relaxed when the weld metal becomes molten liquid. The physically non-existing compressive plastic strains in the FEA model results in the prediction of smaller weld distortions[27].

Besides the problems associated with heating-induced plastic strains, the major obstacles in the FEA simulations of welding processes are the enormous requirements for computing time and processing memory (central processing unit (CPU)) for simulating structural welding. To predict weld distortion caused by a 6 in weld segment, a CPU time of 72 h would be required on a CRAY Y-MP supercomputer[28]. To include numerical procedures for relaxing the plastic strains in the molten weld pool, an estimated additional CPU time of several hours would be required. These CPU requirements practically prohibit the use of the thermal, elastic and plastic FEA method for the prediction of distortion in a welded structure.

The common FEA simulation models may incorporate several unique features, which are briefly summarized as follows.

1.4.1 Modeling strategy

Although realistic welding problems tend to be truly three dimensional (3D) with complex geometry, transient and nonlinear, the 3D numerical modeling procedure usually requires huge computer resources. Since the computational demands by 3D models have limited the size of welds that can be analyzed, 2D models have long been utilized to reduce computing cost. Although it is a common belief that 3D models formulated with a moving heat source simulating the welding heat input and other nonlinear physical parameters can accurately compute weld residual stresses and distortion, using a hybrid weld and structural model with mixed 2D and 3D elements can be a realistic alternative modeling approach in many practical situations[23,24].

For 2D modeling and analysis, some geometrical conditions have to be met and some assumptions have also to be made before a 2D model can be generated and used to simplify a 3D model. According to the geometrical conditions and the assumptions adopted, there are three general types of 2D FEA model that are frequently applied to predict weld residual stresses and distortion.

1. *Axisymmetric model.* When both geometry and loading have a common axis of symmetry, the axisymmetric condition exists. This model is usually applied in the analysis of a spot or projection welding process.
2. *Plane-stress model.* When the plate thickness is small or the temperature and stress changes in the thickness direction are negligible, the plane-stress condition exists. In most structural welding projects, it involves thin-plate weldments in 3D frames. Therefore, it is reasonable to apply the plane-stress weld model in a 3D structural analysis.
3. *Generalized plane-strain model.* It assumes the existence of a plane, which contains all displacement vectors that have a constant strain value normal to the plane (i.e. a cross-sectional plane remains a plane when it deforms). For 2D modeling analysis of a weld cross-section, the generalized plane-strain condition must be specified in lieu of the general plane-strain condition that restricts displacements normal to the cross-sectional plane. When the general plane-strain condition is applied, the entire cross-section would yield because of complete rigidity of the cross-section[26].

Another modeling issue commonly asked by the researchers is the coupling of the thermal and mechanical analyses in the numerical procedure. In the coupled analysis, thermal and mechanical behaviors are analyzed sequentially in the time increments incorporating the effect of the mechanical work in the thermal evolution process. This coupled analysis is CPU intensive. Because the effect mechanical work is insignificantly small in comparison with the welding heat input in most of the welding processes, uncoupled thermomechanical analysis is a common practice. In addition, when using the uncoupled analysis approach the thermal evolution results predicted by the welding analysis can be independently verified prior to the mechanical analysis[26].

1.4.2 Heat input model

The welding heat source has been modeled with different degrees of simplicity depending upon the analysis requirements. For a generalized plane-strain welding model of a 2D cross-section, the heat source is scanned across the cross-section at the travel speed with the linear heat input energy consistent with the welding heat input. In this case, heat diffuses into the material on both sides of the weld without any heat dissipation along the weld axis. This 2D analysis results in significant inconsistency in the cooling rates due to ignorance of heat diffusion along weld axis. This is particularly true for a slow travel speed. When the welding travel speed is high, direct heat deposition to the joint circumvents the material's diffusivity effect, which would reduce the inaccuracy of the 2D heat flow model. In order to compensate for the heat flow inconsistency in a 2D model; a ramp-heat model has been studied[29]. This ramp-heat model assigns more heat to the later part of the scanning period to adjust both heating and cooling rates. If only the final states of weld stresses (residual stresses) and deformation (distortion) are of interest, they are generally insensitive to the heat input model because these final states are primarily influenced by the cooling temperatures below materials softening temperature. These temperatures occur at greater distance from the moving heat source and are usually insensitive to the heat input model.

The other issue associated with welding of a thick joint is simulation of multiple weld passes in the 2D model. The root and cap weld passes usually dominate the formation of weld residual stresses and distortion. The fill passes have less influence on the final states of weld stresses and distortion. Therefore, the fill passes may be lumped into individual weld layers without compromising the accuracy of the prediction[29].

In the 2D plane-stress model for a thin plate or the 3D model of a plate with a finite thickness a moving heat source is usually incorporated in the analysis. Several different heat source distribution models have been studied. The traditional Adam's point (thick-plate) or line (thin-plate) heat source models are often used in the theoretical welding heat analysis[30–33]. A singularity condition occurs at the heat source owing to the assumption of heat concentration at a point or along a line through the plate thickness.

Several heat source models with flux distribution over a surface area have been studied. Pavelic *et al.*[34] proposed a Gaussian distribution of flux deposited on the surface. Tsai[35] modified the model proposed by Pavelic *et al.* with a skewed Gaussian distribution of flux in which the arc column is distorted backwards to reflect the effect of welding speed and to adjust the cooling evolution in the analysis. Anderson[36] used a piecewise constant flux over the weld pool. Goldak *et al.*[37,38] proposed a non-axisymmetric 3D heat source model in which the power density is dispensed with a Gaussian distribution in a double ellipsoid representing the weld pool. All these heat source models require calibration with the weld pool boundary (i.e. the peak temperature at this boundary is the material's melting temperature) as an intimate correction factor. The double-ellipsoidal heat input function has been commonly used by researchers in the modeling analysis requiring accurate prediction of thermal evolution in or near the weld pool[39,40]. However, for weld distortion analysis, such a detailed heat source model is usually unnecessary. FEA analysis using normal heat distribution function has also been conducted for weld residual stresses and distortion studies by researchers[41–43].

1.4.3 Strain relaxation model due to melting

When melting occurs in the weld pool, molten metal does not retain the plastic strains accumulated during the heat cycle. Upon solidification and reducing the temperature to regain the material's strength and rigidity, the weld shrinks by one to several times the yield strain magnitude depending upon the joint stiffness, as demonstrated by the one-dimensional bar–spring model. However, a numerical procedure incorporating the strain relaxation phenomenon may be unnecessary in the analysis. Weld shrinkage strains may be directly superimposed on the inherent shrinkage strains predicted with a model without considering the relaxation phenomenon. The magnitude of the weld shrinkage strain can be determined by adding about once the yield strain magnitude to the predicted total strain in the weld using an iterative procedure. As illustrated by the analogous examples presented earlier in this chapter, any shrinkage strain magnitude more than one yield strain beyond the total strain is irrelevant to the force and moment equilibrium conditions as required in the analysis procedure.

1.4.4 Metallurgical evolution model

The metallurgical evolution may develop stresses due to strain incompatibility during solid-state phase transformation[36,44]. This strain incompatibility results in stresses between the grains of different phases, which may have a significant influence on the state of residual stresses in the weld metal and heat-affected zone in joining certain types of material such as martensitic weldments. However, this phase transformation stress is usually ineffective in influencing the final state of weld distortion. The actual weld shrinkage strain is already beyond the effective shrinkage strain magnitude in the weld zone; therefore, weld distortion is usually not affected by the transformation stress.

1.4.5 Large-scale distortion analysis model

To date, analysis of weld residual stresses and distortion using either a 2D or a 3D FEA model has been primarily applied to study individual weld joints. For practical industrial applications, the large structural FEA model with adequately meshed elements is usually cost prohibitive for a detailed thermal elastic–plastic analysis with moving heat source. Attempts have been made by many researchers to develop new techniques to handle the large-scale structure problem. Paul *et al.*[45] studied distortion of long steady welds using an Eulerian formulation. The ESI Group[46] developed a decoupled 3D weld modeling and 3D structural response and implemented them in a FEA program, called SYSWELD. This hybrid modeling technique has also been applied for large structures using a 2D weld model and 3D structural model for thin-plate space frames.

To study the structural behaviors of welded components during their fabrication process, a large-scale computational algorithm is usually required[47]. This algorithm may be realized on the basis of the physical fact that thermal strains above several times the yield strain of material are almost ineffective in causing weld residual stresses and distortion. Therefore, simple but more efficient computational heat flow schemes may be implemented to reduce the computational requirements. Using the adaptive meshing technique[48,49] in the numerical analyses can also drastically reduce the computational requirements. This computation algorithm can predict distortion of large structures. It can also be used to study the effect of the scaling factors between a mock-up model and a real structure model[50].

Since the inherent shrinkage strain is the primary cause for weld distortion, these plastic strains may be mapped onto a structural model to interact with the structure rigidity if the plastic strains can be determined by simulation of an individual weld using either a 2D or a 3D model. The longitudinal plastic strain component has been mapped to a butt or T-joint to predict buckling behavior in large ship panels[51–53]. More recent work by Jung and Tsai[54] studied the significance of each individual plastic strain components on angular distortion of fillet welded T-joints. This plasticity-based distortion analysis verified a unique relationship between cumulative plastic strains and angular distortion. It is believed that modeling of the shrinkage phenomenon of welds alone may be sufficiently accurate in predicting the weld distortion in large structures. User interface software has been developed to map the plastic strains from the weld model automatically to a structural model[14].

1.4.6 Practical applications of modeling analysis results

To incorporate variations in the field practice of welding assemblage, the heuristic rules expressed in algorithmic form (analytical), as experimental data (empirical), and in a realm of experience (skills or know-how) are needed. This heuristic knowledge is generally stored in the mind of an 'expert' in the field. Modeling analysis is an effective supplement or an alternative to the empirical experience of experts in developing the heuristic knowledge base in the form of an electronic expert system for practical applications.

In production environments, heuristic knowledge on weld residual stresses and distortion control may be implemented via an expert system that is a computer program using the artificial intelligence (AI) architecture. The AI architecture integrates components such as knowledge (algorithmic and empirical), heuristics and decision making. It also produces 'intelligent' behaviors by operating on the knowledge of a human or a virtual (i.e. modeling and simulation) expert in a well-defined application domain. The expert guidelines may be used to provide consultation services for structural designers and welding fabricators to determine quickly the optimum relationship between structural geometry, joint detail or jigging planning and the resulting weld residual stresses and distortion. The plasticity-based algorithms may be used as the heuristic engine of the expert system.

As a commentary on the analysis approaches, regardless of the complexity or completeness in modeling the physical behaviors of the welding simulation models, the purpose of modeling analysis should not be considered only as a predictive tool trying to quantify weld residual stresses and distortion. This common misconception may be counterproductive in solving the weld residual stresses and distortion problems. Precise prediction of quantitative measures is often a very difficult task to accomplish, especially for large-scale structures or welding fabrication in any production environments. It is recommended that all modeling efforts should be regarded as a good numerical approach for generating heuristic knowledge. A realistic solution to practical problems in weld residual stresses and distortion is to make intelligent decisions by a learned human expert with the assistance of a virtual expert with AI.

1.5 Conclusions

Although the inherent shrinkage model has been demonstrated to be accurate, convenient and cost effective in the prediction of weld residual stresses and distortion, several fundamental issues remain unsolved to date. The inherent shrinkage model needs calibration constants for different applications. These calibration constants will depend on the thermal characteristics of the welding process and the geometric details of the joint, as well as the jigging constraints. For example, the submerged arc welding process will result in very distinct thermal characteristics in the weldment from that caused by the shielded metal arc welding process. Welding thick joints will result in complete different calibration constants from welding thin sheet metals. To prescribe the inherent shrinkage strains in the weld joint, the equivalent thermal strains may be defined using the artificial temperature fields. A standard procedure to determine the effective shrinkage strain temperatures will be required. Finally, an inherent shrinkage strain database for various production situations, joint designs and welding procedures needs to be developed. The FEA modeling, simulation and analysis procedures can be useful in determining the inherent shrinkage strains.

1.6 References

1. Tsai, C.L., 'Find the root causes of welding induced distortion by numerical modeling method', *Proc. Int. Welding/Joining Conference*, Korea, 2002, Korea Welding Society, Gyeongju, Korea, 2002, pp. 681–687.
2. Blodgett, O.W., 'Types and causes of distortion in welded steel and corrective measures', *Weld. J.*, **39**(7), 692–697, 1960.
3. Spraragen, W., and Ettinger, W.G., 'Shrinkage Distortion in Welding', *Weld. J.*, **16**(7), 29–39, 1937.
4. Guyot, F., 'A note on the shrinkage and distortion of welded joints', *Weld. J.*, **26**(9), 519-s–529-s, 1947.
5. Watanabe, M., and Satoh, K., 'Effect of welding conditions on the shrinkage distortion in welded structures', *Weld. J.*, **40**(8), 377-s–384-s, 1961.
6. Ueda, Y., Fukuda, K., and Tanigawa, M., 'New measuring method of three dimensional residual stresses based on theory of inherent strain', *Trans. JWRI*, **8**(2), 249–256, 1979.
7. Ueda, Y., Kim, Y.C., and Yuan, M.G., 'A predictive method of welding residual stress using source of residual stress (report I) characteristics of inherent strain (source of residual stress)', *Trans. JWRI*, **18**(1), 135–141, 1989.
8. Ueda, Y., and Yuan, M.G., 'Prediction of residual stresses in butt-welded plates using inherent strains', *Trans. ASME, J. Engng Mater. Technol.*, **115**(10), 417–423, 1993.
9. Yuan, M.G., and Ueda, Y., 'Prediction of residual stresses in welded T- and J-joints using inherent shrinkage strains', *Trans. ASME, J. Engng Mater. Technol.*, **118**(4), 229–234, 1996.
10. Luo, Y., Murakawa, H., and Ueda, Y., 'Prediction of welding deformation and residual stress by elastic FEM based on inherent strain (report I)', *Trans. JWRI*, **26**(2), 49–57, 1997.
11. Tsai, C.L., and Shim, Y.L., 'Determination of welding-induced shrinkage strains for distortion prediction of welded frame structures', Research Report submitted to General Motors Technical Center, The Ohio State University, Columbus, Ohio, USA, 1993.

12. Tsai, C.L., Cheng, W.T., and Lee, H.T., 'Development of weld distortion prediction model using springs and its application to the GM tubular joint', Research Report submitted to the GM Technical Center, The Ohio State University, Columbus, Ohio, USA, 1995.
13. Han, M.S., 'Fundamental studies on welding induced distortion in thin plate', PhD Dissertation (advisor C.L. Tsai), The Ohio State University, Columbus, Ohio, USA, 2002.
14. Jung, G.H., 'Plasticity-based distortion analysis for fillet welded thin plate T-joints', PhD Dissertation (advisor C.L. Tsai), The Ohio State University, Columbus, Ohio, USA, 2003.
15. Tsai, C.L., 'Residual stresses and distortion', WE821 Course Notes, The Ohio State University, Columbus, Ohio, USA, 2004.
16. Okerblom, H.O., *Calculations of Deformations of Welded Metal Structures*, MASHGIZ, Moscow, 1955.
17. Papzoglou, V.J., and Masubuchi, K., 'Analysis and control of distortion in welded aluminum structures', *Weld. J.*, **57**(9), 251-s–262-s, 1978.
18. Hou, C.A., 'Modeling and control of welding distortion in tubular frame structures', PhD Dissertation (advisor C.L. Tsai), The Ohio State University, Columbus, Ohio, USA, 1986.
19. Tsai, C.L., Kim, D.S., Jaeger, J., Shim, Y., Feng, Z., and Lee, S., 'Determination of residual stresses in thick section weldments', *Weld. J.*, **71**(9), 305-s–312-s, 1992.
20. Rybicki, E.F., Nagel, G.L., Stonesifer, R.B., and Miller, E.G., 'A semi-empirical model for transverse weld deflections of square tubular automotive beams', *Weld. J.*, **72**(8), 317-s–380-s, 1993.
21. Feng, Z.L., and Tsai, C.L., 'Welding residual stresses in splicing heavy section shapes', SAE 48th Earthmoving Industry Conf. and Exposition, Session: Manufacturing Processes, Peoria, Illinois, USA, 9–10 April 1997, SAE Paper 971585, Society of Automotive Engineers, New York, 1997.
22. Dong, Y., Hong, J.K., Tsai, C.L., and Dong P., 'Finite element modeling of residual stresses in pipe girth welds', *Weld. J.*, **76**(10), 442-s–449-s, 1997.
23. Tsai, C.L., Lee, B.N., and Cheng, W.T. 'Design optimization of body braze joints', *Proc. Sheet Metal Welding Conference 98*, Troy, Michigan, USA, 13–16 October 1998, American Welding Society, Detroit Section, Troy, Michigan, 1998, Paper 4-5.
24. Tsai, C.L., Park, S.C., and Cheng, W.T., 'Welding distortion of a thin-plate panel structure', *Weld. J.*, **78**(5), 156-s–165-s, 1999.
25. Tsai, C.L., Feng, Z., Kim, D.S., Jaeger, J., Shim, Y., and Lee, S., 'Design analysis for welding of heavy W-shapes', *Weld. J.*, **80**(2), 35–41, 2001.
26. Lee, S.G., 'Modeling of residual stress in thick section weldments', PhD Dissertation (advisor C.L. Tsai), The Ohio State University, Columbus, Ohio, USA, 1992.
27. Hong, J.K., Tsai, C.L., and Dong, P., 'Assessment of numerical procedures for residual stress analysis of multipass welds', *Weld. J.*, **77**(9), 372-s–382-s, 1998.
28. Brown, S.B., and Song, H., 'Implications of three-dimensional numerical simulations of welding of large structures', *Weld. J.*, **71**(2), 55-s–62-s, 1992.
29. Shim, Y., Feng, S., Lee, D., Kim D., Jaeger, J. and Tsai, C.L., 'Modeling of welding residual stresses', *Proc. 112th ASME Winter Annual Meeting*, Atlanta, Georgia, USA, 1–6 December 1991, American Society of Mechanical Engineers, New York, 1991, pp. 29–41.

30. Rosenthal, D., and Schmerber, R., 'Thermal study of welding: experimental verification of theoretical formulas', *Weld. J.*, **17**(4), 2-s–8-s, 1938.
31. Rosenthal, D., 'Mathematical theory of heat distribution during welding', *Weld. J.*, **20**(5), 220-s–234-s, 1941.
32. Rosenthal, D., 'The theory of moving sources of heat and its application to metal treatments', *Trans. ASME*, **68**, 849–866, 1946.
33. Wells, A.A., 'Heat flow in welding', *Weld. J.*, **31**(5), 263-s–267-s, 1952.
34. Pavelic, V., Tanbakuchi, R., Uyehara, O.A., and Meyers, P.S., 'Experimental and computed temperature histories in gas tungsten-arc welding of thin plates', *Weld. J.*, **48**(7), 295-s–305-s, 1969.
35. Tsai, C.L., 'Finite source theory', *Modeling of Casting and Welding Process II*, American Institute of Mining, Metallurgical and Petroleum Engineers, New York, 1983, pp. 329–341.
36. Anderson, B.A., 'Thermal stresses in a submerged-arc welded joint considering phase transformations', *Trans. ASME, J. Engng Mater. Technol.*, **100**, 356–362, 1978.
37. Goldak, J., Chakrevarti, A., and Bibby, M., 'A new finite element model for welding heat source', *Metall. Trans. B*, **15B**(7), 299–305, 1984.
38. Goldak, J., Bibby, M., Moore, J., House, R., and Patel, B., 'Computer modeling of heat flow in welds model for welding heat source', *Metall. Trans. B*, **17B**, 587–600, 1986.
39. Feng, Z.L., Cheng, W.T., and Chen Y.S., 'Development of new modeling procedures for 3D welding residual stress and distortion assessment', EWI CRP Report SR9818, Edison Welding Institute, Columbus, Ohio, USA, November 1998.
40. Bae, D.H., Kim, C.H., Hong, J.K., and Tsai, C.L., 'Numerical analysis of welding residual stress using heat source models for the multi-pass weldment', *KSME Int. J.*, **16**(9), 1054–1064, 2002.
41. Friedman, G., 'Thermomechanical analysis of the welding process using the finite element method', *J. Pressure Vessel Technol.*, **97**, 206–213, 1975.
42. Tsai, C.L., Lee, S.G., and Shim, Y.L., 'Modeling techniques for welding-induced residual stress predictions', *Proc. Int. Conf. on Modeling and Control of Joining Processes* (Ed. T. Zacharia), Orlando, Florida, 1993, American Welding Society, Miami, Florida, 1993, pp. 462–469.
43. Tsai, C.L., Cheng, W.T., and Lee, H.T., 'Modeling strategy for control welding-induced distortion', *Modeling of Casting, Welding and Advanced Solidification Processes VII*, 10–15 September 1995, The Minerals, Metals and Materials Society of AIME, Warrendale, Pennsylvania, 1995, pp. 335–345.
44. Feng, Z., David, S.A., Zacharia, T., and Tsai, C.L., 'Quantification of thermomechanical conditions for weld solidification cracking', *Sci. Technol. Weld. Joining*, **2**(1), 11–19, 1997.
45. Paul, S., Michaleris, P., and Shanghvi, J., 'Optimization of thermo-elasto-plastic finite element analysis using an Eulerian formulation', *Int. J. Numer Meth. Engng*, **56**, 1125–1150, 2003.
46. ESI Group, *SYSWELD: Engineering Simulation Solution for Heat Treatment, Welding, and Welding Assembly*, ESI Group, Paris, 2003.
47. Argyris, J.H., Szimmat, J., and William, K.J., 'Computational aspects of welding stress analysis', *Comput. Meth. Appl. Mech. Engng*, **33**, 635–666, 1982.

48. Prasad, N.S., and Sankaranrayanan, T.K., 'Estimation of residual stresses in weldments using adaptive grids', *Comput. Structs*, **60**(6), 1037–1045, 1996.
49. Runnemalm, H., and Hyun, S., 'Three-dimensional welding analysis using an adaptive mesh scheme', *Comput. Meth. Appl. Mech. Engng*, **189**(2), 515–523, 2000.
50. Tsai, C.L., Cheng, X.D., Zhao, Y.F., and Jung, G.H., 'Plasticity modeling to predict welding distortion in thin-shell tubular structures', *Advances in Computational Engineering and Sciences*, Vol. 2, (eds. S.N. Atluri and F.W. Brust), Tech Science Press, Irvine, California, 2000, pp. 1913–1918.
51. Michaleris, D.A., Dantzig, J.A., and Tortorelli, D.A., 'Minimization of welding residual stress and distortion in large structures', *Weld. J.*, **78**(11), 361-s–366-s, 1999.
52. Michaleris, D.A., Tortorelli, D.A., and Vidal, C.A., 'Analysis and optimization of weakly coupled thermo-elasto-plastic systems with applications to weldment design', *Int. J. Numer. Meth. in Engng*, **38**(8), 2471–2500, 1995.
53. Michaleris, D.A., and DeBiccari, A., 'Prediction of welding distortion', *Weld. J.*, **76**(4), 172-s–180-s, 1997.
54. Jung, G.H., and Tsai, C.L., 'Plasticity-based distortion analysis for fillet welded thin-plate T-joints', *Weld. J.*, **83**(6), 1–6, 2004.

2
Welding heat transfer

J. ZHOU and H. L. TSAI, University of Missouri-Rolla, USA

2.1 Introduction

Heat transfer during welding can strongly affect phase transformations and thus the metallurgical structure and mechanical properties of the weld. Also, the complicated heat and mass transfer and fluid flow phenomena in welding can result in the formation of residual stresses in the joint region and distortion of the welded structure. Understanding the fundamentals of heat flow and the basic features of heat transfer, such as the cooling rates of the weld metal, the distribution of heat between the weld metal and heat-affected zone provides insight into ways in which the welding process can be optimized. It also may suggest new directions for improvements in welding technology and how welding defects can be minimized.

In this chapter, the fundamentals of heat transfer in various welding processes, such as arc welding, laser welding and some new welding technologies including hybrid laser–arc welding and dual-beam laser welding are discussed. The complex mathematical models that describe the heat transfer process in the aforementioned welding processes are included. With the advent of modern computer technology, the complicated differential equations in the models can be solved numerically. The advanced numerical technologies for solving these equations are discussed and some simulation results are provided. Finally, the related publications and references and future trends of the welding heat transfer research will be discussed.

2.2 Arc welding

Welding generally involves the application or development of localized heat near the intended joint. The term *arc welding* applies to a large and diversified group of welding processes that use an electric arc as the source of heat to melt and join metals, as illustrated in Fig. 2.1[1]. The welding arc is struck between the workpiece and the tip of an electrode. The electrode

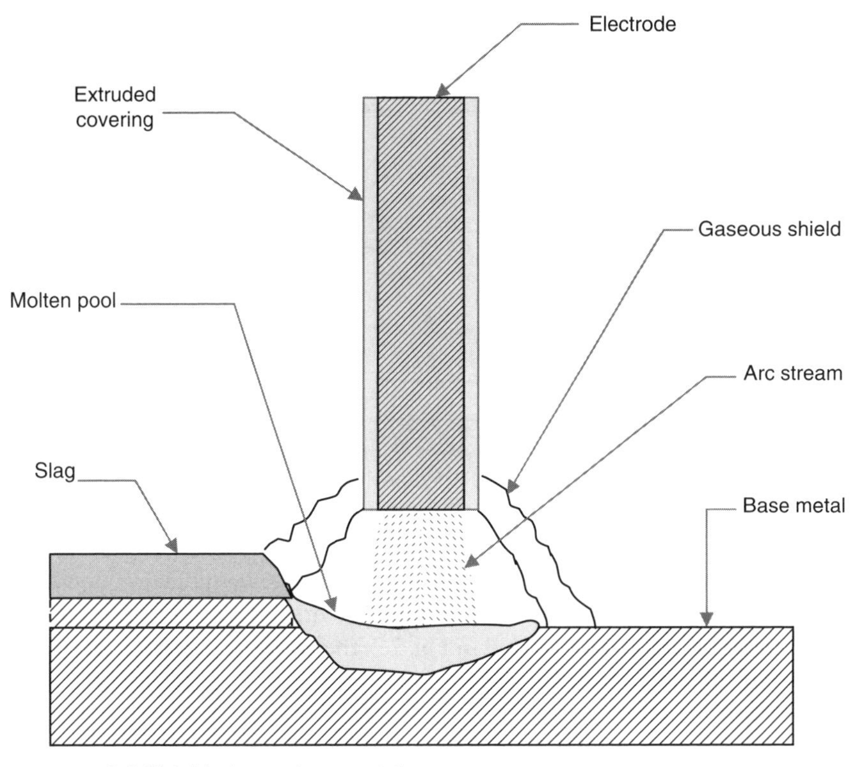

2.1 Shielded metal arc welding process.

will be either a consumable wire or rod or a nonconsumable carbon or tungsten rod which carries the welding current. When a nonconsumable is used, filler metal can be supplied by a separate rod or wire if needed. A consumable electrode is designed not only to conduct the current that sustains the arc but also to melt and supply filler metal to the joint. An electric arc consists of a relatively high current discharge sustained through a thermally ionized gaseous column called plasma. The power of an arc may be expressed in electrical units as the product of the current passing through the arc and the voltage drop across the arc.

2.2.1 Conduction heat transfer

In arc welding, the heat in the weld pool is transported by means of convection and conduction. A rigorous solution of the complete heat flow equation considering heat transfer by both conduction and convection is complicated. As a first step, it is often useful to discuss a simplified solution considering only conduction heat transfer. This simplification is attractive

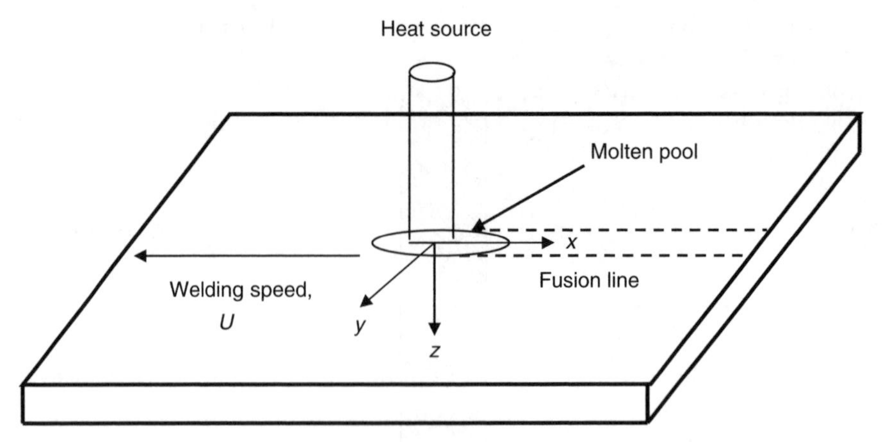

2.2 Coordinate system for conduction heat transfer analysis in welding.

since analytical solutions can be obtained for the heat conduction equation in many situations, and these solutions can provide interesting insight about the fusion process. As illustrated in Fig. 2.2, the workpiece is stationary and the origin of the coordinate system moves with the heat source at a constant speed U in the negative x direction. The transfer of heat in the weldment is governed primarily by the time-dependent conduction of heat, which is expressed by the following equation[2]:

$$\frac{\partial}{\partial x}\left(k(T)\frac{\partial T}{\partial x}\right)+\frac{\partial}{\partial y}\left(k(T)\frac{\partial T}{\partial y}\right)+\frac{\partial}{\partial z}\left(k(T)\frac{\partial T}{\partial z}\right)$$
$$=\rho\frac{\partial}{\partial T}[C_{\mathrm{M}}(T)T]+U\rho\frac{\partial}{\partial x}[C_{\mathrm{M}}(T)T] \qquad [2.1]$$

where x is the coordinate in the welding direction, y is the coordinate transverse to the weld, z is the coordinate normal to the weldment surface, T is the temperature (°C) in the weldment, $k(T)$ is the thermal conductivity (J/cm s) of the metal, ρ is the density (g/cm^3) of the metal and $C_{\mathrm{M}}(T)$ is the specific heat (J/g °C) of the metal.

Rosenthal[2] used the following assumptions to derive analytical solutions for the heat transfer equation during welding.

1. Steady-state heat flow.
2. A point heat source.
3. Constant thermal properties.
4. Negligible heat of fusion.
5. No heat losses from the workpiece surface.
6. No convection in the weld pool.

When the thickness of the workpiece is small, the heat transfer in the thickness direction is assumed negligible and heat flow is assumed two dimensional. Rosenthal[2] derived the following equation to calculate the two-dimensional temperature distributions during welding of thin sheets of infinite width:

$$T - T_0 = \frac{Q}{2\pi kg} \exp\left(\frac{Ux}{2\alpha}\right) K_0\left(\frac{U(x^2 + y^2)^{1/2}}{2\alpha}\right)$$

$$\alpha = \frac{k}{\rho C_M} \qquad [2.2]$$

where T_0 is initial workpiece temperature before welding, g is the workpiece thickness, U is the welding speed, K_0 is the modified Bessel function of second kind and zero order, as shown in Fig. 2.3, and Q is the heat flux in terms of the welding current I, the welding voltage V and the efficiency η of the arc as in

$$Q = \eta VI \qquad [2.3]$$

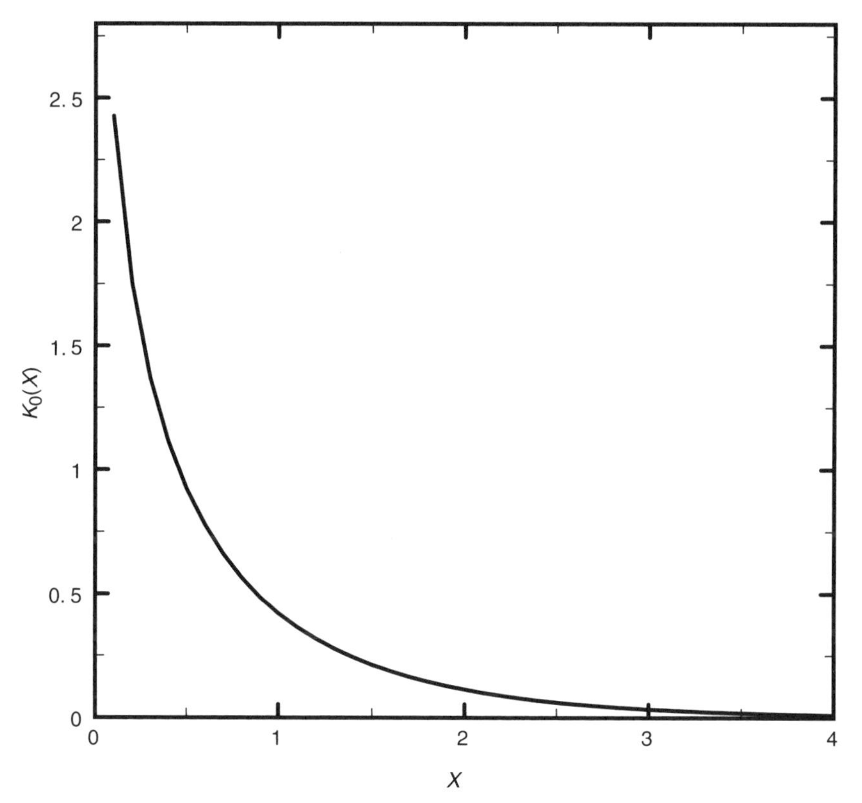

2.3 Modified Bessel function of second kind and zero order.

Table 2.1 Arc efficiencies

Process	η
Manual metal arc	0.7–0.85
Tungsten–inert gas (argon)	0.22–0.48
Metal-inert gas (argon)	0.66–0.75
Submerged arc	0.90–0.99

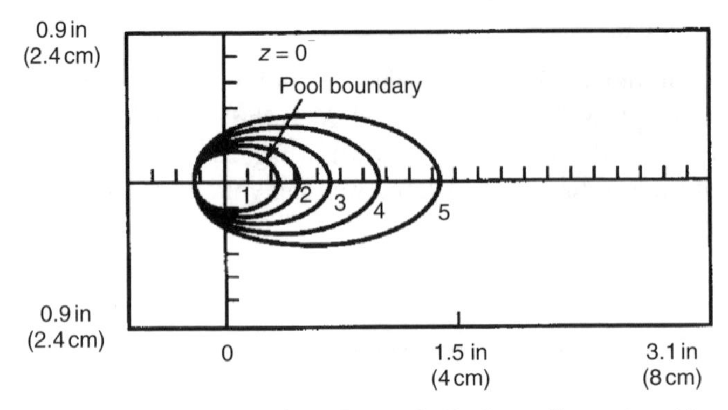

2.4 Calculated results from Rosenthal's three-dimensional heat flow equation (welding speed, 2.54 mm/s; heat input, 3200 W; material, 1018 steel): curve 1, 5018 °F (1530 °C); curve 2, 3668 °F (1100 °C); curve 3, 2588 °F (780 °C); curve 4, 1886 °F (550 °C); curve 5, 1357 °F (400 °C).

The arc efficiency depends on the welding process and some typical values are given in Table 2.1[3]. The 1–10% loss in efficiency of the submerged arc process is related to that part of the flux which does not fuse. The lower efficiency of the tungsten–inert gas (TIG) process is associated with heat losses through the electrode holder.

The analytical solution derived by Rosenthal for three-dimensional heat conduction in a semi-infinite workpiece during welding is as follows:

$$T - T_0 = \frac{Q}{2\pi k(x^2 + y^2 + z^2)^{1/2}} \exp\left(\frac{-U((x^2 + y^2 + z^2)^{1/2} - x)}{2\alpha} \right) \qquad [2.4]$$

This equation implies that on the transverse-cross section of the weld all isotherms, including the fusion boundary and the outer boundary of the heat-affected zone, are semicircular in shape. Despite the restrictions imposed by the aforementioned simplifying assumptions, these equations have been widely used in the welding community mainly because of their simplicity. Examples of results based on Rosenthal's two- and three-dimensional heat flow equations are presented in Fig. 2.4[4]. More discussion has been given by Kou[4] and Jenney and O'Brien[5].

2.2.2 Convection heat transfer

In both gas–tungsten arc welding (GTAW) and gas–metal arc welding (GMAW), fluid flow in the weld pool can significantly affect the heat transfer, weld penetration depth, segregation and porosity[6]. Szekely and coworkers[7] made the most significant breakthrough in computer simulation of weld pool convection. Then, numerous investigations followed his method[8–14]. Tsai and Kou[15] used the finite-difference method and pool-surface-fitting orthogonal curvilinear coordinates to calculate heat transfer and fluid flow in a weld pool for GTAW, focusing on the effect of the electromagnetic force on weld pool surface deformation. Kim and Na[16], on the other hand, used the boundary-fitted coordinates to calculate the pool surface shape, as well as heat transfer and fluid flow in the weld pool. Fan *et al.*[17] developed a numerical model for partially or fully penetrated weld pools in stationary GTAW.

The process of GMAW is much more complicated than that of GTAW, as GMAW involves the impingement of droplets onto the weld pool, creating the mixing of mass, momentum, thermal energy and species in the weld pool. Tsao and Wu[18] presented a two-dimensional stationary weld pool convection model, in which the weld pool surface was assumed to be flat and mass transfer was not considered. Wang and Tsai[19] investigated the dynamic impingement of droplets on to the weld pool and the solidification process in spot GMAW. They[20] continued to investigate the Marangoni effect caused by surface-active elements on weld pool mixing and weld penetration. Also, the effect of droplet size and drop frequency on weld pool penetration was studied[21]. Using boundary-fitted coordinates, Kim and Na[22] presented a three-dimensional quasi-steady heat and fluid flow analysis for the moving heat source of the GMAW process with a free surface. Ushio and Wu[23] used a boundary-fitted nonorthogonal coordinate system to handle the largely deformed GMA weld pool surface and predicted the area and configuration of weld reinforcement. By using the volume-of-fluid (VOF) technique[24] and the continuum formulation[25], Wang[26] developed a mathematical model for three-dimensional moving GMAW. The fluid flow, heat transfer and species transfer are calculated when droplets carrying mass, momentum, thermal energy and species periodically impinge on to the weld pool. So far, calculations of convective heat transfer in the weld pool are mainly performed through the numerical solution of the equations of conservation of mass, heat and momentum. These equations and the associated boundary conditions for solving these equations will be discussed in the following.

Governing equations

Figure 2.5 is a schematic sketch of a moving GMAW for a plain plate. The three-dimensional x–y–z coordinate system is fixed to the stationary base

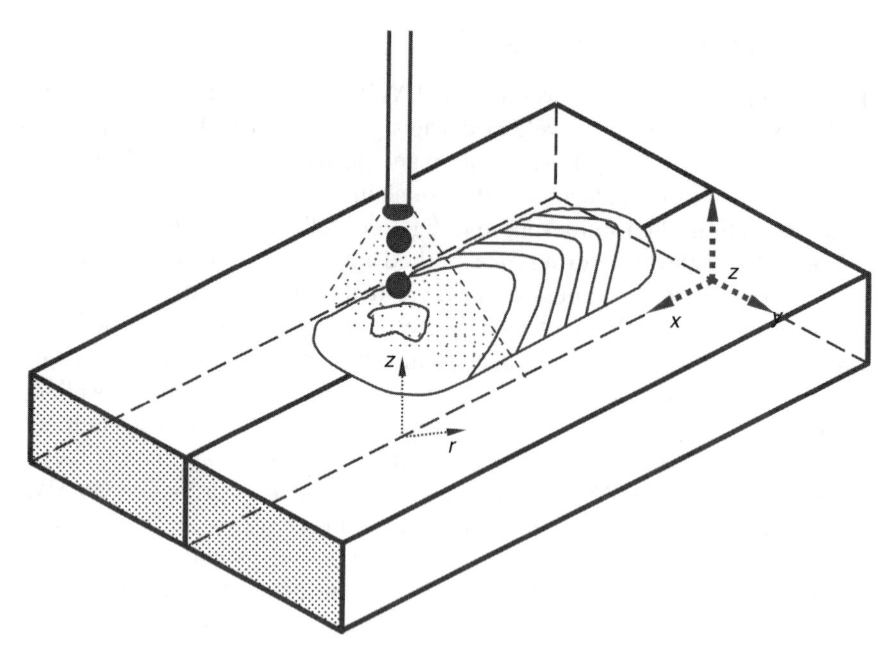

2.5 Schematic sketch of a moving GMAW system; *x–y–z* is a fixed coordinate system and *r–z* is a local coordinate system moving along with the arc center.

metal, while a two-dimensional *r–z* cylindrical coordinate system is moving with the arc center. In GMAW, the total arc energy is split into two parts: one to melt the electrode and generate droplets, and the other to heat the base metal directly. Hence, in addition to thermal energy, mass and momentum carried by the droplets, arc heat flux is simultaneously impacting on the base metal. Droplets containing sulfur at different concentrations from that of the base metal periodically impinge onto the base metal in the negative z direction, while they move at the same velocity along the y direction as the arc. The mathematical formulation given below is valid for both the base metal and the liquid droplets. Once a droplet reaches the free surface, it is immediately considered to be part of the base metal and then the exchange of momentum, energy and species between the droplet and the weld pool occurs. The differential equations governing the conservation of mass, momentum, energy and species based on continuum formulation are as follows[26]: the equation for the conservation of mass is

$$\frac{\partial}{\partial t}(\rho) + \nabla \cdot (\rho \boldsymbol{V}) = 0 \qquad [2.5]$$

the equations for the conservation of momentum are

$$\frac{\partial}{\partial t}(\rho u)+\nabla\cdot(\rho \boldsymbol{V} u)=\nabla\cdot\left(\mu_1\frac{\rho}{\rho_1}\nabla u\right)-\frac{\partial p}{\partial x}-\frac{\mu_1}{K}\frac{\rho}{\rho_1}(u-u_s)$$
$$-\frac{C\rho^2}{K^{1/2}\rho_1}|u-u_s|(u-u_s)$$
$$-\nabla\cdot(\rho f_s f_1 \boldsymbol{V}_r u_r)+\nabla\cdot\left[\mu_1 u\nabla\left(\frac{\rho}{\rho_1}\right)\right]+\boldsymbol{J}\times\boldsymbol{B}|_x \qquad [2.6]$$

$$\frac{\partial}{\partial t}(\rho v)+\nabla\cdot(\rho \boldsymbol{V} v)=\nabla\cdot\left(\mu_1\frac{\rho}{\rho_1}\nabla v\right)-\frac{\partial p}{\partial y}-\frac{\mu_1}{K}\frac{\rho}{\rho_1}(v-v_s)$$
$$-\frac{C\rho^2}{K^{1/2}\rho_1}|v-v_s|(v-v_s)$$
$$-\nabla\cdot(\rho f_s f_1 \boldsymbol{V}_r v_r)+\nabla\cdot\left[\mu_1 v\nabla\left(\frac{\rho}{\rho_1}\right)\right]+\boldsymbol{J}\times\boldsymbol{B}|_y \qquad [2.7]$$

$$\frac{\partial}{\partial t}(\rho w)+\nabla\cdot(\rho \boldsymbol{V} w)=\rho g+\nabla\cdot\left(\mu_1\frac{\rho}{\rho_1}\nabla w\right)-\frac{\partial p}{\partial z}-\frac{\mu_1}{K}\frac{\rho}{\rho_1}(w-w_s)$$
$$-\frac{C\rho^2}{K^{1/2}\rho_1}|w-w_s|(w-w_s)-\nabla\cdot(\rho f_s f_1 \boldsymbol{V}_r w_r)$$
$$+\nabla\cdot\left[\mu_1 w\nabla\left(\frac{\rho}{\rho_1}\right)\right]+\rho g[\beta_T(T-T_0)+\beta_s(f_1^\alpha-f_{1,0}^\alpha)]\boldsymbol{J}\times\boldsymbol{B}|_z \qquad [2.8]$$

the equation for the conservation of energy is

$$\frac{\partial}{\partial t}(\rho h)+\nabla\cdot(\rho \boldsymbol{V} h)=\nabla\cdot\left(\frac{k}{c_s}\nabla h\right)+\nabla\cdot\left(\frac{k}{c_s}\nabla(h_s-h)\right)$$
$$-\nabla\cdot[\rho(\boldsymbol{V}-\boldsymbol{V}_s)(h_1-h)] \qquad [2.9]$$

and the equation for the conservation of species is

$$\frac{\partial}{\partial t}(\rho f^\alpha)+\nabla\cdot(\rho \boldsymbol{V} f^\alpha)=\nabla\cdot(\rho D\nabla f^\alpha)+\nabla\cdot[\rho D\nabla(f_1^\alpha-f^\alpha)]$$
$$-\nabla\cdot[\rho(\boldsymbol{V}-\boldsymbol{V}_s)(f_1^\alpha-f^\alpha)] \qquad [2.10]$$

where u, v and w are the velocities in the x, y and z directions respectively and $\boldsymbol{V}_r$ ($= V_1 - V_s$) is the relative velocity vector between the liquid phase and the solid phase. The subscripts s and l refer to the solid and liquid phases respectively; p is the pressure; μ is the viscosity; f is the mass fraction; K is the permeability, which is a measure of the ease with which fluids pass through the porous mushy zone; C is the inertial coefficient; β_T is the thermal expansion coefficient; g is the gravitational acceleration; T is the temperature; the subscript 0 represents the initial condition; $\boldsymbol{J}$ is the electric current density vector; $\boldsymbol{B}$ is the magnetic flux vector; h is the enthalpy; k is the thermal conductivity; c is the specific heat.

The third and fourth terms on the right-hand side of Eqns [2.6], [2.7] and [2.8] represent the first and second drag forces respectively for the flow in the mushy zone. The fifth term represents an interaction between the solid and liquid phases. These three terms are zero except in the mush zone, in which case neither f_s nor f_l is zero. The sixth term represents the effect of shrinkage and is identical with zero when the density difference between the solid and liquid phases is not considered. The second-to-last term on the right-hand side of Eqn [2.8] is the buoyancy force term which is based on the Boussinesq approximation for natural convection caused by both thermal and solutal convection. The last terms of Eqns [2.6], [2.7] and [2.8] are electromagnetic force (Lorentz force) terms. The first two terms on the right-hand side of Eqn [2.9] represent the net Fourier diffusion flux. The last represents the energy flux associated with the relative phase motion. The electromagnetic force is assumed to be independent of the properties of fluid flow in the weld pool, and the x, y and z components are calculated first, as discussed next, before the velocity is calculated.

In Eqns [2.5]–[2.10], the continuum density, specific heat, thermal conductivity, mass diffusivity, solid mass fraction, liquid mass fraction, velocity, enthalpy and mass fraction of constitute are defined as follows:

$$\begin{aligned} &\rho = g_s\rho_s + g_l\rho_l, \quad c = f_s c_s + f_l c_l, \quad k = g_s k_s + g_l k_l \\ &D = f_s D_s + f_l D_l, \quad f_s = \frac{g_s\rho_s}{\rho}, \quad f_l = \frac{g_l\rho_l}{\rho} \\ &\boldsymbol{V} = f_s\boldsymbol{V}_s + f_l\boldsymbol{V}_l, \quad h = h_s f_s + h_l f_l, \quad f^\alpha = f_s f_s^\alpha + f_l f_l^\alpha \end{aligned} \qquad [2.11]$$

where g_s and g_l are the volume fractions of the solid and liquid phases respectively. If the phase specific heats are assumed constant, the phase enthalpies for the solid and the liquid can be expressed as

$$h_s = c_s T, \quad h_l = c_l T + (c_s - c_l)T_s + H \qquad [2.12]$$

where H is the latent heat of fusion of the alloy.

The permeability function analogous to fluid flow in porous media is assumed employing the Carman–Kozeny equation[27,28]:

$$K = \frac{g_l^3}{c_l(1-g_l)^2}, \quad c_l = \frac{180}{d^2} \qquad [2.13]$$

where d is proportional to the dendrite dimension, which is assumed to be a constant and is on the order of 10^{-2} cm The inertial coefficient C can be calculated as follows[29]:

$$C = 0.13 g_l^{-3/2} \quad [2.14]$$

Tracking of solid–liquid interface

The solid–liquid-phase change is handled by the continuum formulation[24]. The third, fourth and fifth terms on the right-hand side of Eqns [2.6] and [2.7], and similar terms in Eqn [2.8], vanish at the solid region because $u = u_s = v = v_s = w = w_s = 0$ and $f_l = 0$ for the solid phase. For the liquid region, since K goes to infinity because $g_l = 1$ in Eqn [2.13] and $f_s = 0$, all these terms also vanish. These terms are only valid in the mushy zone where $0 < f_l < 1$ and $0 < f_s < 1$. Therefore, the liquid region, mushy zone and solid region can be handled by the same equations. Also, in GMAW, as the arc heat flux is rather concentrated and solidification time is very short (compared with casting), it is expected that the mushy zone in the base metal is very small, and the solid-phase velocity is assumed to be zero in the mushy zone. During the fusion and solidification process, latent heat is absorbed or released in the mushy zone. By using enthalpy, conduction in the solid region, and conduction and convection in the liquid region and mushy zone, the absorption and release of latent heat are all handled by the same equation, namely Eqn [2.9].

The algorithm of the VOF is used to track the dynamic geometry of the free surface[23]. The fluid configuration is defined by a VOF function $F(x, y, z, t)$, which tracks the location of the free surface. This function represents the VOF per unit volume and satisfies the following conservation equation:

$$\frac{\mathrm{d}F}{\mathrm{d}t} = \frac{\partial F}{\partial t} + (\boldsymbol{V} \cdot \nabla) F = 0 \quad [2.15]$$

When averaged over the cells of a computing mesh, the average value of F in a cell is equal to the fractional volume of the cell occupied by the fluid. A unit value of F corresponds to a cell full of fluid, whereas a zero value indicates a cell containing no fluid. Cells with F values between zero and one are partially filled with fluid and identified as surface cells.

Boundary conditions

The boundary conditions for the solution of previous equations (Eqns [2.5]–[2.10]) are divided as follows.

1. Normal to the free surface boundary.
2. Tangential to the free surface boundary.
3. Top surface boundary.
4. Symmetrical boundary.
5. Other boundaries including a convective boundary.

The last two kinds of boundary condition are simple and will not be discussed here. Wang[26] has given details of these. The first three types of boundary conditions are complicated and are discussed in detail below.

1. *Normal to the local free surface.* For the free surface, i.e. the liquid metal–arc plasma interface, the following pressure conditions must be satisfied[23]:

$$p = p_v + \gamma\kappa \qquad [2.16]$$

where p is the pressure at the free surface in a direction normal to the local free surface and p_v is the vapor pressure or any other applied external pressure acting on the free surface, which is the plasma arc pressure in this study. The plasma arc pressure is assumed to have a radial distribution in the following form[23]:

$$p_v = P_{max} \exp\left(-\frac{r^2}{2\sigma_p^2}\right) \qquad [2.17]$$

where P_{max} is the maximum arc pressure at the arc center, r is the distance from the arc center ($x = x_a$, $y = 0$), and σ_p is the arc pressure distribution parameter. In Eqn [2.16], κ is the free-surface curvature given by[24]

$$\kappa = -\left(\nabla \cdot \frac{\boldsymbol{n}}{|\boldsymbol{n}|}\right) = \frac{1}{|\boldsymbol{n}|}\left[\left(\frac{\boldsymbol{n}}{|\boldsymbol{n}|} \cdot \nabla\right)|\boldsymbol{n}| - (\nabla \cdot \boldsymbol{n})\right] \qquad [2.18]$$

where $\boldsymbol{n}$ is a normal vector to the local surface, which is the gradient of the VOF function

$$\boldsymbol{n} = \nabla F \qquad [2.19]$$

2. *Tangential to the local free surface.* The temperature- and sulfur-concentration-dependent Marangoni shear stress at the free surface in a direction tangential to the local free surface is given by

$$\tau_s = \mu_1 \frac{\partial(\nabla \cdot \boldsymbol{s})}{\partial \boldsymbol{n}} = \frac{\partial \gamma}{\partial T}\frac{\partial T}{\partial \boldsymbol{s}} + \frac{\partial \gamma}{\partial f^\alpha}\frac{\partial f^\alpha}{\partial \boldsymbol{s}} \qquad [2.20]$$

where $\boldsymbol{s}$ is a tangential vector to the local surface. The surface tension γ for a pseudobinary Fe–S system as a function of temperature T and the sulfur concentration f^α is given by[30]

$$\gamma = 1.943 - 4.3 \times 10^{-4}(T - 1723) - RT \times 1.3 \times 10^{-8} \ln\left[1 + 0.00318 f^\alpha \exp\left(\frac{1.66 \times 10^8}{RT}\right)\right] \qquad [2.21]$$

where R is the gas constant. Figure 2.6 shows the surface tension and its gradients as functions of temperature and sulfur concentration.

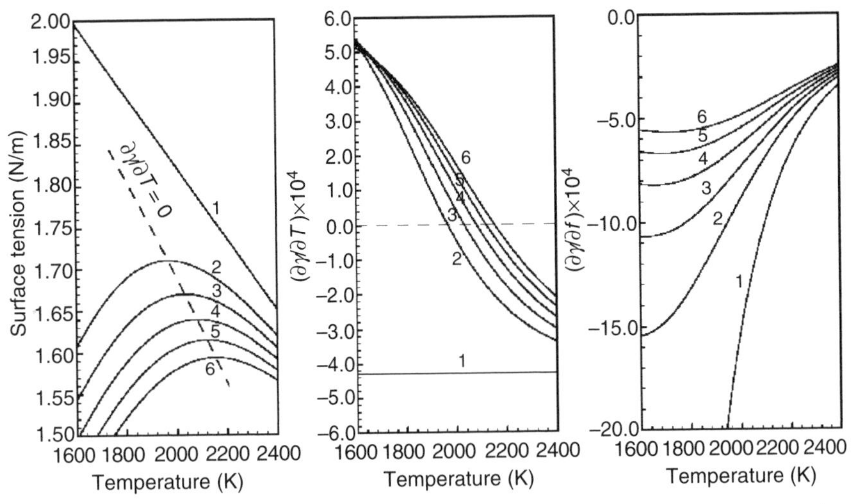

2.6 Surface tension and its gradients as functions of temperature and sulfur concentration for the pseudo-binary Fe-S system: curves 1, 0 ppm sulfur; curves 2, 100 ppm sulfur; curves 3, 150 ppm sulfur; curves 4, 200 ppm sulfur; curves 5, 250 ppm sulfur; curves 6, 300 ppm sulfur.

3. *Top surface.* At the moving arc center, in addition to droplet impingement, the arc heat flux is also impacting on the base metal. As the arc heat flux is relatively concentrated, it is assumed that the heat flux is perpendicular to the base metal (i.e. neglecting the inclination nature of current and heat flux). Hence, the temperature and concentration boundary conditions at the top surface of the base metal are

$$k\frac{\partial T}{\partial z} = \frac{\eta(1-\eta_{\mathrm{d}})Iu_{\mathrm{w}}}{2\pi\sigma_q^2}\exp\left(-\frac{r^2}{2\sigma_q^2}\right) - q_{\mathrm{conv}} - q_{\mathrm{radi}} - q_{\mathrm{evap}} \qquad [2.22]$$

$$\frac{\partial f^{\alpha}}{\partial z} = 0 \qquad [2.23]$$

where I is the welding current, η is the arc thermal efficiency, η_{d} is the ratio of droplet thermal energy to the total arc energy, u_{w} is the arc voltage and σ_q is the arc heat flux distribution parameter. The heat losses due to convection, radiation and evaporation can be written as

$$q_{\mathrm{conv}} = h_{\mathrm{c}}(T - T_{\infty}) \qquad [2.24]$$

$$q_{\mathrm{radi}} = \sigma\varepsilon(T^4 - T_{\infty}^4) \qquad [2.25]$$

$$q_{\mathrm{evap}} = WH_{\mathrm{v}} \qquad [2.26]$$

where H_{v} is the latent heat for liquid–vapor-phase change and W is the melt mass evaporation rate. For a metal such as steel, W can be written as[31]

$$\log(W) = A_v + \log P_{atm} - 0.5 \log T \tag{2.27}$$

$$\log P_{atm} = 6.121 - \frac{18836}{T} \tag{2.28}$$

Electromagnetic force

In each of Eqns [2.6]–[2.8], there is a term caused by the electromagnetic force that should be calculated first before the calculation of velocity. Assuming that the electric field is in a quasi-steady state and the electrical conductivity is constant, the scalar electric potential ϕ satisfies the following Maxwell equation in the local r–z coordinate system[10]:

$$\nabla^2\phi = \frac{1}{r}\frac{\partial}{\partial r}\left(r\frac{\partial \phi}{\partial r}\right) + \frac{\partial^2 \phi}{\partial z^2} = 0 \tag{2.29}$$

Assuming that the current is in the negative z direction, the required boundary conditions for the solution of Eqn [2.29] are:

$$-\sigma_e \frac{\partial \phi}{\partial z} = \frac{I}{2\pi\sigma_c^2}\exp\left(-\frac{r^2}{2\sigma_c^2}\right) \text{ at the top surface} \tag{2.30}$$

$$\frac{\partial \phi}{\partial z} = 0 \text{ at } z = 0 \tag{2.31}$$

$$\frac{\partial \phi}{\partial r} = 0 \text{ at } r = 0 \tag{2.32}$$

$$\phi = 0 \text{ at } r = 10\sigma_c \tag{2.33}$$

where σ_e is the electrical conductivity and σ_c is the arc current distribution parameter. After the distribution of electrical potential is solved, current density in the r and z directions can be calculated via

$$J_r = -\sigma_e \frac{\partial \phi}{\partial r} \tag{2.34}$$

$$J_z = -\sigma_e \frac{\partial \phi}{\partial z} \tag{2.35}$$

The self-induced azimuthal magnetic field is derived from Ampere's law through[10]

$$B_\theta = \frac{\mu_0}{r}\int_0^r J_z r \, dr \tag{2.36}$$

where μ_0 is the magnetic permeability. Finally, the three components of electromagnetic force in Eqns [2.6]–[2.8] are calculated via

$$\boldsymbol{J} \times \boldsymbol{B}\big|_x = -B_\theta J_z \frac{x - x_\mathrm{a}}{r} \qquad [2.37]$$

$$\boldsymbol{J} \times \boldsymbol{B}\big|_y = -B_\theta J_z \frac{y}{r} \qquad [2.38]$$

$$\boldsymbol{J} \times \boldsymbol{B}\big|_z = B_\theta J_r \qquad [2.39]$$

Driving forces for the weld pool convection

The fluid flow in the weld pool during the arc welding process is mainly driven by the buoyancy force, the electromagnetic force (Lorentz force), the surface-tension-gradient-driven force, the arc pressure and the stirring force resulting from the falling momentum of droplets.

1. The electromagnetic force is caused by the converging current field, along with the magnetic field that it induces. They can be found in the first from last term in Eqn [2.6], the first from last term in Eqn [2.7] and the second from last term in Eqn [2.8] respectively.
2. The buoyancy force results from three contributions. The first is caused by the density difference between the liquid metal and surrounding air. Since the density of air is much less than metal, this part can be regarded as a pure gravity force. In GTAW, the surface can be considered flat[32] when the arc current is below 200 A but, in GMAW, when droplets depress the free surface, the gravity can translate potential energy into kinetic energy and also translate kinetic into potential energy; then a wave is produced. The second is due to the temperature gradient in the weld pool. Since there is nearly no difference between the results considered with this force or not, it is concluded that this force is much smaller than the electromagnetic force and can be precluded from the analysis of GMAW just as in GTAW[7]. The third is caused by the difference of species concentration in the weld pool. It is also much smaller than the electromagnetic force and can be precluded from the analysis of GMAW.
3. The surface-tension-gradient-driven force is called the *Marangoni stress*. In welding, as shown in Eqn [2.16], this stress may arise as a result of variations in the concentration of the surface-active agents and the temperature gradient. It cannot be considered small when there is a great gradient in temperature and concentration. For GMAW, when droplets just reach the surface, this gradient in temperature and concentration is dramatically great.
4. The plasma arc pressure is effective near the arc center and cannot be ignored when the current is above 200 A. It depresses the weld pool near the arc center.

5. The stirring force resulting from the impinging process of the droplets is the most complicated of all the forces since it involves the mixing of mass, momentum, thermal energy and species between droplets and weld pool.

These five driving forces interact with each other during the welding process and they are included either in the governing equations or as boundary conditions in the mathematical model of fluid flow and heat transfer in the weld pool. The weld pool dynamics, penetration depth and heat transfer in the weld pool are strongly influenced by these forces, as shown in Fig. 2.7 and Fig. 2.8.

2.2.3 Metal transfer

The subject of this section is liquid-metal transfer in arc welding, namely the detachment of metal from the electrode, transfer of droplets across the arc, and flow in the weld pool. Consumable electrode arc welding processes are used extensively because filler metal is deposited more efficiently and at higher rates than is possible with other welding processes. To be most effective, the filler metal needs to be transferred from the electrode with small losses due to spatter. In GMAW, the heat transfer and fluid flow in the welding are strongly affected by the liquid-metal transfer process. In the following, the physics of liquid-metal transfer in arc welding are investigated by studying the droplet formation and detachment, the droplet impingement and the ripple formation processes using the aforementioned mathematical models.

Droplet generation and transfer

For electrode melting and droplets generation, two major models have been proposed; one is the static-force balance theory (SFBT)[33,34] and the other is the magnetic pinch instability theory (MPIT)[32,35]. The SFBT considers the balance between gravity, electromagnetic force, plasma drag force and surface tension acting on a pendant drop. The MPIT considers the perturbation of the surface of an infinite cylindrical liquid column due to a radially inward force of the magnetic pinch. Nemchinsky[36] developed a steady-state model to describe the equilibrium shape of a pendant droplet, accounting for surface tension and magnetic pinch force. Simpson and Zhu[37] developed a one-dimensional dynamic model to predict the formation of the droplet. Haidar[38] and Haidar and Lowke[39] developed a time-dependent two-dimensional model for the prediction of droplet formation including arc plasma. Fan and Kovacevic[40] developed a model that includes the droplet formation, detachment and impingement on the welding pool.

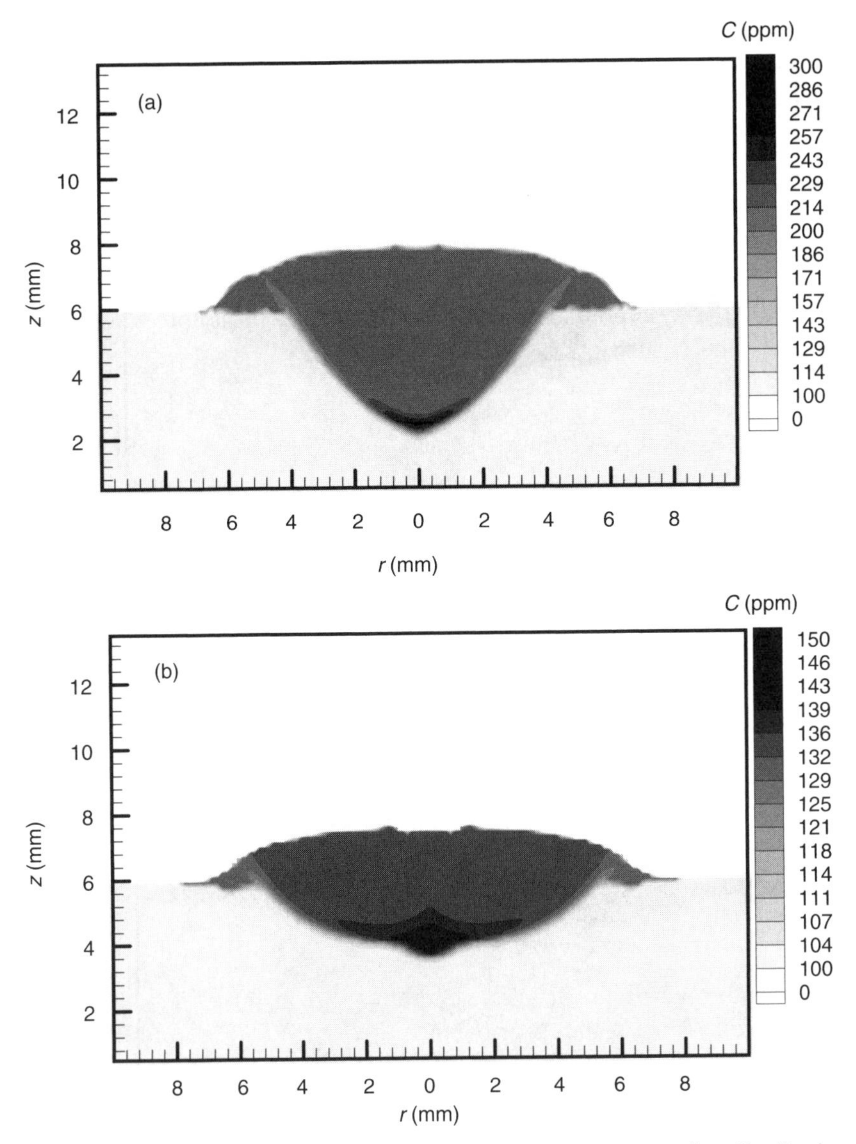

2.7 Final shape of the weld pool and sulfur concentration distribution when the droplet sulfur concentration is (a) 300 ppm and (b) 150 ppm (t = 6.000 s)[20].

However, a distribution of current density was assumed, and the arc plasma was not included in the model. Zhu[41] assumed that the pressure of the surface cells is equal to the pressure of the adjacent arc plasma cell since the surface tension has been converted to a body force:

$$p_{\text{liquid metal}} = p_{\text{arc}} \qquad [2.40]$$

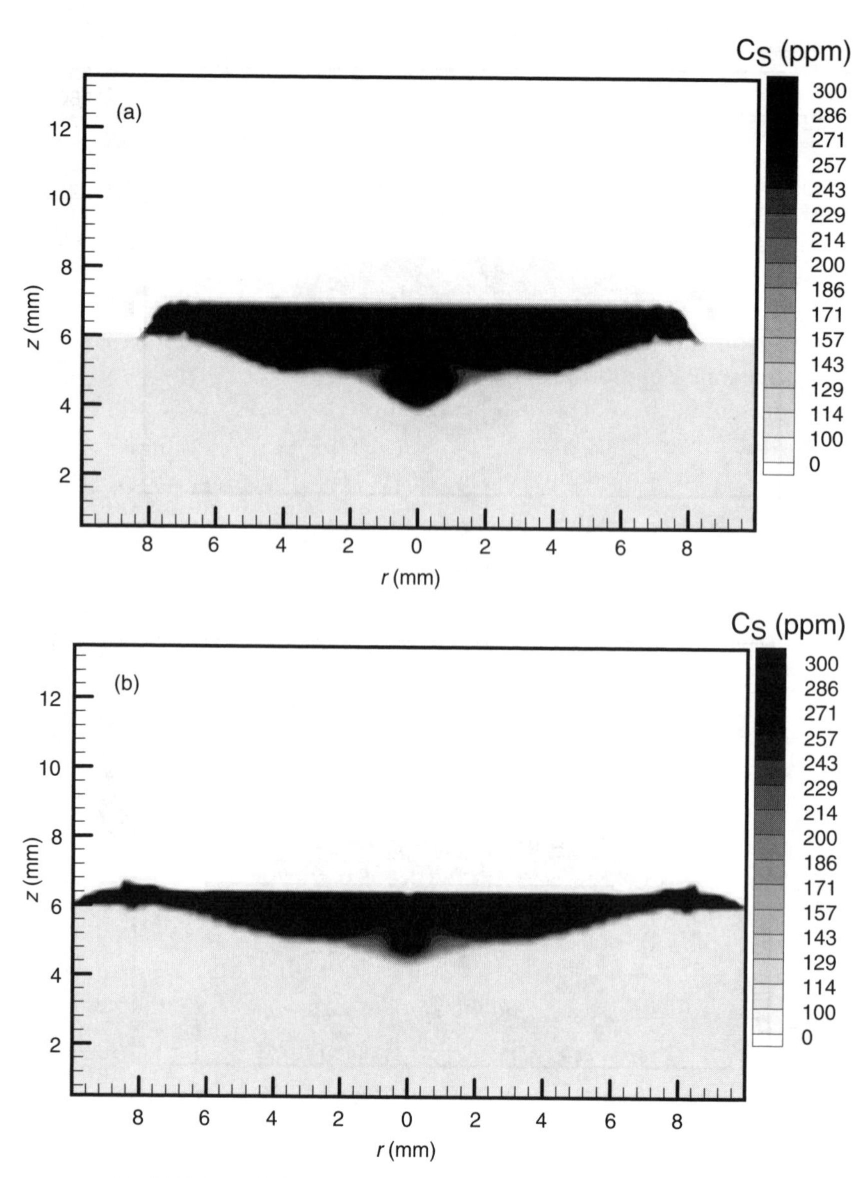

2.8 Weld bead shape and sulfur concentration distribution (a) when surface tension is neglected and (b) when both surface tension force and electromagnetic force are neglected ($t = 6.000$ s)[20].

The friction drag force on the surface is given as follows[42]:

$$\tau_0 = \mu\left(\frac{\partial V_r}{\partial n}\right)_s \quad [2.41]$$

where μ is the viscosity of arc plasma, V_r is the relative velocity between plasma and liquid metal, n is the normal direction and the subscript s means at the surface.

At the plasma–electrode interface, there exists an anode sheath region[43]. In this region, the mixture of plasma and metal vapor may depart from local thermodynamic equilibrium (LTE); thus it no longer complies with the model presented above. The thickness of this region is very thin (about 0.02 mm)[44]. According to Lowke *et al.*[44], the sheath does not have a major influence on the temperature of the arc column, but it will have a major effect on the temperature of the electrode. Since the sheath region is very thin, it is treated as a special interface and takes into account the thermal effects on the electrode. For the surface of the anode, the energy balance equation [2.9] is modified to include an additional source term S_a as the following[40, 42, 45]:

$$S_a = \frac{k_{eff}(T_{arc} - T_{anode})}{\delta} + j_a\phi_w + j_a\frac{5K_BT_e}{2e} - \varepsilon K_BT^4 + F_R - q_{ev}L \quad [2.42]$$

The first term on the right-hand side of Eqn [2.42] is the contribution due to thermal conduction from the plasma to the anode. The second term represents the electron heating associated with the work function (ϕ_w) of the anode material; the magnitude of this term is in the order of $10^8\,W/m^2$. The third term accounts for thermal energy carried by the electrons into the anode; this term is also in the order of $10^8\,W/m^2$. The fourth term is black-body radiation loss from the anode surface; it is in the order of $10^5\,W/m^2$. The fifth term is the radiation heat from arc to electrode; it is in the order of $10^5\,W/m^2$. The final term is heat loss due to the evaporation of electrode materials. The symbol δ is the width of the anode sheath region; the maximum experimentally observed thickness of the anode fall region is 0.1 mm[46]; T_{arc} is the arc temperature; T_{anode} is the anode temperature; k_{eff} is the effective thermal conductivity at the anode–plasma interface; ε is the emissivity of the surface; K_B is the Stefan–Boltzmann constant; T_e is the electron temperature and, since LTE is assumed in the model, it is equal to the local temperature; J_a is the square root of j_r^2 and j_z^2; L is the latent heat of vaporization; q_{ev} is the mass rate of evaporation of metal vapor from the droplet. For metals such as steel, q_{ev} can be written as[31]

$$\log(q_{ev}) = A_v + \log P_{atm} - 0.5\log T \quad [2.43]$$

$$\log P_{atm} = 6.121 - \frac{18\,836}{T} \qquad [2.44]$$

Figure 2.9 shows the formation and detachment of the first droplet[41]. As shown in Fig. 2.9(a), with increasing time, the electrode is heated by Joule heating and the high-temperature arc plasma. When the temperature of metal reaches the liquidus temperature, the electrode melts and a droplet is formed on the electrode tip and detaches. Under the aggregate efforts of gravity, electromagnetic force, surface tension, viscous drag force and arc pressure, necking starts to form at approximately $t = 19.0$ ms when the total melting volume reaches a certain value. As the volume of melting metal increases, at time $t = 24.0$ ms, the detaching forces (gravity, electromagnetic force and arc drag force) surpass the surface tension, leading to the detachment of the first droplet. Figure 2.9(b) shows the corresponding velocity distributions for liquid metal. It is seen that the droplet velocity is less than 0.4 m/s. By controlling the 'current wave', Zhu[41] has demonstrated that 'one droplet per pulse' can be achieved. This can significantly reduce the spatter phenomena.

Droplet impingement and weld pool dynamics

After the droplet is detached from the electrode, the droplet will fall across the arc plasma and impinge on the weld pool. As soon as the droplet touches the weld pool, the mass, momentum and thermal energy carried by the droplet will mix with and merge into the base metal; very complicated heat and mass transfer phenomena will occur and weld pool dynamics and the solidification process can be greatly affected by this impinging process[19]. The calculation of its velocity, temperature and concentration distributions follows the formulas in Eqns [2.6]–[2.10].

Figure 2.10 shows a sequence of a typical periodic droplet impinging process onto the weld pool[26]. As shown, at $t = 2.938$ s, resulting from the previous droplet, the fluid flows outward in two directions at about $x = 33$ mm, one along the top of the base metal in the welding direction and the other toward the left-hand side at a much higher speed. The left-hand fluid is downward and along the inclined solid-phase base metal, and it then flows upward after reaching the bottom of the weld pool. This creates a clockwise vortex between $x = 28$ mm and $x = 32$ mm. As the former droplet carries higher thermal energy, a skewed hot spot is formed in the weld pool near the arc center. It is noted that the flow of high-temperature fluid downward and toward the left-hand side has a 'cutting' effect to melt the base metal so that the weld pool 'moves' to the right-hand side. When this 'hot fluid' turns left near the weld pool surface and flows to the left-hand side (between $x = 25.0$ mm and $x = 30.0$ mm), it collides with another fluid from

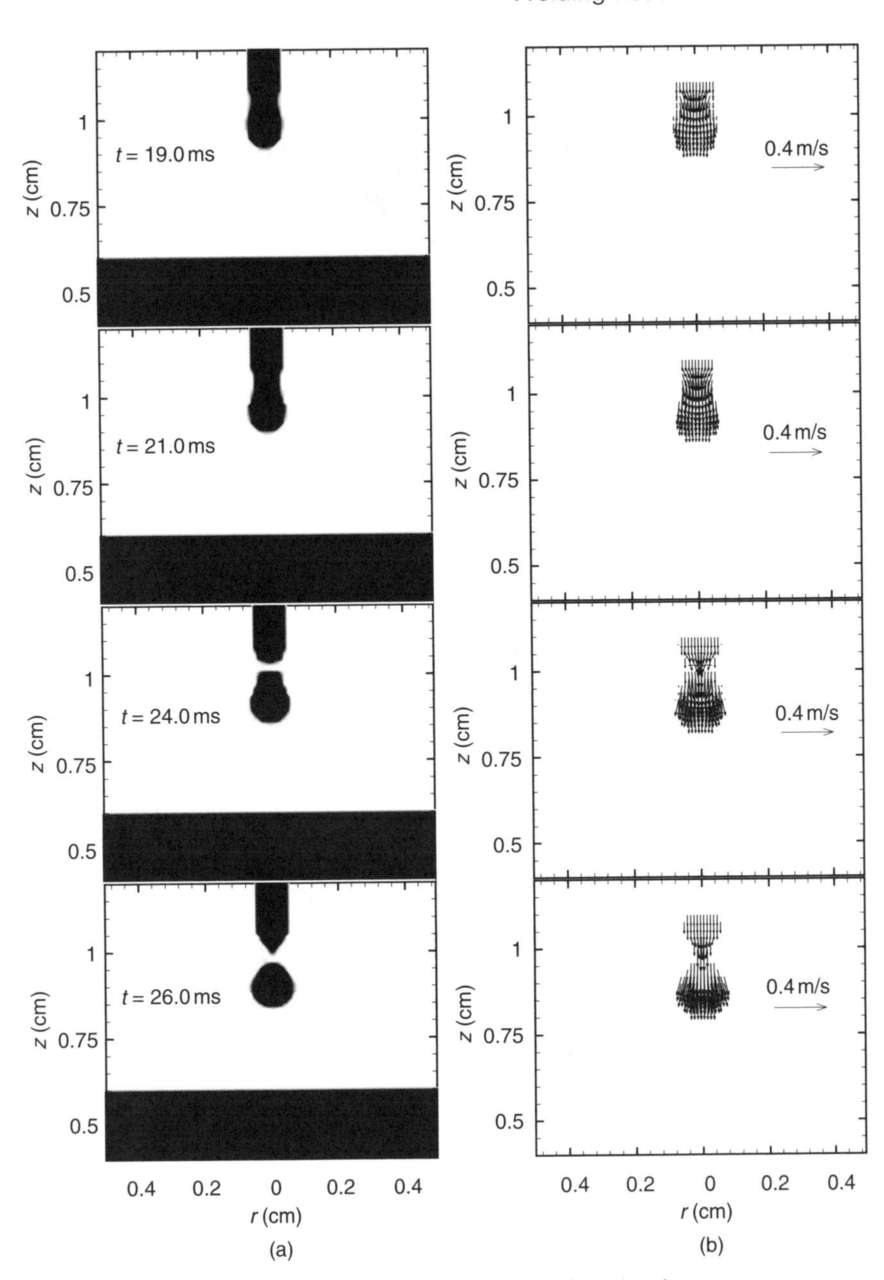

2.9 The formation and detachment of the first droplet.

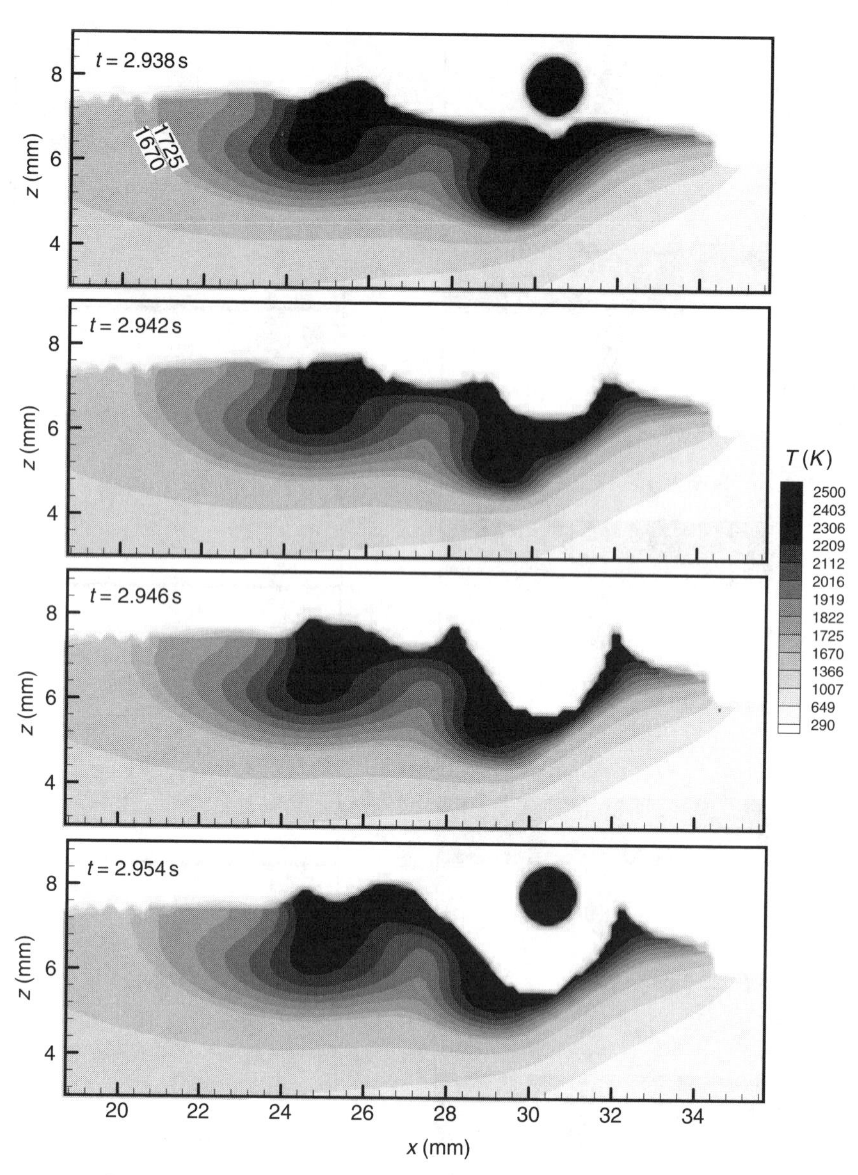

2.10 A typical sequence showing the impinging process, weld pool dynamics and temperature distributions.

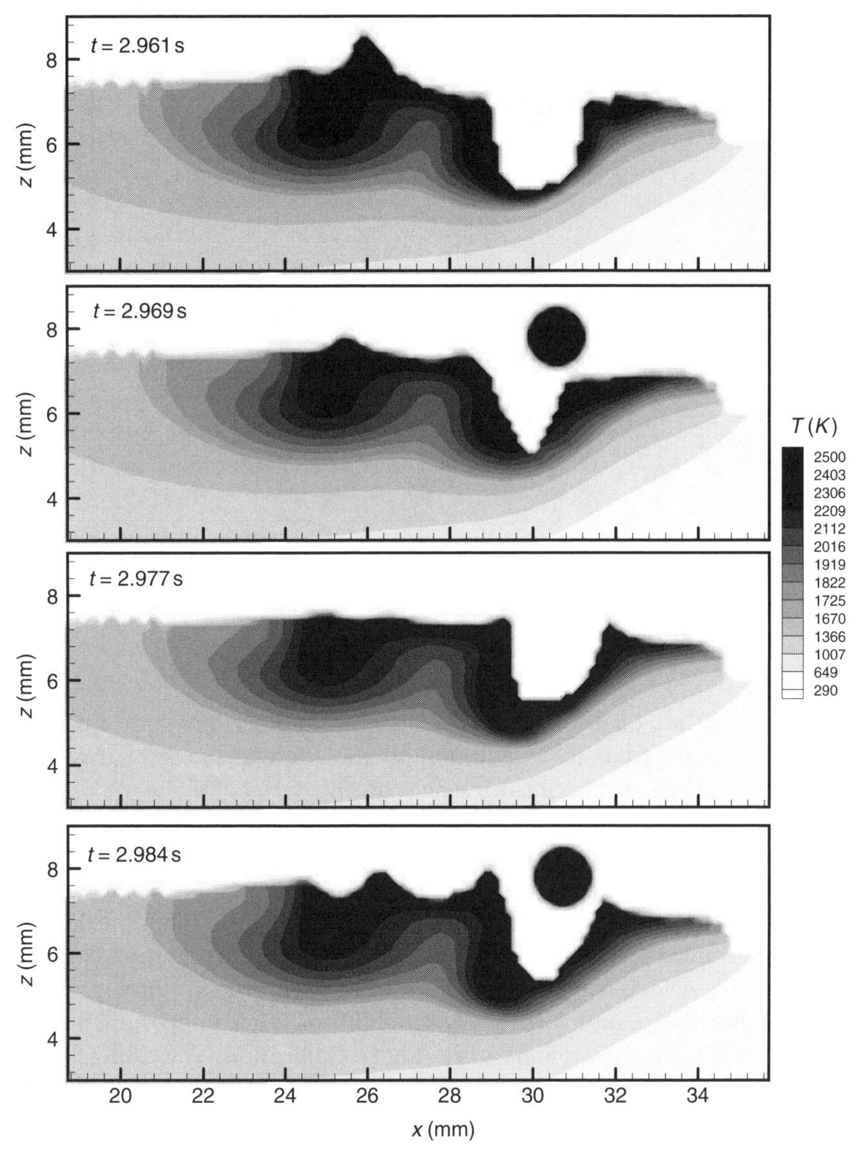

2.10 Continued

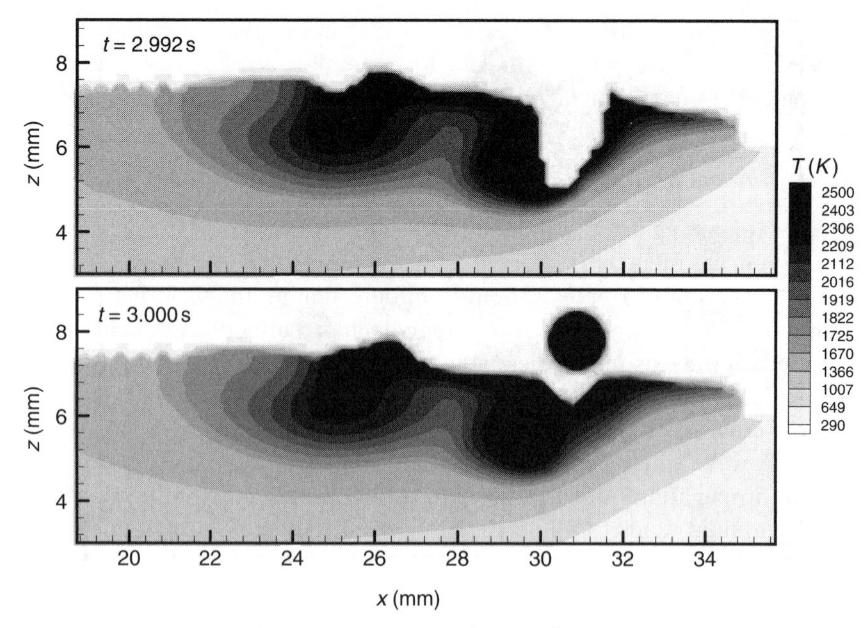

2.10 Continued

the left-hand side of the weld pool, leading to another 'hot spot' near the surface. It is noted that, in addition to the thermal energy carried by droplets, thermal arc flux also impacts on the surface of the weld pool, resulting in a relatively high surface temperature near the arc center. The fluid collision creates a 'bump' with the highest liquid level at about x = 25 mm. These two hot spots form the shape of a 'skewed W', which moves as the welding proceeds in the right-hand direction. The shape of the liquidus line implies that at a location along the horizontal direction (say, z = 5.5 mm) the weld is subjected to a cycle of high temperature (possibly melted), low temperature (possibly resolidified), high temperature (possibly remelted) and low temperature (final solidification) as welding moves in the right-hand direction. This heat–cool phenomenon can be important if one wants to consider weld microstructure evolution and residue stress in the weld.

At t = 2.942 s, the droplet has already plunged into the weld pool at the arc center. The just flattened surface is deformed by the impinging momentum, while the temperature near the arc center is higher than that of its surroundings. A 'crater' is formed. The impinging momentum destroys the existing vortex under the arc center and enhances the outer flow at the surface near the arc center (between x = 25.0 mm and x = 30.0 mm). When the 'crater' becomes deeper and wider, the surface fluid velocity near the left-hand side of the crater increases, pushing the liquid level there upward.

As a result, the kinetic energy of the droplet has converted into potential energy of the weld pool fluid. The formation of a crater facilitates the cutting effect on base metal, influencing weld penetration and weld mixing.

Ripple formation

Ripples remaining in the solidified weld bead, as shown in Fig. 2.11, are very common in gas metal arc welding[26]. Understanding the mechanisms leading to the formation of ripples is helpful in determining the weld bead shape and understanding undercut formation which can affect weld quality.

Based on the existing studies[26], the mechanisms leading to the formation of a ripple can be summarized as follows. First, owing to droplet impingement, a crater is created, pushing the fluid upward and away from the arc center. A wave with its peak height greater than the liquid level is also generated, propagating outward. Second, owing to the hydrostatic force, the high-level fluid tends to fill up the crater and to decrease the liquid level. Third, as welding proceeds in the right-hand direction, solidification also moves in the same direction. As high-level fluid near the tail edge is solidified before moving down and flowing back in the crater direction, a ripple is created. Hence, the formation of ripples is related to the opening and closing of the crater and the resulting up-and-down nature of the weld pool fluid level. The time required for the up-and-down cycle for fluid level near

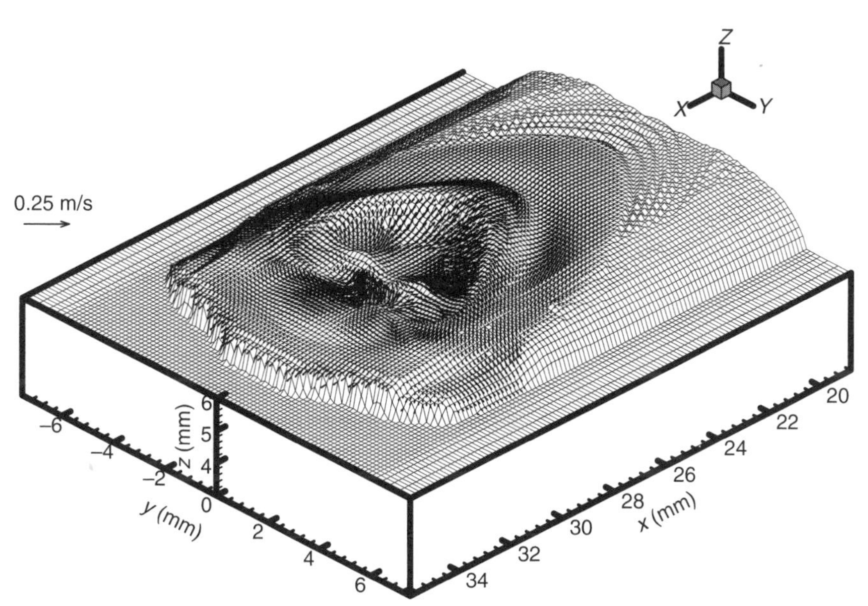

2.11 Partial view of the three-dimensional mesh system, weld bead shape and velocity distributions.

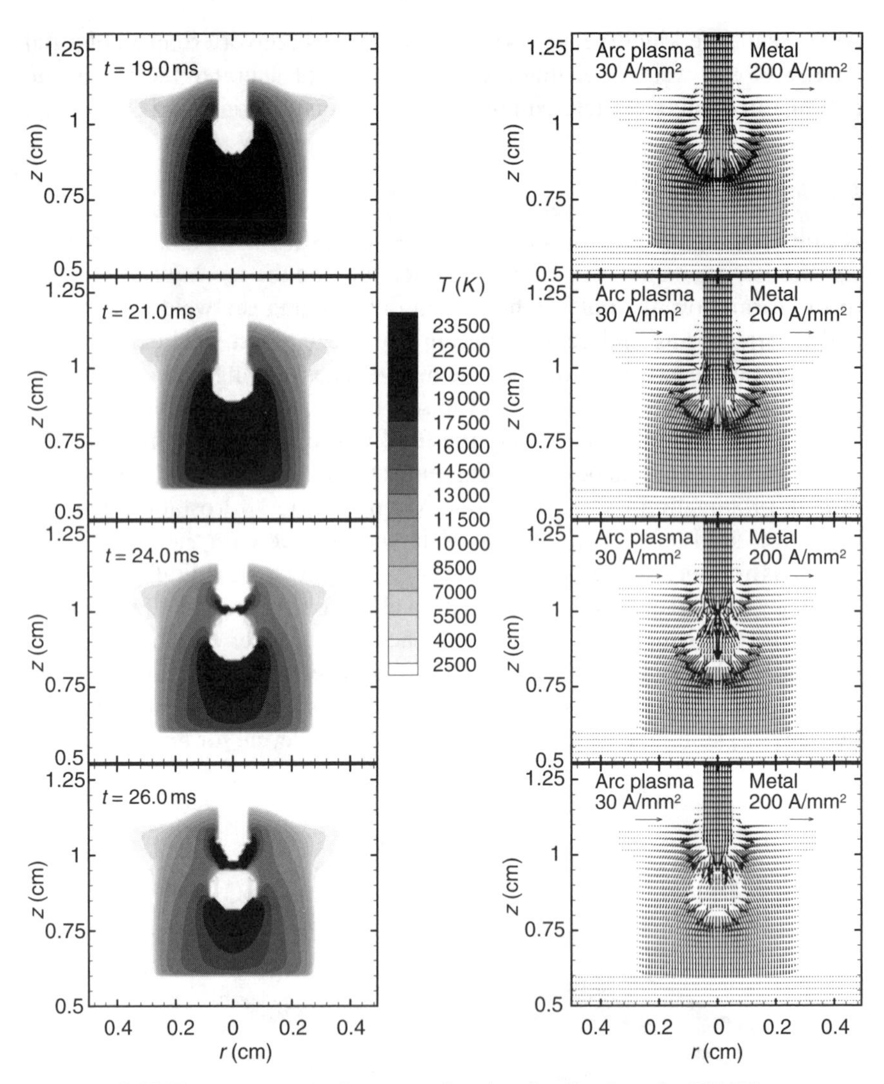

2.12 Temperature and current density distributions in GMAW.

the tail edge and the 'pitch' of the ripples depends on many welding parameters, including droplet size (electrode diameter), droplet momentum, drop frequency (wire feed speed), welding speed, welding power (current and voltage) and others. It is noted that all these welding parameters and conditions are coupled together.

2.2.4 Arc plasma

In the GMAW process, the arc plasma can affect the droplet formation and detachment and provide additional heat and pressure to the weld pool, which will affect the heat transfer and the final weld bead shape. Hence, it is necessary to study the characteristics of the arc plasma. Modeling the heat transfer and fluid flow in arc plasma during an arc welding process has been well documented. Choo *et al.*[11] dealt with the arc plasma between a tungsten electrode (cathode) and a water-cooled copper plate (anode). Waszink and Graat[33] studied the arc characteristics in GMAW of aluminum using argon as the shielding gas. Haidar[38] developed a unified model for arc plasma and electrode in GMAW.

Figure 2.12 shows the arc temperature and current distributions[41]. As shown, an arc plasma mantle covers the tip of the electrode, the droplet and the top surface of the weld pool. The maximum temperature occurs just under the tip of the electrode before the droplet detaches from the electrode. After detachment, the droplet is stuck between the electrode and the workpiece, distorting the arc plasma. Before droplet detachment, the maximum current density is just under the electrode tip. Because the major heating source of arc plasma is the Ohmic heat by the arc current, a high-temperature region corresponds to a high-current-density region. After droplet detachment, the droplet is stuck in the pathway of the arc current, distorting the distribution of current density, and two high-current-density areas are formed corresponding to two high-temperature regions. Since the current density under the electrode tip is higher than that of the droplet, the temperature is higher under the electrode tip. As the current flows out of the electrode tip, the distribution is similar to a fan shape, and the current converges to the base metal on a circle of 3.6 mm diameter. When the droplet is stuck between the electrode and base metal, almost half of the current flows out of the electrode tip, converges to the droplet and flows out from the droplet in another fan shape.

2.3 Laser welding

Laser welding has gained in popularity in recent years because of its high intensity, small heat-affected zone, higher welding speed and precision, etc. Laser welding represents a delicate balance between heating and cooling within a spatially localized volume overlapping two or more solids. Maintenance of the balance between heat input and heat output depends on the absorption of laser radiation and dissipation of heat inside the workpiece. The absorption by laser radiation toward the workpiece is often interrupted because of the evolution of hot gas from the laser focus. Under certain conditions this hot gas may turn into plasma that can severely attenuate the laser beam owing to absorption and scattering. The complexity of the interactions involved in this process and the subsequent behavior of liquids generated in a solid by incident laser radiation preclude a completely rigorous simulation of laser welding. However, heat and mass transfer calculations are useful in quantifying the laser welding process and in suggesting how changes in process parameters may influence weld properties.

The two fundamental modes of laser welding are, first, conduction welding and, second, keyhole or penetration welding[47]. The basic difference between these two modes is that the surface of the weld pool remains unbroken during conduction welding and opens up to allow the laser beam to enter the melt pool in keyhole welding. The schematic sketches of these two welding modes are shown in Fig. 2.13. The transition from the conduction mode to a keyhole mode depends on the peak laser intensity. Normally, when the laser intensity is smaller than 10^6 W/cm^2, a conduction mode welding is formed. When the laser intensity is greater than 10^6 W/cm^2, a keyhole mode welding is started.

2.3.1 Conduction-mode welding

Conduction-mode welding offers less perturbation to the system because laser radiation does not penetrate into the material being welded; as a result, conduction welds are less susceptible to gas entrapment during welding. The energy equation for solving the temperature distributions in a thick workpiece during bead-on-plate conduction-mode welding is given in Eqn [2.1], which is due to Rosenthal. Equations [2.3] and [2.4] give the temperature distributions in the workpiece during arc welding assuming a point source. However, in laser welding, the shape of the temperature field should be more accurate if the point heat source was replaced by a diffuse source. A solution to this, using a diffuse (Gaussian) heat source, has been given by Ashby and Easterling[48]. The diffuse source solution was actually developed for the case of a laser beam traversing the surface of a sample, as illustrated in Fig. 2.14. With reference to the figure, consider a laser beam of power q and radius r_B tracking in the x direction with a velocity v. A point (y, z) below the surface of the workpiece is subject to a thermal cycle $T(y, z, t)$, where t refers to time. The temperature field $T(z, t)$ at a point below the center of the tracking beam (y = 0) is given by[2]

$$T(z,t) = T_0 + \frac{Aq/v}{2\pi\lambda[t(t+t_0)]^{1/2}} \exp\left(-\frac{(z+z_0)^2}{4at}\right) \qquad [2.45]$$

where λ is the thermal conductivity and a is the thermal diffusivity. The term A refers in laser treatments to the efficiency of absorption by the sample of the energy of a laser (light) beam. For steels, A is usually about 0.7, depending on the coating material used. In welding, it can be assumed that A = 1. The constant t_0 measures the time for heat to diffuse over a distance equal to the beam radius as given by

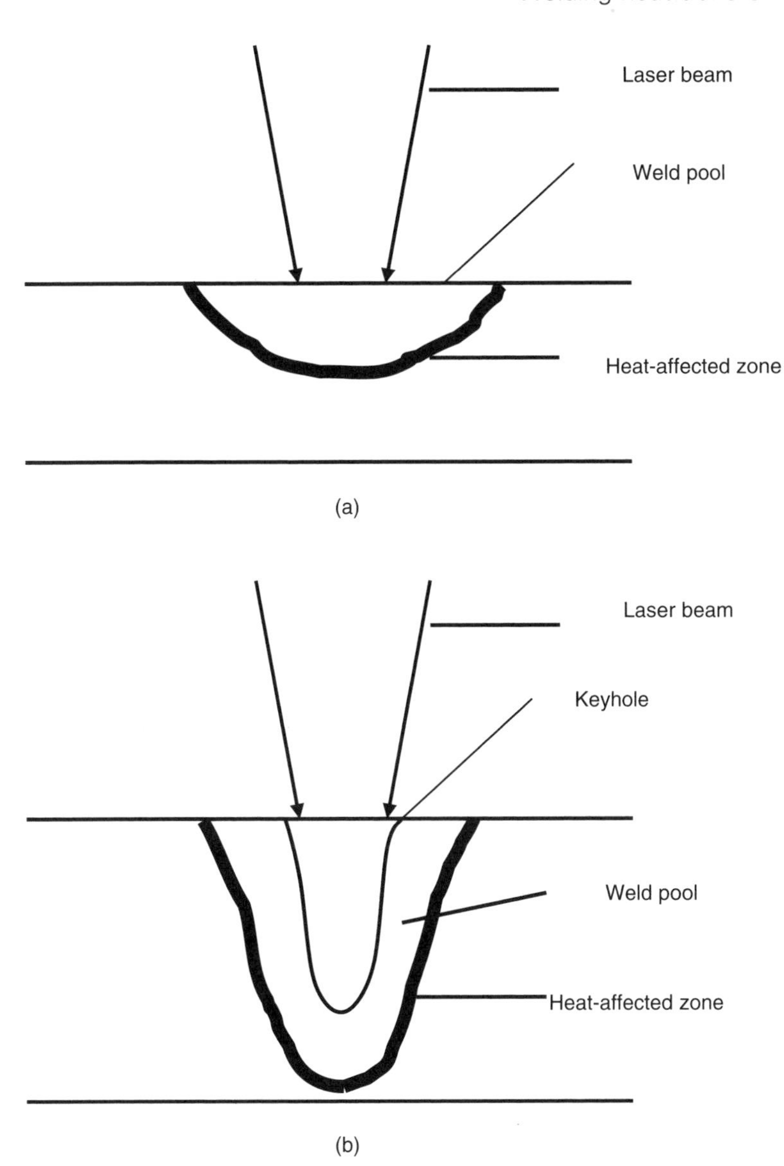

2.13 Schematic sketch of (a) conduction-mode and (b) keyhole-mode welding.

$$t_0 = \frac{r_B^2}{4a} \qquad [2.46]$$

The length z_0 measures the distance over which heat can diffuse during the beam interaction time r_B/v and it should be pointed out that this parameter is a mathematical 'adjustment' factor and as such should not strictly

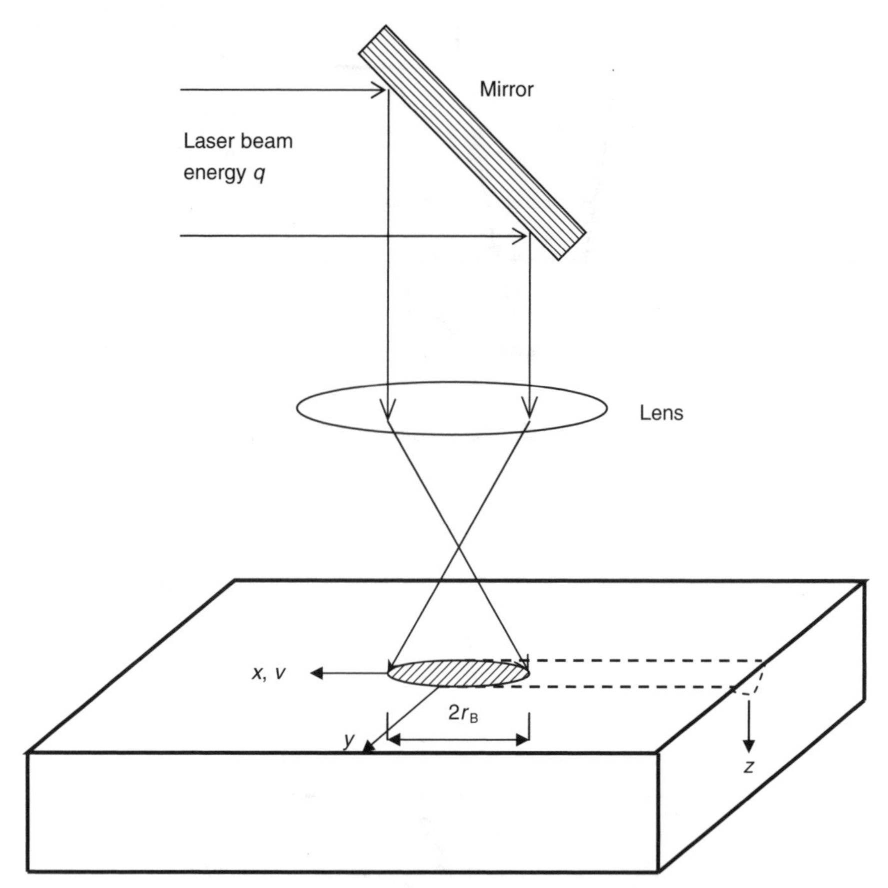

2.14 Welding configuration in terms of a diffuse (Gaussian) heat source of energy q and a constant velocity v.

be assigned any physical meaning. More details have been given by Rosenthal[2].

2.3.2 Keyhole-mode welding

Laser welding is in general keyhole-mode welding. Figure 2.15 shows a schematic sketch of keyhole laser welding. The formation of a keyhole increases the coupling of laser energy into the workpiece resulting in a weld with high depth-to-width ratio and a narrow heat-affected zone compared with conventional arc welding and conduction-mode laser welding. As the weld quality depends on the heat transfer and molten metal flow during the formation and collapse of the keyhole, it is necessary to understand the detailed transport phenomena associated with the keyhole formation and collapse.

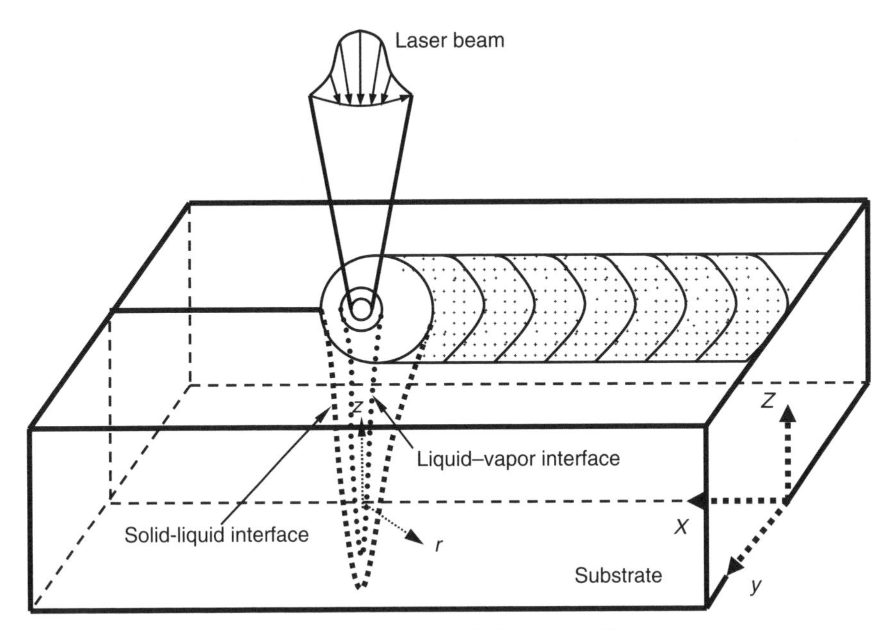

2.15 A schematic sketch of keyhole laser welding.

Over the years, a number of researchers have developed models to simulate keyhole-mode welding. Semak *et al.*[49] divided those models into two categories; one is based on the vaporization approach and the other is based on the recoil pressure approach. The vaporization approach can be further divided into two types. In the first type[50–52], it is assumed that there is a balance between the sum of recoil pressure and vapor pressure in the keyhole and the sum of surface tension and hydrostatic pressure. The first two keep the keyhole open, while the last two tend to close the keyhole. This type of model fails to describe the process of keyhole formation and predicts lower values of melt velocity than experimental results do[53]. The second type[54] assumes that the keyhole is stabilized and held open when the vaporization rate is equal to the mass flow rate of molten metal coming into the keyhole due to hydrostatic pressure. The predicted behavior for a longer time process using this type of model does not coincide with experimental data of keyhole dynamics[53].

The model based on the recoil pressure approach assumes that the recoil pressure acts as a driving force for melt motion and keyhole formation. The recoil pressure, acting toward the metal, is a kind of reaction force as opposed to the action force caused by the sudden 'burst' of vapor from the metal when a laser beam impinges onto it. Also, the laser-induced plasma is considered to be an important factor affecting keyhole behavior and is a limiting factor for penetration. Klemens[55] proposed a model in which the motion of the melt was a significant factor for keyhole support. Allmen and Blatter[56] proposed that a Knudsen layer was formed above the molten region and extended beyond the keyhole surface for a few mean free paths. Solana *et al.*[57] further developed a mathematical model for the ablation process. Clucas *et al.*[58] developed a mathematical model for the keyhole welding process by suggesting a pressure and energy balance at the keyhole walls. Ducharme *et al.*[59] developed an integrated keyhole and weld pool model for the welding of thin metal sheets. Sudnik *et al.*[60] further analyzed the three driving forces of melt flow, including the force resulting from temperature-dependent surface tension, the friction force of metal vapor escaping from the capillary and the movement of capillary relative to the workpiece. Zhou[61] developed a comprehensive model of three-dimensional keyhole laser welding that includes the simulations of temperature field, pressure balance, melt flow, free surface, laser-induced plasma and multiple reflections.

Governing equations

In keyhole laser welding, since the energy transport phenomena in keyhole plasma is different from that in the workpiece, two sets of governing equations are discussed to handle the transport phenomena in plasma and workpiece respectively. For the workpiece (including solid and liquid metal), the differential equations governing the conservation of mass, momentum and energy in the laser welding process are in the same form as listed in Eqns [2.5]–[2.10]. In laser welding, the electromagnetic force terms in Eqns [2.6], [2.7] and [2.8] are neglected. As to the keyhole plasma, the governing equations and the associated boundary conditions to solve the equations are discussed in the following section.

Laser-induced plasma

Laser-induced plasma was presented as a result of ablation vapor from the keyhole wall, which was heated sufficiently to become partially ionized under the laser radiation. Kapadia *et al.*[62] thought the high vapor temperature required for plasma ignition was due to the small droplet radius and the presence of surface tension around the droplet. Farson and Kim[63] simulated the laser-induced evaporation and plume formation at the same time. The simulation included material melting and evaporation phenomena, a Knudsen layer and the formation of a partially ionized plume consisting of shielding gas and metal vapor. It was pointed out that a stable plume could be formed when the material surface irradiance was reduced. For the welding process at a lower flow rate of argon assistant gas, the inverse

Bremsstrahlung absorption in the plasma and the heating of the keyhole walls by the heat transfer from plasma lump could not be neglected[64].

The energy equation for calculating the temperature distributions in plasma can be written in the following form[65, 66]:

$$\frac{\partial}{\partial t}(\rho_v h_v) = \nabla\cdot\left(\frac{k_v}{c_v}\nabla h_v\right) + \nabla\cdot(-\boldsymbol{q}_r) + k_{pl} I_{laser}(1-\alpha_{iB,1}) + \sum_{m r=1}^{n} k_{pl} I_{laser}(1-\alpha_{iB,1})(1-\alpha_{Fr})(1-\alpha_{iB,mr}) \quad [2.47]$$

where h_v and ρ_v represent the enthalpy and density respectively of the plasma, k_v and c_v represent the thermal conductivity and specific heat of the plasma respectively, $\boldsymbol{q}_r$ stands for the radiation heat flux vector, k_{pl} is the plasma absorption coefficient due to inverse Bremsstrahlung absorption, I represents the incident laser intensity, $\alpha_{iB,1}$ is the absorption fraction in the plasma due to the original laser beam, $\alpha_{iB,mr}$ is the absorption fraction due to the m_r^{th} reflected laser beam and α_{Fr} is the Fresnel absorption coefficient on the keyhole wall.

When an intense laser beam interacts with metal vapor, a significant amount of the laser energy is absorbed by the ionized particles. The radiation, absorption and emission by the vapor plume may couple strongly with the plume hydrodynamics. This coupling, shown on the right-hand side of Eqn [2.47], is included through the plasma laser light absorption and radiation cooling terms. The radiation source term $\nabla\cdot(-\boldsymbol{q}_r)$ is defined via[66]

$$\nabla\cdot\boldsymbol{q}_r = k_a\left(4\pi I_b - \int_{4\pi} I\, d\Omega\right) \quad [2.48]$$

where k_a, I_b and Ω denote the Planck mean absorption coefficient, black-body emission intensity and solid angle respectively. For the laser-induced plasma inside the keyhole, the scattering effect is not significant compared with the absorbing and emitting effects. So it will not lead to large errors to assume the plasma is an absorbing-emitting medium. The radiation transport equation[66] has to be solved for the total directional radiative intensity I:

$$(\boldsymbol{s}\cdot\nabla)I(\boldsymbol{r},\boldsymbol{s}) = k_a[I_b - I(\boldsymbol{r},\boldsymbol{s})] \quad [2.49]$$

where $\boldsymbol{s}$ denotes a unit vector along the direction of the radiation intensity and $\boldsymbol{r}$ denotes the local position vector. The Planck mean absorption coefficient k_a is defined in the following equation[66]:

$$k_a = \left(\frac{128}{27}k\right)^{1/2}\left(\frac{\pi}{m_e}\right)^{3/2}\frac{Z^2 e^6 \bar{g}}{h\sigma c^3}\frac{n_e n_i}{T_v^{7/2}} \quad [2.50]$$

where n_i and n_e represent the particle density of ions and electrons, T_v is the temperature of the plasma, Z stands for the charge of ions, e is the proton charge and m_e is the mass of electrons.

The boundary conditions at the vapor–liquid interface are critical to solving Eqns [2.47] and [2.49]. In the classical kinetic model of evaporation, the vapor particles escaping from a hot liquid surface possess a half-space Maxwellian distribution, corresponding to the liquid surface temperature. This anisotropic velocity distribution is transformed into an isotropic distribution by collisions among the vapor particles within a few mean free paths (typically the order of a few micrometers) from the surface in a discontinuity region known as the Knudsen layer, as illustrated in Fig. 2.16. Beyond the Knudsen layer the vapor reaches a new internal equilibrium at a temperature different from the surface temperature. The vapor temperature outside the Knudsen layer can be used as a temperature boundary. The calculation of the vapor temperature outside the Knudsen layer is given by[67]

$$\frac{T_K}{T_L} = \left\{ \left[1 + \pi \left(\frac{\gamma - 1}{\gamma + 1} \frac{m}{2} \right)^2 \right]^{1/2} - \sqrt{\pi} \frac{\gamma - 1}{\gamma + 1} \frac{m}{2} \right\}^2 \qquad [2.51]$$

where T_K is the temperature outside of the Knudsen layer, T_L is the liquid surface temperature adjacent to the Knudsen layer, m is closely related to the Mach number M_K at the outer edge of the Knudsen layer and is defined as $m = M_K(2/\gamma)^{\frac{1}{2}}$ and γ is the specific heat. So, the temperature and intensity boundary conditions for the liquid–vapor interface are given below as[66]

$$T_v = T_K \qquad [2.52]$$

$$I = \varepsilon_w I + \frac{1 - \varepsilon_w}{\pi} \int_{\boldsymbol{n} \cdot \boldsymbol{\Omega}' < 0} I |\boldsymbol{n} \cdot \boldsymbol{\Omega}| \, d\Omega' \qquad [2.53]$$

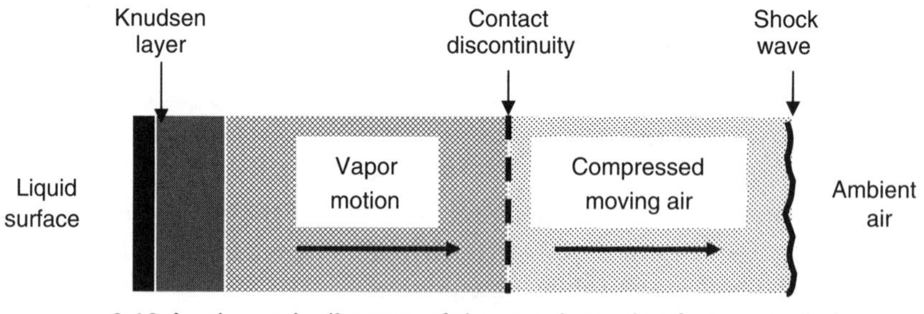

2.16 A schematic diagram of the gas dynamic of vapor and air away from a liquid surface at an elevated temperature.

Keyhole dynamics

The keyhole is basically an unstable structure even in the absence of motion. To understand the keyhole formation and collapse, the force and energy balance at the keyhole wall must be considered. At the keyhole wall, the following pressure conditions must be satisfied:

$$P = P_r + P_\sigma + P_h + P_g \quad [2.54]$$

where P is the pressure at the free surface in a direction normal to the local free surface, P_r is recoil pressure, P_σ is surface tension, P_g is hydrostatic pressure and P_h is hydrodynamic pressure. The pressure terms P_r and P_h tend to keep the keyhole open, whereas P_σ and P_g try to close the keyhole. According to Duley[47], in laser keyhole welding, the values of P_g and P_h are very small compared with P_r and P_σ; so only P_r and P_σ are considered. P_σ is the main restoring force that tries to close the keyhole and it can be calculated in the following form[19]:

$$P_\sigma = \kappa\gamma \quad [2.55]$$

where κ is the free surface curvature given by Eqn [2.18] and γ is the surface tension coefficient. For a pseudo-binary Fe–S system, γ is expressed in the form of Eqn [2.21]. P_r is the evaporation-induced recoil pressure, which is exerted on the surface of the weld pool to open the keyhole. This recoil pressure results from the rapid evaporation from the surface that has been heated to high temperatures. When the liquid–vapor interface temperature reaches the boiling point, evaporation begins to occur. It is known that there exists a very thin layer of several mean free path lengths, called the kinetic Knudsen layer, just outside the liquid–vapor interface. Inside the Knudsen layer, there is a transition from a non-equilibrium state at the liquid wall to an equilibrium state of the metallic vapor. Anisimov[68] and Knight[67] carried out the early investigations on the Knudsen layer. Here, the calculation of recoil pressure follows the work of Semak and Matsunawa[69]:

$$P_r = A\frac{B_0}{T_s^{1/2}}\exp\left(\frac{-U}{T_s}\right) \quad [2.56]$$

where A is a factor dependent on the environmental pressure, and $A = 0.55$ was employed; B_0 is an evaporation constant and was taken to be 3.9×10^{12} kg/m s^2 for iron; T_s is the temperature of the keyhole wall. U is defined as follows[69]:

$$U = \frac{M_a H_v}{N_A k_B} \quad [2.57]$$

where M_a is the atomic mass, H_v is the latent heat of evaporation, N_A is Avogadro's number and k_B is the Boltzmann constant.

The energy balance at the liquid–vapor interface (keyhole wall) must satisfy

$$k\frac{\partial T}{\partial \boldsymbol{n}} = q_{\text{laser}} + q_{\text{rad}} - q_{\text{evap}} - q_{\text{conv}} \tag{2.58}$$

The meaning and calculation of each term is discussed in the following.

1. q_{conv} is the heat loss due to convection. Since there is assumed to be no vapor flow inside the keyhole, q_{conv} can be omitted.
2. q_{evap} is the heat loss due to evaporation. It is defined in the following form:

$$q_{\text{evap}} = WH_v \tag{2.59}$$

where H_v is the latent heat for liquid–vapor phase change and W is the melt mass evaporation rate. For a metal such as steel, W can be written as[69]

$$W = n_l\left(\frac{k_B T_l}{2\pi m_a}\right)^{1/2} \exp\left(-\frac{H_v}{k_B T_l}\right) - \theta_s n_v\left(\frac{k_B T_v}{2\pi m_a}\right)^{1/2} \tag{2.60}$$

where T_l is the liquid surface temperature, m_a is the atomic mass, n_l and n_v are the numbers of atoms per unit volume for liquid and vapor respectively, T_v represents the evaporation temperature and θ_s denotes the probability that a vapor atom returning to the liquid surface from equilibrium conditions at the edge of the discontinuity layer manages to penetrate this layer to finally be absorbed on the liquid surface, which is in the range of 15–20%.

3. q_{laser} is the total heat input to the keyhole wall. The following absorption mechanisms are considered. First, the laser power is directly absorbed at the keyhole wall by Fresnel absorption, but also plasma absorption occurs. The metal vapor reaches temperatures much higher than the metal evaporation temperature, resulting in strong ionization. The resulting plasma absorbs laser power by the effect of inverse Bremsstrahlung absorption. Another absorption mechanism is Fresnel absorption due to multiple reflections of the beam inside the keyhole. So, q_{laser} consists of Fresnel absorption $\boldsymbol{I}_{\alpha,\text{Fr}}$ of the incident intensity which is directly from the laser beam and Fresnel absorption $\boldsymbol{I}_{\alpha,m_r}$ due to multiple reflections of the beam inside the keyhole. They are defined as follows[61]:

$$q_{\text{laser}} = I_{\alpha,\text{Fr}} + I_{\alpha,\text{m}_r} \tag{2.61}$$

$$I_{\alpha,\text{Fr}} = I_{\text{laser}}(1 - \alpha_{i\text{B},1})\alpha_{\text{Fr}}(\varphi_1) \tag{2.62}$$

$$I_{\alpha,m_r} = \sum_{m_r=1}^{n} I_{\text{laser}}(1-\alpha_{\text{iB},1})(1-\alpha_{\text{Fr}})(1-\alpha_{\text{iB},m_r})\alpha_{\text{Fr}}(\varphi_{m_r}) \qquad [2.63]$$

where I_{laser} is the incoming laser intensity. For a Gaussian-like laser beam, it is in the following form:

$$I_{\text{laser}}(x, y, z) = I_0\left(\frac{r_f}{r_{f0}}\right)^2 \exp\left(-\frac{2r^2}{r_f^2}\right) \qquad [2.64]$$

where r_f is the beam radius, r_{f0} is the beam radius at the focal position and I_0 is the peak intensity. α_{Fr} is the Fresnel absorption coefficient and it can be defined by using the following formula[59]:

$$\alpha_{\text{Fr}}(\varphi) = 1 - \frac{1}{2}\left(\frac{1+(1-\varepsilon\cos\varphi)^2}{1+(1+\varepsilon\cos\varphi)^2} + \frac{\varepsilon^2 - 2\varepsilon\cos\varphi + 2\cos^2\varphi}{\varepsilon^2 + 2\varepsilon\cos\varphi + 2\cos^2\varphi}\right) \qquad [2.65]$$

where φ is the angle of incident light with the normal to the keyhole surface and ε is a material-dependent coefficient. Also n is the total amount of incident light from multiple reflections, $\boldsymbol{I}$ is the unit vector along the laser beam radiation direction and $\boldsymbol{n}$ is a unit vector normal to the free surface. In CO_2 laser welding of mild steel, $\varepsilon = 0.2$ is used. $\alpha_{\text{iB},1}$ is the absorption fraction in the plasma due to the original laser beam; α_{iB,m_r} is the absorption fraction due to the mr^{th} reflected laser beam. They can be defined by the following formula[70]:

$$\begin{aligned} \alpha_{\text{iB},1} &= 1 - \exp\left(-\int_0^{s_0} k_{\text{pl}}\,\mathrm{d}s\right) \\ \alpha_{\text{iB},m_r} &= 1 - \exp\left(-\int_0^{s_m} k_{\text{pl}}\,\mathrm{d}s\right) \end{aligned} \qquad [2.66]$$

where $\int_0^{s_0} k_{\text{pl}}\,\mathrm{d}s$ and $\int_0^{s_m} k_{\text{pl}}\,\mathrm{d}s$ are optical thicknesses of the laser transportation path for the first incident reflection and multiple reflections respectively and k_{pl} is the plasma absorption coefficient due to inverse Bremsstrahlung absorption[47]:

$$k_{\text{pl}} = \frac{n_e n_i Z^2 e^6 2\pi}{6\times 3^{1/2} m\varepsilon_0^3 ch\omega^3 m_e^2}\left(\frac{m_e}{2\pi k_B T_e}\right)^{1/2}\left[1-\exp\left(-\frac{\omega}{k_B T_e}\right)\right]\bar{g} \qquad [2.67]$$

where Z is the average ionic charge in the plasma, ω is the angular frequency of the laser radiation, ε_0 is the dielectric constant, k_B is the Boltzmann constant, n_e and n_i are particle densities of electrons and ions respectively, h is Planck's constant, m_e is the electron mass,

T_e is the excitation temperature, c is the speed of light and $\bar{g}$ is the quantum-mechanical Gaunt factor. For the weakly ionized plasma in the keyhole, the Saha equation[47] is used to calculate the densities of plasma species:

$$\frac{n_e n_i}{n_0} = \frac{g_e g_i}{g_0} \frac{(2\pi m_g k_B T_e)^{3/2}}{h^3} \exp\left(-\frac{E_i}{k_B T_e}\right) \quad [2.68]$$

4. q_{rad} is the heat input due to radiation heat transfer by the high-temperature plasma in the keyhole. It is a heat transfer problem of an enclosure containing emitting–absorbing–scattering media. Because the temperature of the keyhole wall is much lower than the temperature of the plasma, only the radiation transfer between the plasma and the keyhole wall is considered. By using the zone method, q_{rad} is calculated from the formula[61]

$$q_{rad} = \varepsilon q_i \quad [2.69]$$

$$q_i = \frac{1}{A} \sum_{\gamma=1}^{\Gamma} \overline{g_\gamma s} \sigma T_\gamma^4 \quad [2.70]$$

where $\overline{g_\gamma s}$ is the gas–surface direct-exchange area, T_γ is the mean temperature in the gas zone, Γ is the number of gas zones and ε is the radiation coefficient. Figure 2.17 illustrates the keyhole formation process in three-dimensional moving keyhole laser welding[61]. The corresponding velocity and temperature distributions are shown in Fig. 2.18 and Fig. 2.19 respectively.

Weld porosity formation and prevention

In laser keyhole welding, pores are frequently observed[71–74], which cause the strength of the welded part to deteriorate. In order to optimize a laser welding process and to ensure high weld quality and strength, it is necessary to understand the porosity formation mechanism and subsequently to find methods to reduce or eliminate pore defects.

Over the years, a number of researchers have experimentally investigated pore formation in laser welding. Katayama *et al.*[75] studied pore formation in welding A5083 alloy and type 304 steel with a high-power yttrium aluminum garnet (YAG) laser. They reported that many pores were formed owing to the evaporation of metal from the bottom tip of the keyhole, and the vapor was trapped by the solidifying front. Seto and coworkers[76] took X-ray transmission images of the keyhole dynamics during laser welding by using a high-speed video camera. They concluded that the shape of the

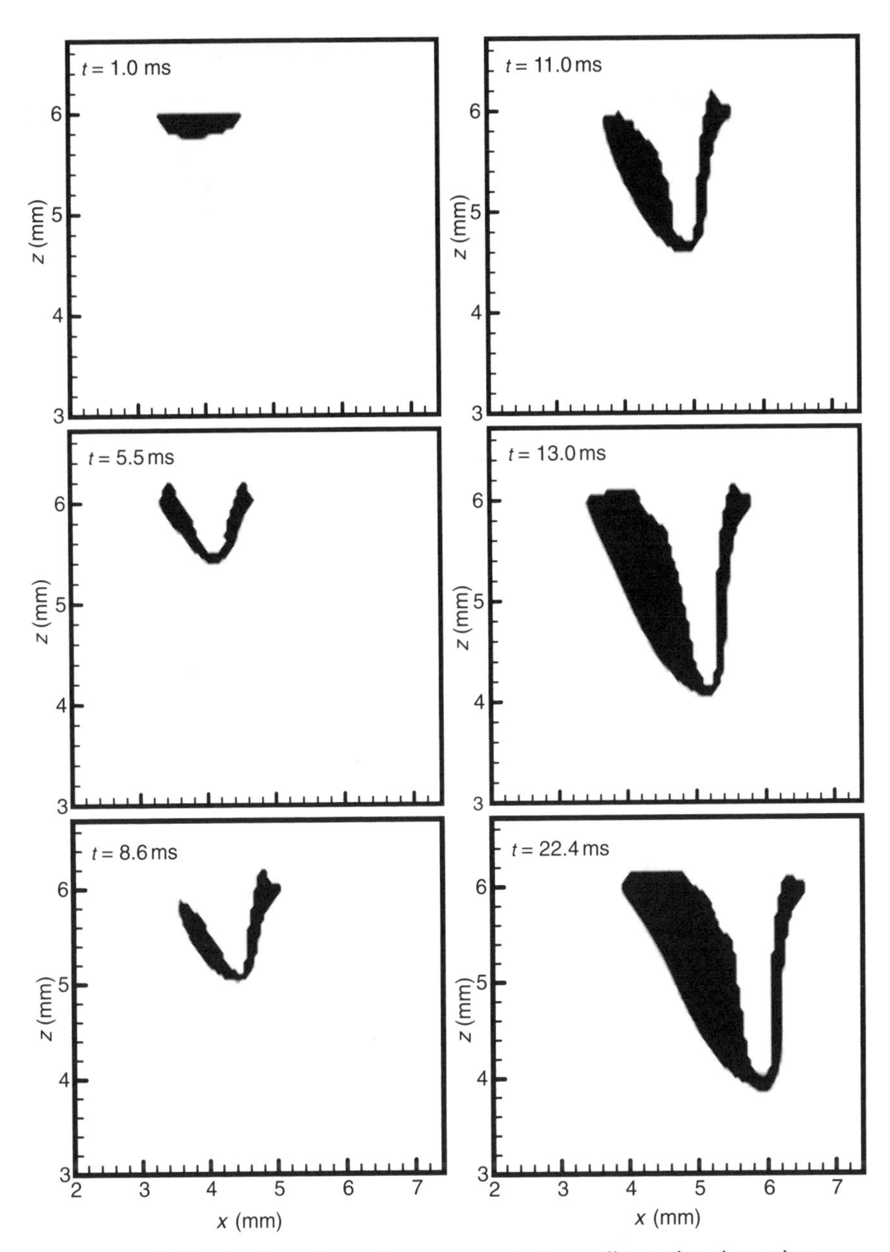

2.17 The keyhole formation process in three-dimensional moving keyhole laser welding.

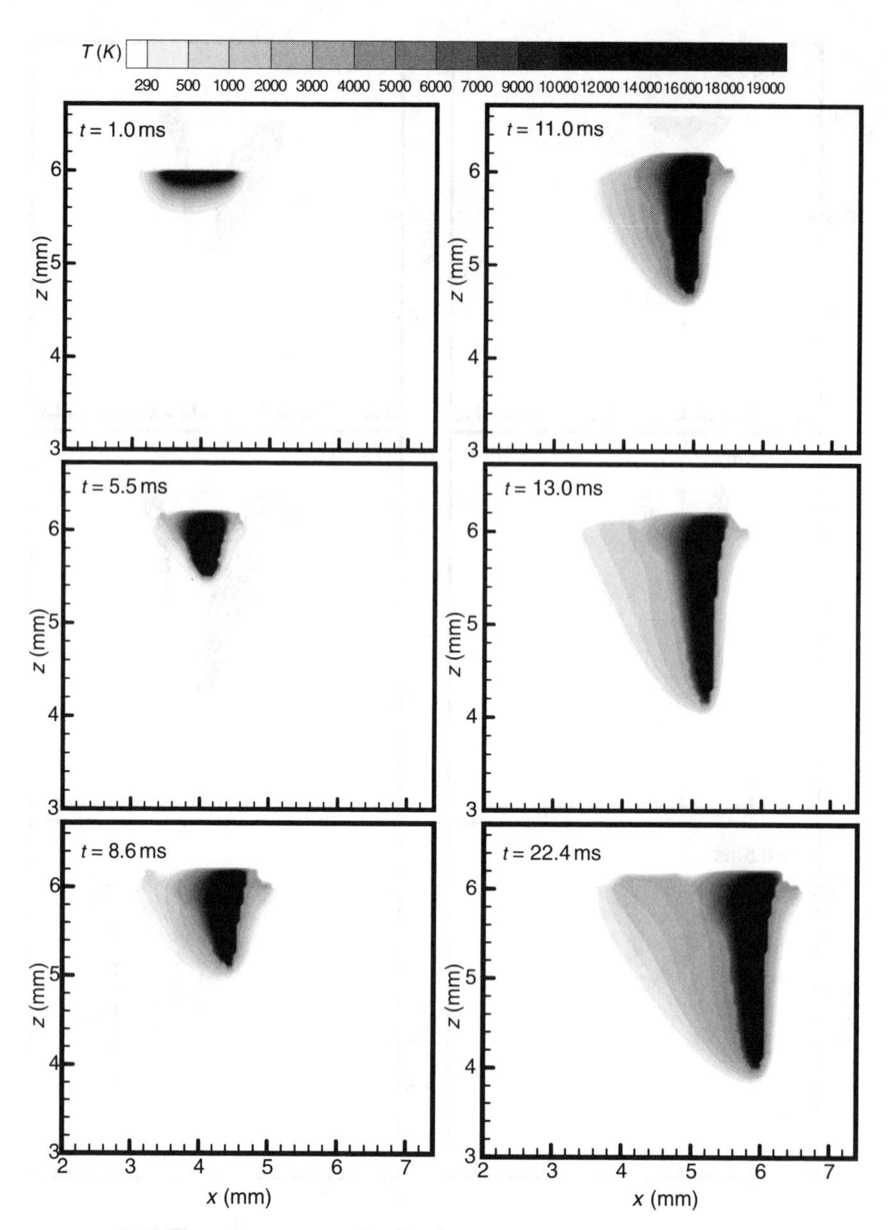

2.18 The temperature distributions corresponding to Fig. 2.17.

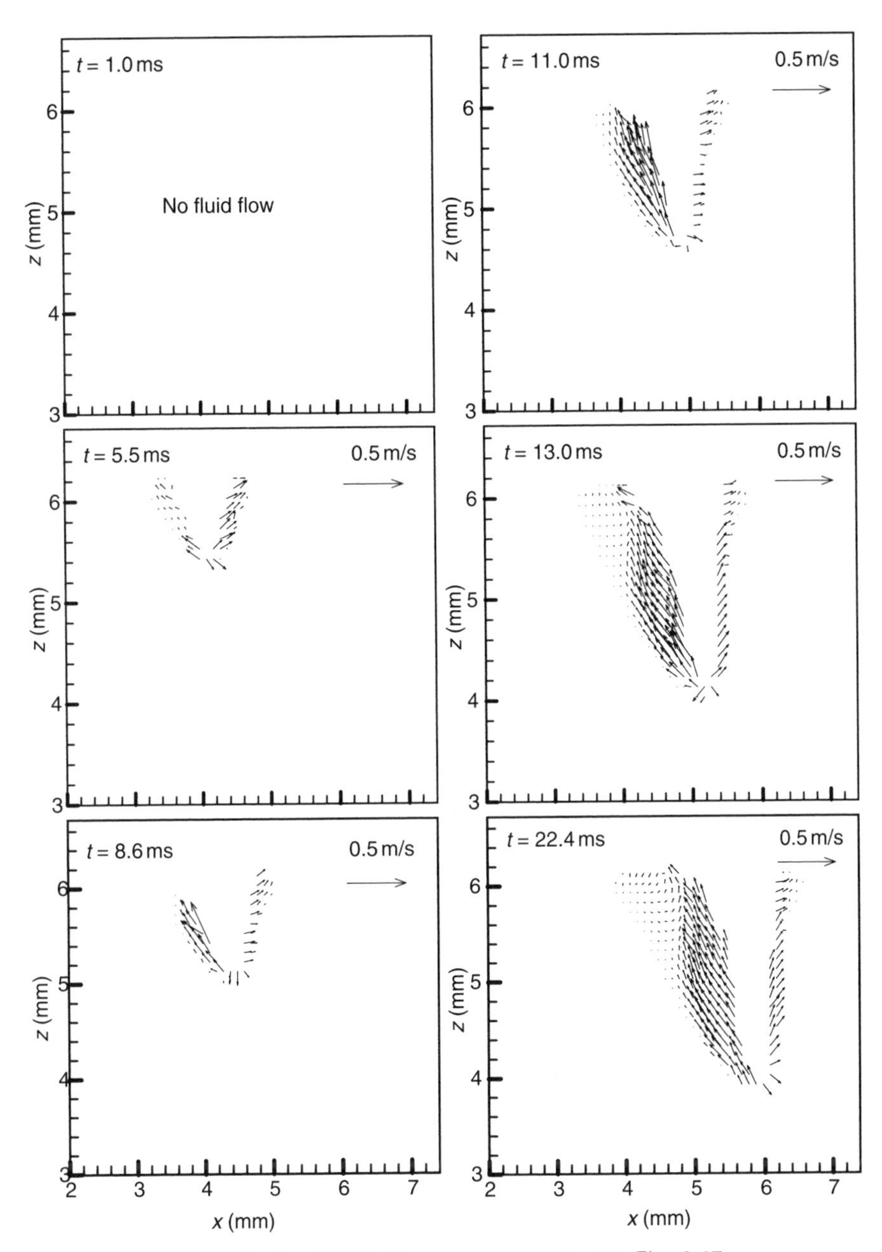

2.19 The velocity distributions corresponding to Fig. 2.17.

keyhole front had a great influence on the pore formation. Katayama *et al.*[77] observed the fusion and solidification behavior of a molten puddle during laser spot welding of type 316S steel. They found that the formation of pores had a close correlation with the collapse of the keyhole right after the irradiation termination. Once the laser beam was terminated, the melt in the upper part of the keyhole flowed downward to fill the keyhole; meanwhile this part of the melt rapidly solidified, which prevented the melt from flowing to fill the keyhole. Then a pore was formed in the weld.

Zhou and coworkers [78–80] developed mathematical models to investigate systematically the porosity formation mechanisms and its prevention method in pulsed laser welding. Figure 2.20 shows a typical porosity formation process in pulsed laser welding[78]. The corresponding temperature distributions are given in Fig. 2.21. As shown, the formation of porosity is found to be determined by two competing mechanisms: one is the solidification rate and the other is the speed at which molten metal backfills the keyhole after laser energy is terminated. If the backfill speed of the liquid on the upper part of the keyhole is not large enough to make the liquid completely fill back the keyhole before its solidification, pores will appear in the final weld. Also, the depth-to-width ratio of the keyhole has a great influence on pore formation. The greater the ratio, the larger is the size of the pore. When the ratio is small enough, no pores occur.

Based on the investigations on porosity formation mechanisms, a technique to control the laser power tailing[77,79] has been proposed (Fig. 2.22). Figure 2.23 shows a porosity elimination process and its corresponding temperature distributions in workpiece by using the controlled power tailing as shown in Fig. 2.22[79]. As shown, this technique can postpone the solidification speed and, as a result, prevent the porosity formation. However, this method fails for a 'deep' keyhole. Another technique to apply external electromagnetic force to increase the backfill speed of the molten metal is proposed[80]. Figure 2.24 shows the keyhole collapse and molten metal solidification processes by using this technique, and its corresponding temperature distribution is shown in Fig. 2.25. This method is shown to be capable of eliminating the porosity formation in laser welding effectively, but parametric studies are needed to determine the desired strength of the electromagnetic force and its duration.

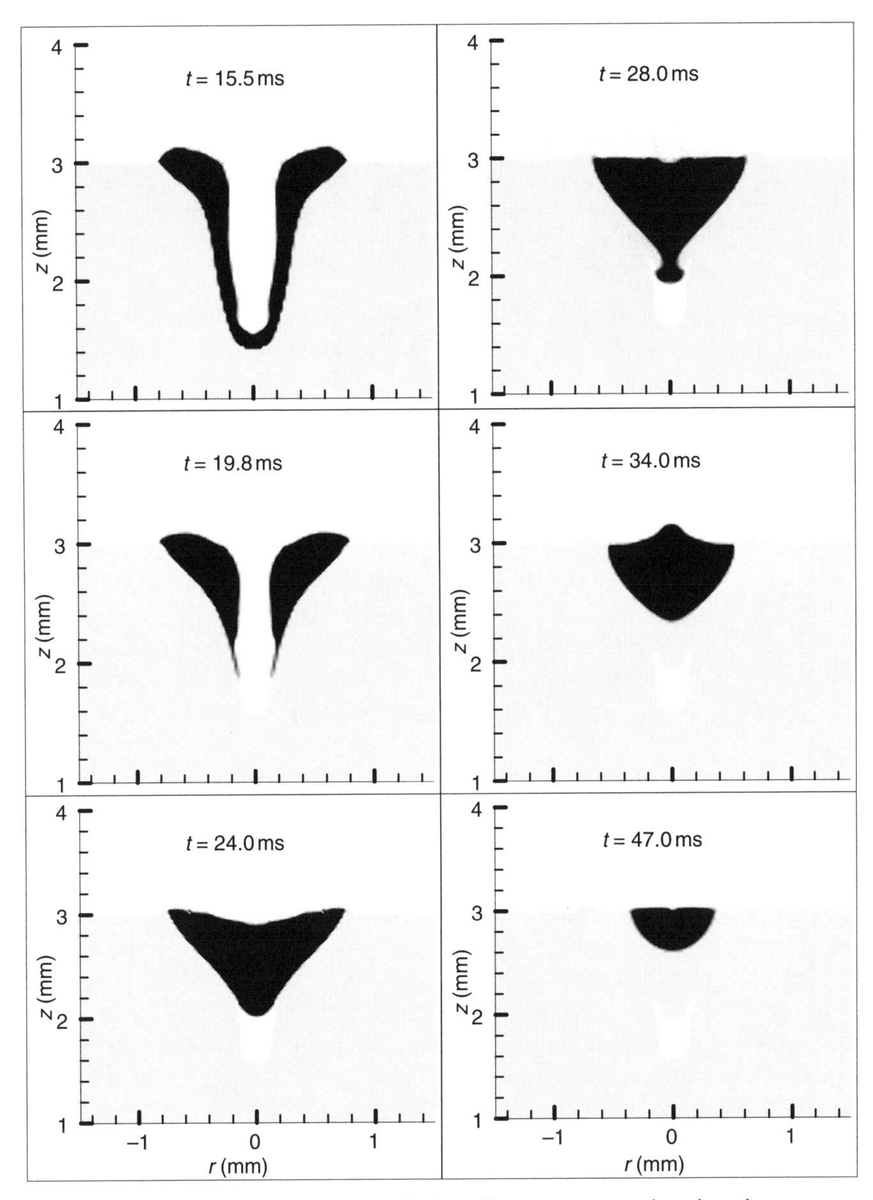

2.20 A sequence of the keyhole collapse process showing the possibility of forming a pore in the weld.

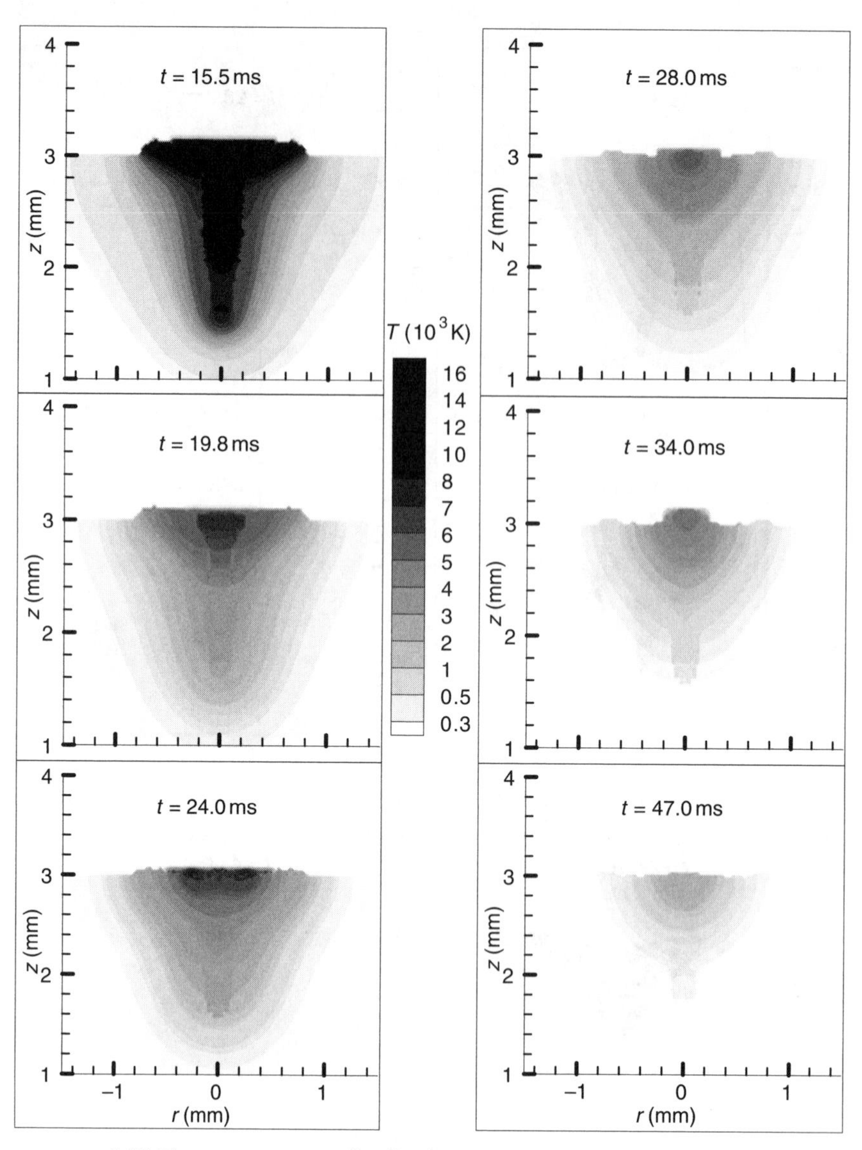

2.21 The temperature distributions corresponding to Fig. 2.20.

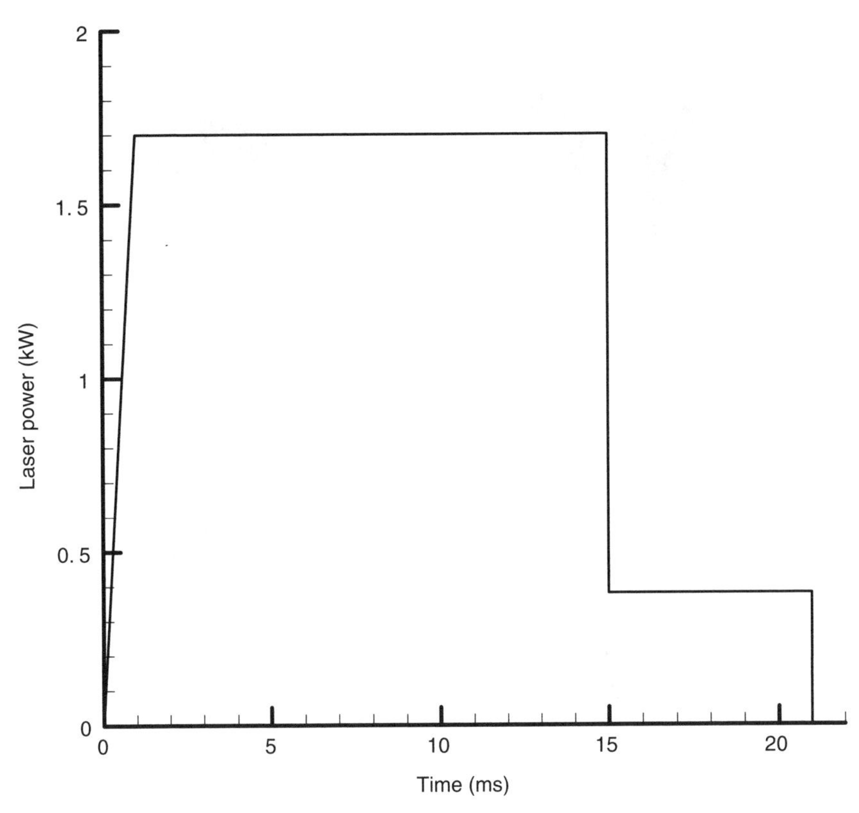

2.22 Laser pulse profiles used to prevent pore formation for laser keyhole welding.

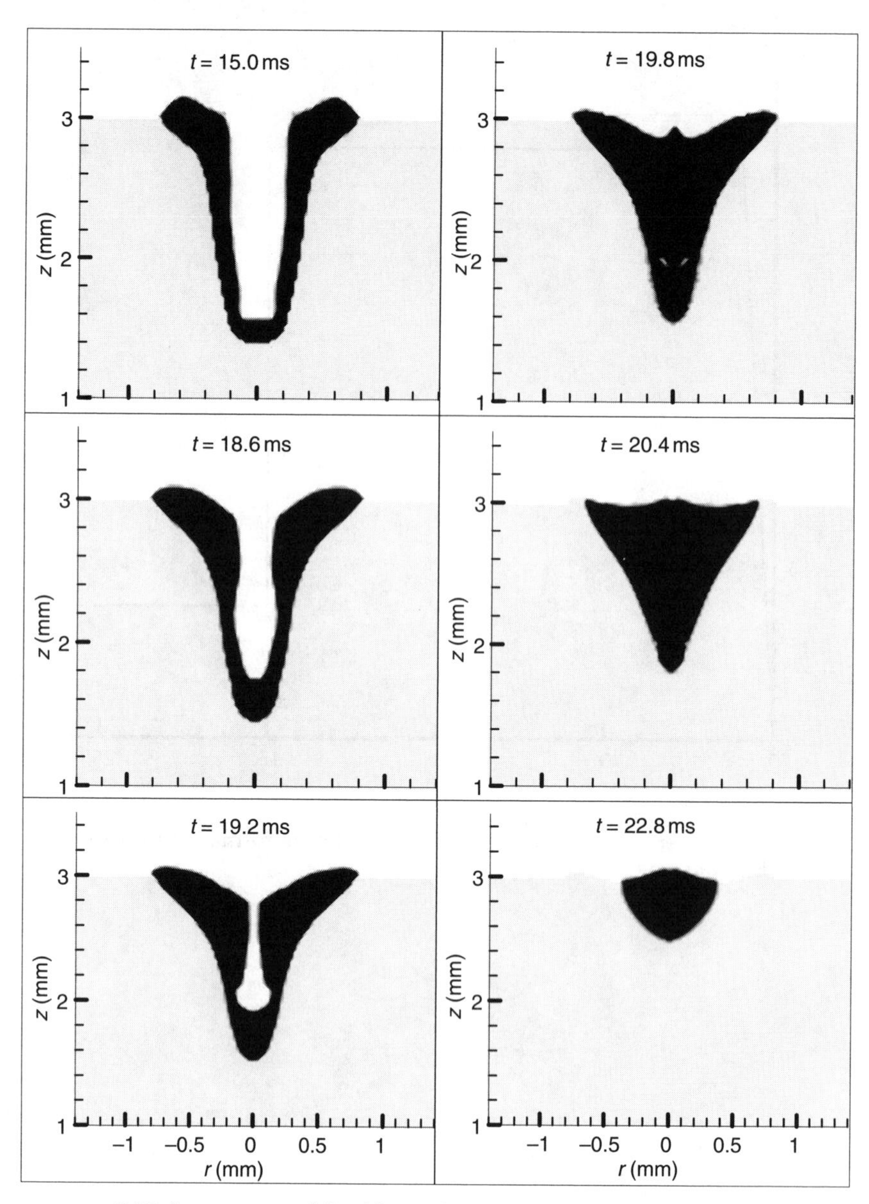

2.23 A sequence of liquid metal evolution during pore elimination for keyhole welding by using the controlled pulse profile in Fig. 2.22.

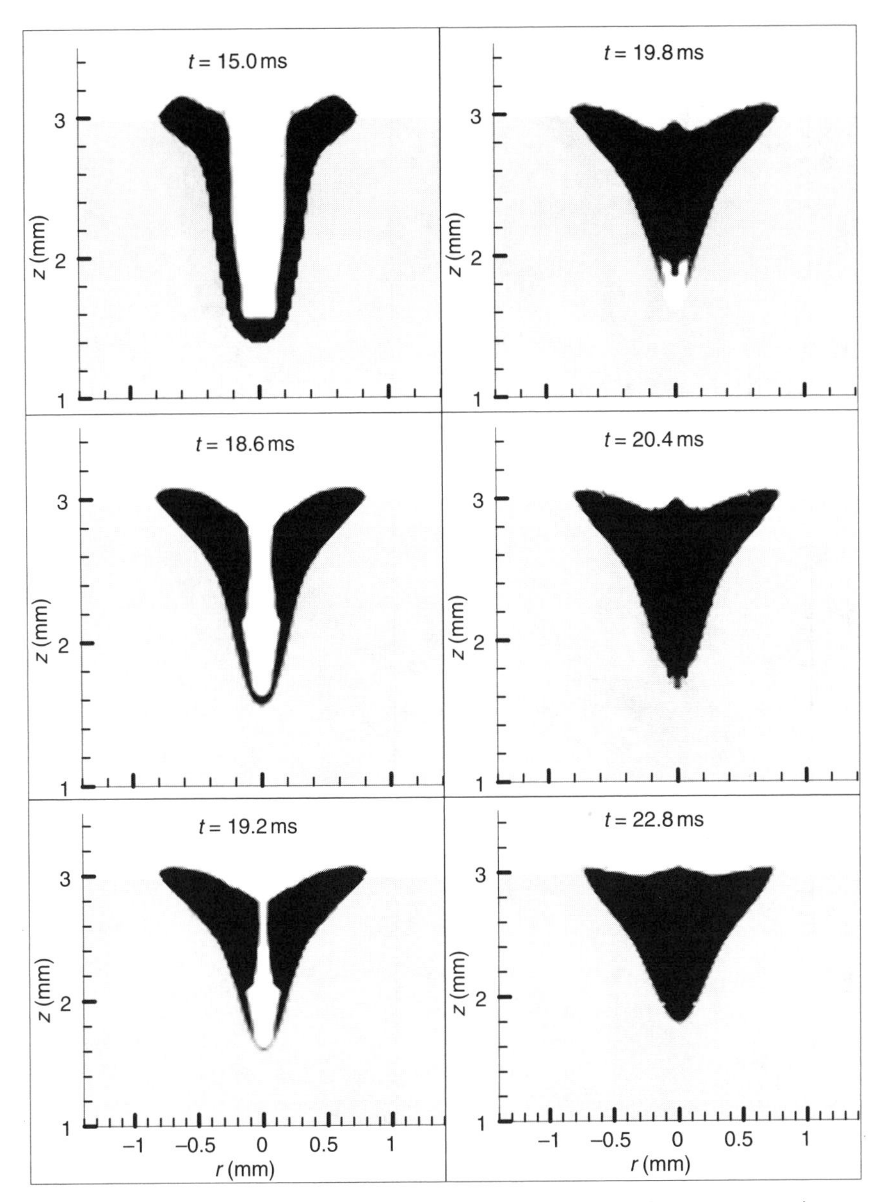

2.24 Pore prevention for keyhole laser welding using an external electromagnetic force.

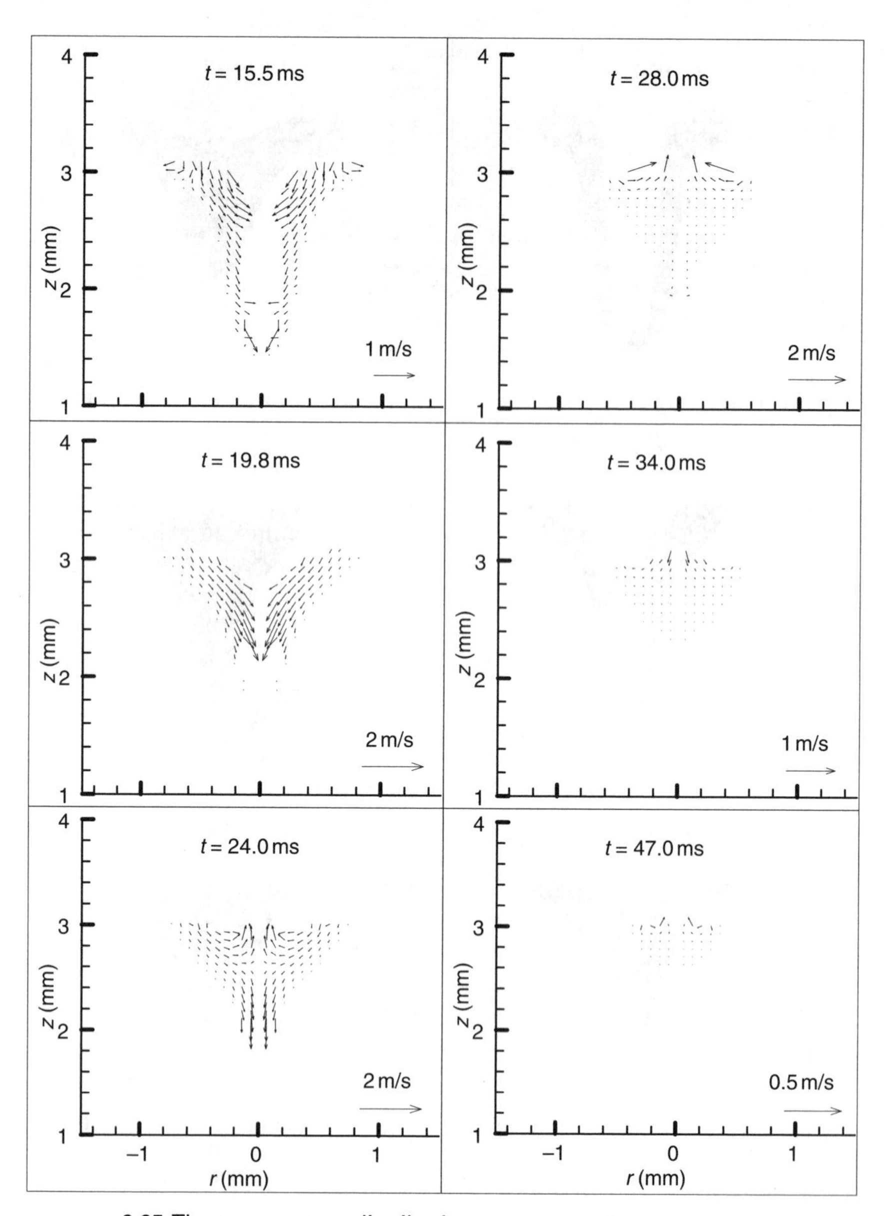

2.25 The temperature distributions corresponding to Fig. 2.24.

2.4 New welding processes

In recent years, new welding technologies, such as hybrid laser–arc welding and dual-beam laser welding are receiving increasing interest in both academic and industrial fields. Research shows that the welding efficiency and weld quality can be improved by these new welding processes. In the following, the hybrid laser–(metal–inert gas (MIG)) welding process and the dual-beam welding process will be discussed respectively.

2.4.1 Hybrid laser–(metal–inert gas) welding

In order to overcome the disadvantages of laser welding and conventional arc welding, an idea to combine these two processes was proposed to become the hybrid laser–arc welding technique, as illustrated in Fig. 2.26. As shown, in hybrid laser–arc welding, a laser beam (a CO_2 or neodymium-doped yttrium aluminum garnet (Nd: YAG) laser) and an arc (tungsten–inert gas (TIG), plasma or gas–metal) are coupled and they mutually influence and assist each other. The arc, in addition to the laser beam,

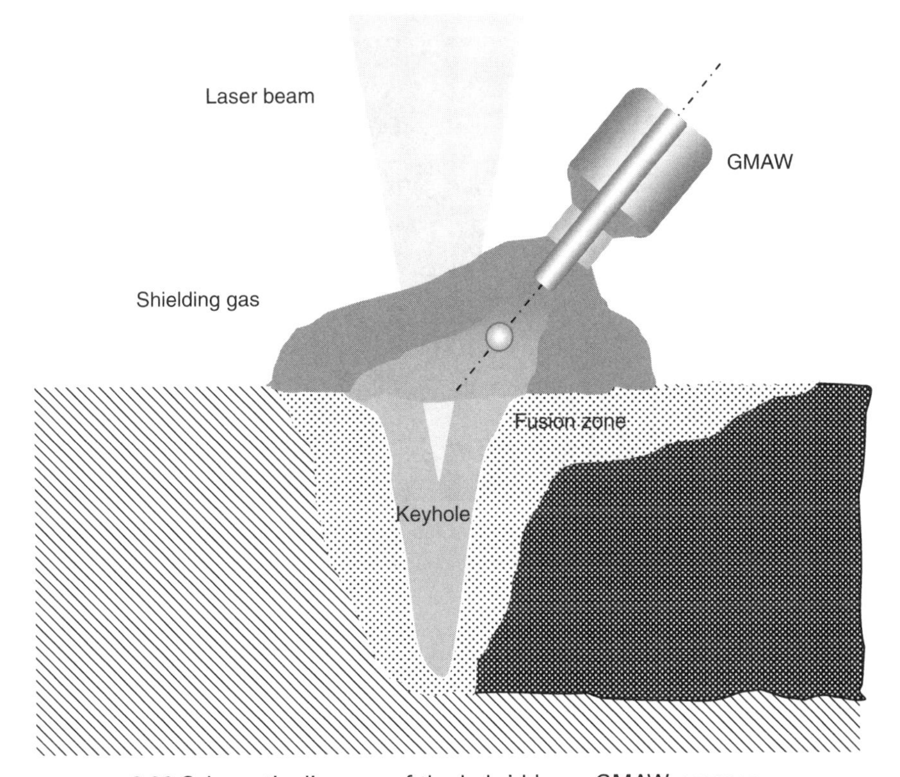

2.26 Schematic diagram of the hybrid laser–GMAW process.

(a)

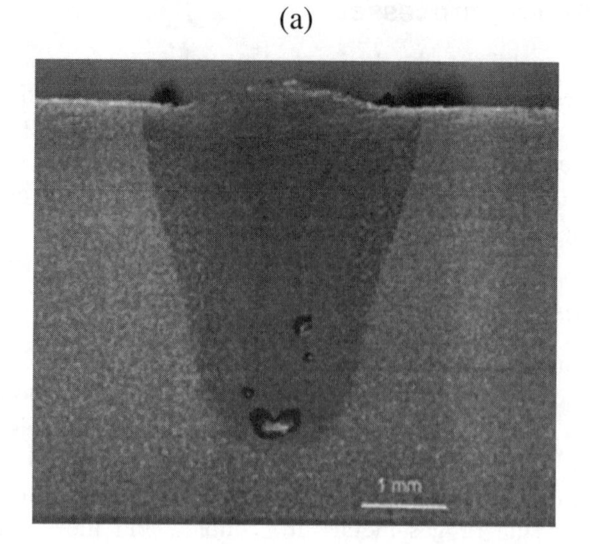

(b)

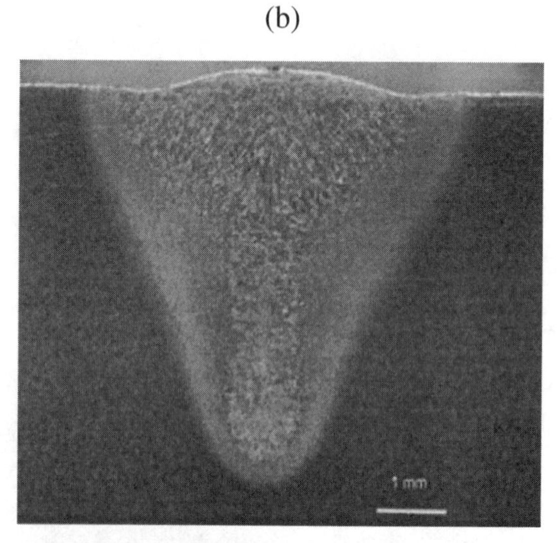

2.27 Comparison between (a) a laser weld and (b) a hybrid laser–arc weld in 250 grade mild steel.

supplies heat to the weld metal in the upper weld region. In summary, the advantages of hybrid welding over laser welding include higher process stability, higher bridgeability, deeper penetration depth, lower capital investment costs, less porosity and cracking, and greater flexibility. The advantages of hybrid welding over MIG welding include higher welding speed, deeper penetration depth at higher welding speeds, lower thermal input, higher tensile strength and narrower weld seams. Figure 2.27 shows some typical weld beads of hybrid laser–arc welding[81].

Research and development of hybrid laser–arc welding have been conducted since 1978[82–84]. After the 1980s for an extended period of time, little progress was achieved for improving the quality of laser welds and no further research and development were undertaken. Recently, researchers have turned their attention to this topic again. Dilthey[85] investigated the prospects in hybrid laser–arc welding by using different kinds of lasers (CO_2 laser and Nd:YAG laser) and arc processes (TIG, plasma and gas–metal). Staufer *et al.*[86] and Graf and Staufer[87] studied the application of hybrid laser–arc welding in Volkswagen vehicles. They concluded that, by merging the two processes, hybrid laser–arc welding could offer synergies for wide fields of application in the automotive industries. Minami *et al.*[88] tried to apply CO_2 laser and hybrid MIG arc welding on thick stainless steel. They found that the penetration depth is influenced by the distance from the laser beam axis to the MIG electrode, and shield gas flow. The weld defects could be reduced by suitable gas components and arc current waveform. Nielsen *et al.*[89] investigated the hybrid CO_2 laser–(metal–active gas (MAG)) welding of 12 mm C–Mn steel, in which the gap size, the process parameters and the hybrid welding speed were compared with those of the pure laser welding process. Engstrom *et al.*[90] studied the laser hybrid welding of high-strength steels, in which a CO_2 laser and a MIG electrode are used. They pointed out that this process is capable of handling different kinds of joint preparation, thereby expanding the industrial production capabilities. Jokinen *et al.*[91] investigated the welding of thick austenitic stainless steel using a Nd:YAG laser and GMAW. They studied the effect of distance between the arc point and laser beam. Nilsson *et al.*[92] investigated the hybrid laser–arc process for different industrial joint preparations. Ishide *et al.*[93] developed various hybrid TIG–YAG and MIG–YAG welding methods, in which the layout of laser and arc, and the optimization of focusing performance were studied. They found that the arc stability could be improved by the hybrid conditions. Ono *et al.*[94] studied the application of hybrid laser–arc welding on zinc-coated steel sheets. They found hybrid laser–arc welding is characterized by large gap tolerance in lap welding, high strength of welded joint and excellent lap weldability. Naito *et al.*[95] experimentally studied the keyhole behavior and liquid flow in the molten pool in laser–arc hybrid welding. In the experiment, they found that the deepest weld bead could be produced when the YAG laser beam was shot on the molten pool which is stabilized by the TIG arc. Kutsuna and Chen[96] investigated the interaction of laser plasma and MAG arc plasma in hybrid laser–MAG welding. Their experimental results indicated that the interaction of laser beam and plasma is dependent on the laser-to-arc distance.

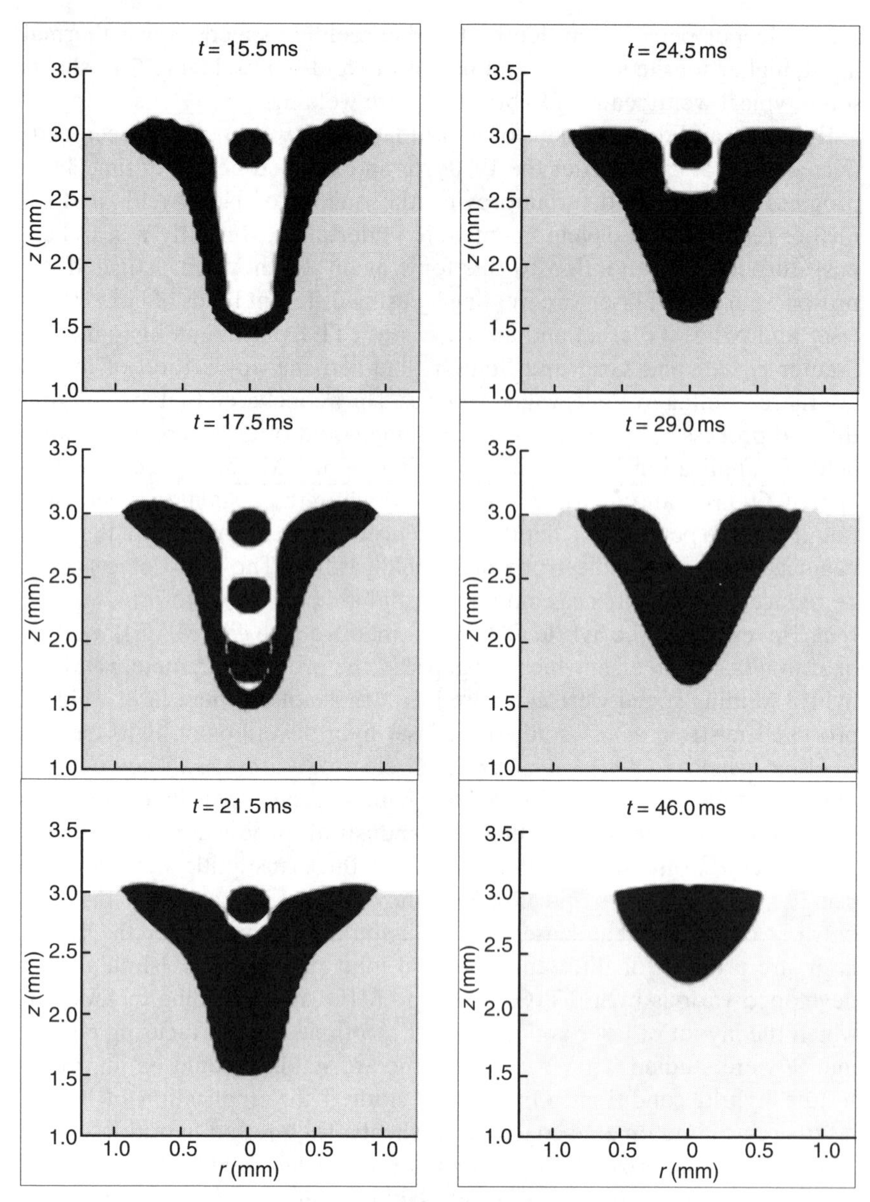

2.28 A sequence of the keyhole collapse process and solidification process during hybrid laser–arc welding.

Zhou *et al.*[97,98] used mathematical models to simulate the heat and mass transfer and fluid flow phenomena in both pulsed and three-dimensional moving hybrid laser–MIG welding. As shown in Fig. 2.28[98], hybrid laser–MIG welding can prevent pores that are easily found in laser welding.

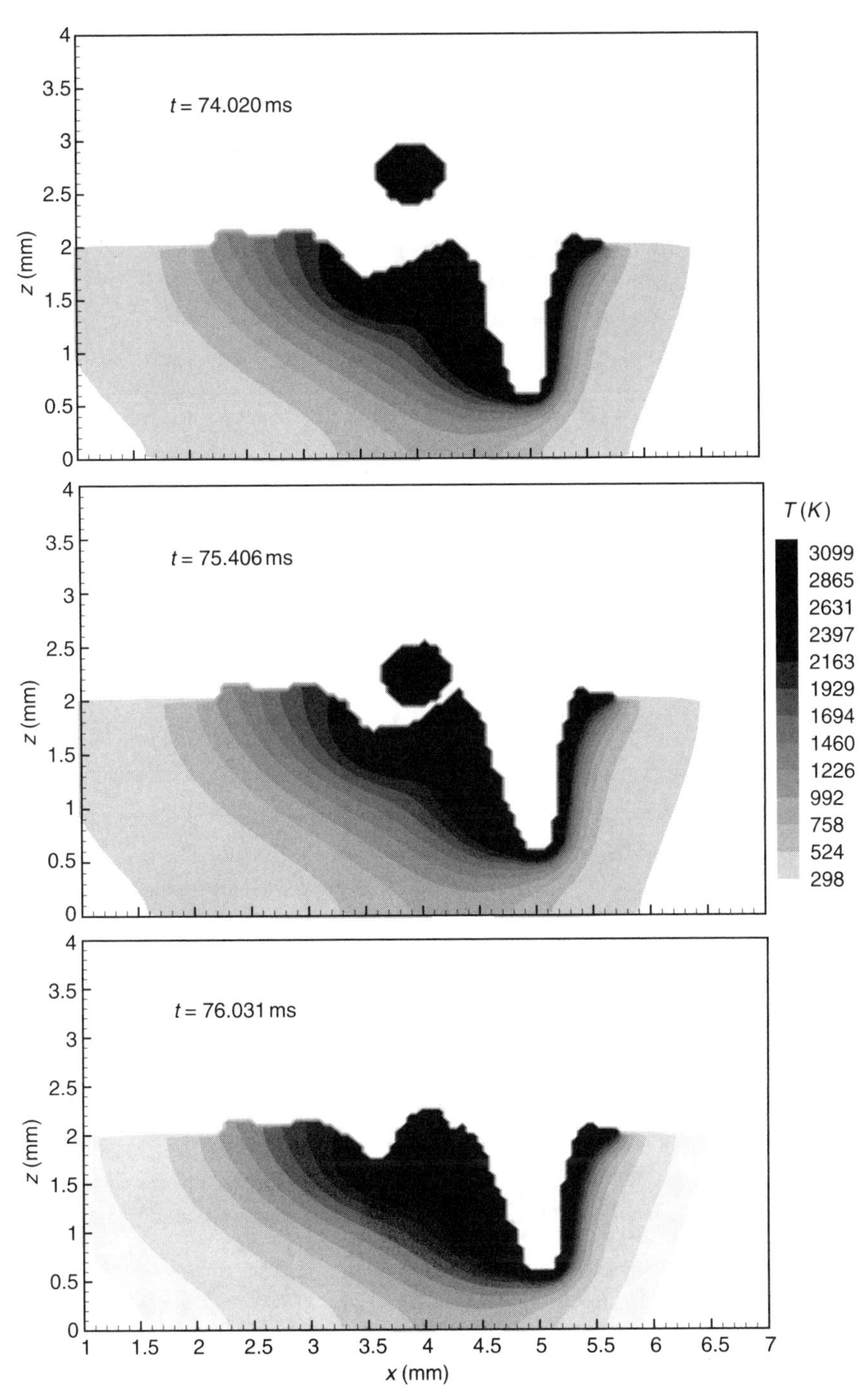

2.29 A sequence of temperature evolution during hybrid laser–MIG welding.

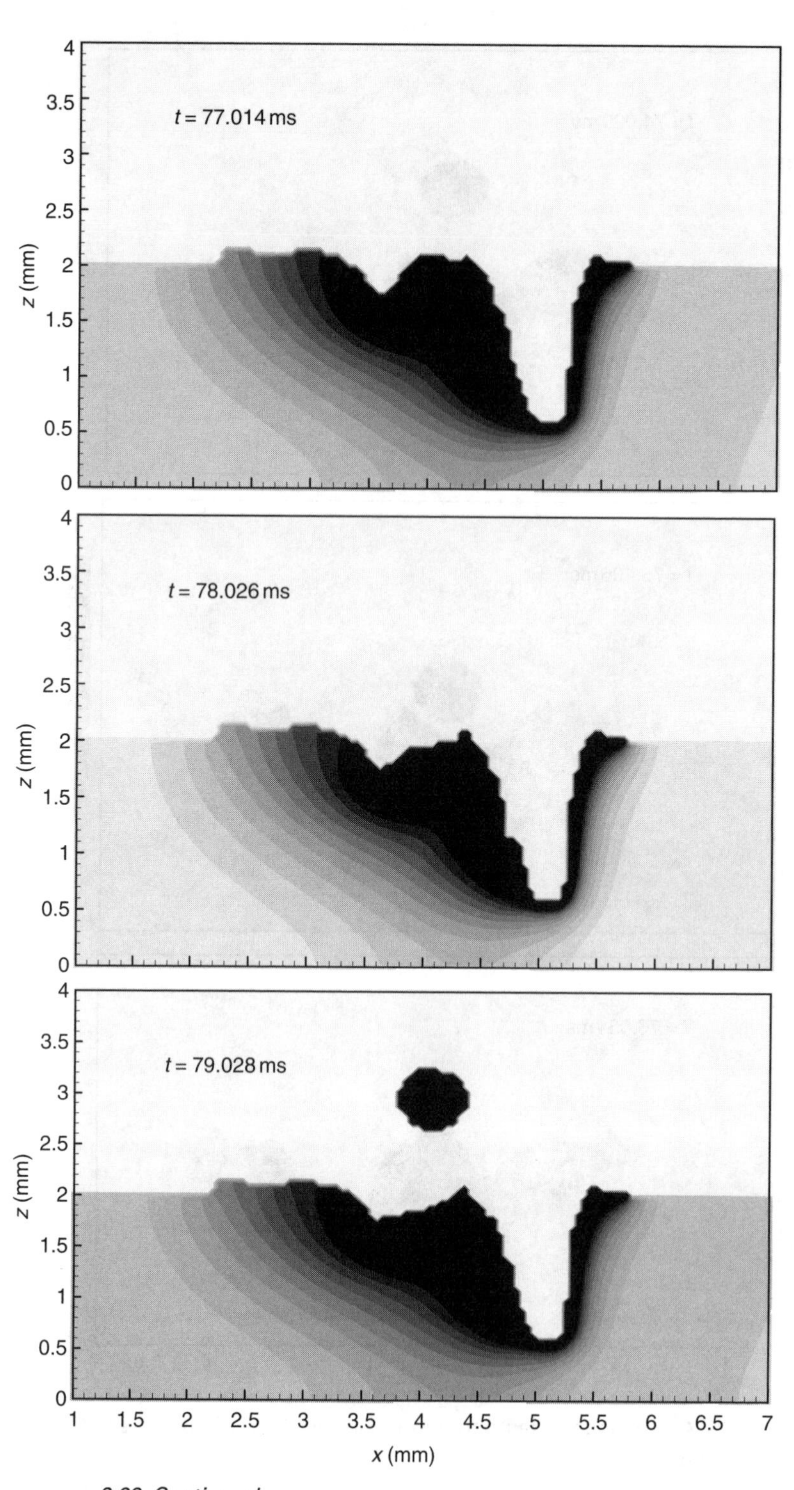

2.29 Continued

In the hybrid welding process, the mixing and heat transfer process in the weld pool are also found to be greatly affected by the droplet size, droplet frequency, etc.[98]. Hence, the microstructure and final weld quality can be improved. Figure 2.29 shows the temperature distributions in a moving three-dimensional hybrid laser–MIG process[97]. The heat transfer process is

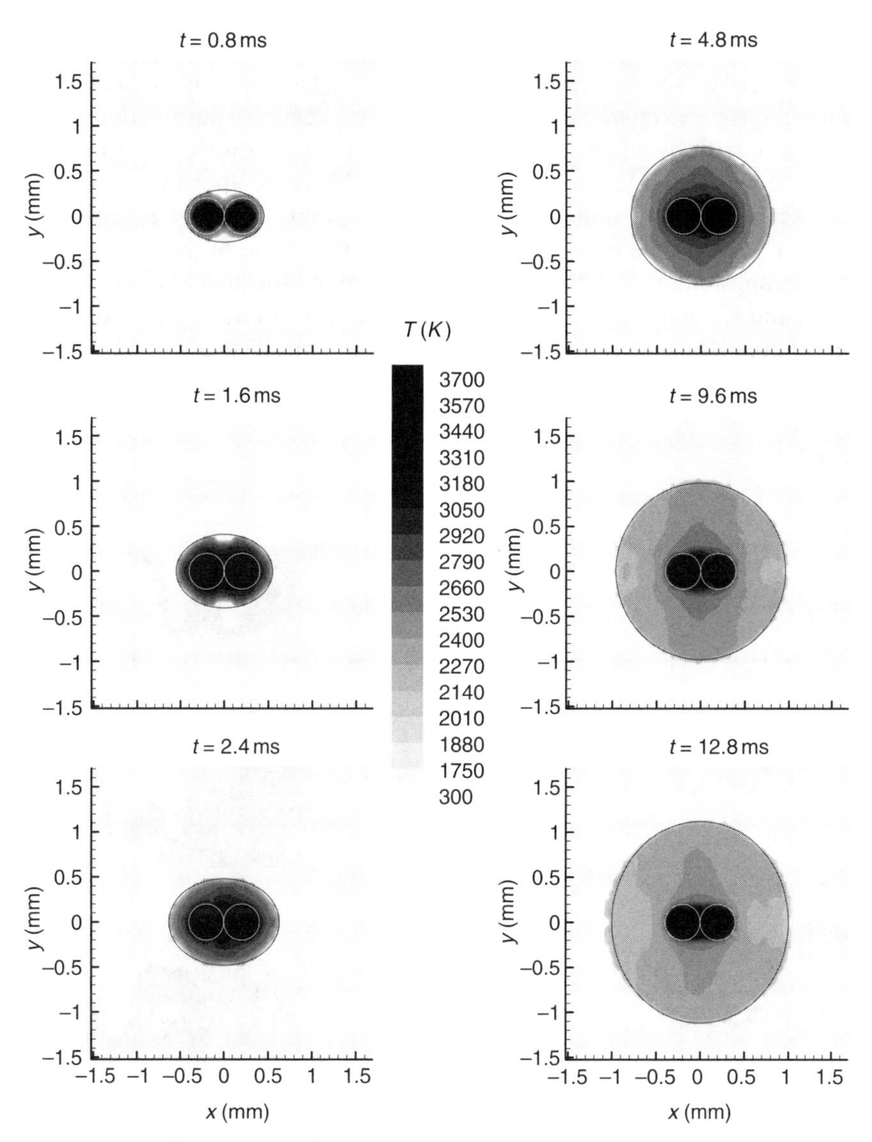

2.30 A sequence of temperature evolution and weld shape variation during dual-beam laser welding.

greatly affected by the laser-to-arc distance, welding speed, etc. More details have been given by Zhou *et al.*[97].

2.4.2 Dual-beam welding

Dual-beam laser welding is reported to increase the stability of the keyhole, which leads to a better weld quality[99–103]. Dual-beam laser welding is also found to be able significantly to stabilize the welding process, to reduce porosity and blowhole formation and to improve surface quality of the weld with fewer surface defects such as undercut, surface roughness, spatter and underfill. Iwase *et al.*[103] observed a larger keyhole during dual-beam Nd:YAG laser welding by real time X-ray observations. An elongated keyhole was found to be the reason that enables the vapor to escape from the keyhole freely and thus results in a stable welding process[101,103]. Fabbro[104] and Coste *et al.*[105,106] reported that the shape of the weld pool in dual-beam welding is different from that observed for single-beam welding. Chen and Kannatey-Asibu[107] studied the heat transfer during a conduction-mode dual-beam laser welding process. Hou[108] and Capello *et al.*[109] simu-

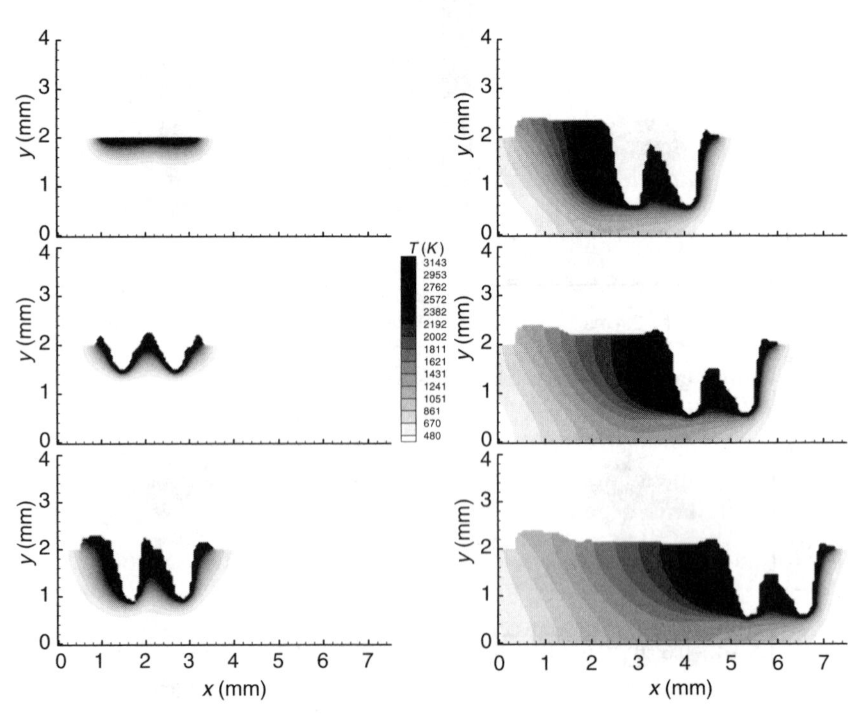

2.31 A sequence of temperature evolution and weld shape variation during three-dimensional moving dual-beam laser welding.

lated a welding process involving two laser beams as a system in which one beam served as a minor heat source and the other was a major heat source. The fluid flow was not considered in their models and the beam distance was much larger than the beam size.

Hu and Tsai[110] developed a stationary three-dimensional model to simulate the dual-beam laser welding process that calculates the heat transfer and fluid flow in the weld pool. Figure 2.30 shows the temperature distribution of the melt pool surface in stationary dual-beam laser welding. The oval weld pool shape agrees with the observations by Coste *et al.*[105,106] and Fabbro[104] in their experiments. The formation of the oval shape is due to the keyhole interaction of the two laser beams which causes the stronger fluid flow in the direction perpendicular to the connection line of the two beam centers. Zhou and Tsai[111] studied the heat transfer and fluid flow phenomena in a three-dimensional moving dual-beam laser welding process. The keyhole dynamics, the temperature distributions in the workpiece and the weld pool dynamics are found to be dependent on the laser beam distance and beam configurations as shown in Fig. 2.31. Detailed information has been provided by Hu and Tsai[110] and Zhou and Tsai[111].

2.5 Numerical considerations

In the welding process, the transport phenomena of the metal and arc plasma–laser-induced plasma is coupled in a transient manner. As the thermophysical properties of the plasma and the metal are very different, it is numerically inefficient to solve simultaneously the entire domain including the metal and the plasma. Hence, only the electric potential equation [2.29] and its associated boundary conditions were solved for the entire domain, while a 'two-domain' method has been used to solve sequentially the arc/keyhole plasma domain and the metal domain. In other words, the primary variables are calculated separately in the metal domain and the arc/keyhole plasma domain. Iterations are required to assure convergence of each domain and the coupling between the two domains.

The governing differential equations (Eqns [2.5]–[2.10] and [2.29]) and all related supplemental and boundary conditions were solved through the following iterative scheme.

1. Equation [2.29] and the associated boundary conditions were solved and the electromagnetic forces were calculated.
2. The momentum equations [2.6]–[2.8] and the associated boundary conditions were solved iteratively to obtain velocity and pressure distributions.
3. The energy equation [2.9] was solved for the enthalpy and temperature distributions. Iteration was performed between steps 2 and 3 until con-

verged solutions of velocity, pressure and temperature distributions were achieved.
4. The free-surface equation [2.15] was solved to obtain the new free surface and domain of the weld pool.
5. Advance to the next time step and back to step 2 until the desired time is reached.

The momentum differential equations are cast into the general format suggested by Patankar[112]:

$$\frac{\partial(\rho \boldsymbol{V})}{\partial t} + \nabla \cdot (\rho \boldsymbol{V}\boldsymbol{V}) = \nabla \cdot \boldsymbol{\tau} + \boldsymbol{S}_{\phi} \qquad [2.71]$$

where $\boldsymbol{\tau}$ is the viscous stress tensor and $\boldsymbol{S}_{\phi}$ is the source term, which includes the pressure gradient, the Darcy function, the relative phase motion, the plasma drag force and the body forces (electromagnetic force, gravitational force and buoyancy force) in the momentum equations. Equation [2.71] is solved in finite-difference form with a two-step projection method involving the time discretization of momentum (Eqns [2.6]–[2.8]). The equation for the first step is

$$\frac{\tilde{\boldsymbol{V}} - \boldsymbol{V}^{n}}{\delta t} = -\nabla \cdot (\boldsymbol{V}\boldsymbol{V})^{n} + \frac{1}{\rho^{n}} \nabla \cdot \boldsymbol{\tau}^{n} + \frac{1}{\rho^{n}} \boldsymbol{S}_{\phi}^{n} \qquad [2.72]$$

where the velocity field $\tilde{\boldsymbol{V}}$ is explicitly computed from incremental changes in the field $\boldsymbol{V}^{n}$ resulting from advection, viscosity and the source term. In the second step, the velocity field $\tilde{\boldsymbol{V}}$ is projected onto a zero-divergence vector field $\boldsymbol{V}^{n+1}$ using the following two equations:

$$\frac{\boldsymbol{V}^{n+1} - \tilde{\boldsymbol{V}}}{\delta t} = -\frac{1}{\rho^{n}} \nabla p^{n+1} \qquad [2.73]$$

$$\nabla \cdot \boldsymbol{V}^{n+1} = 0 \qquad [2.74]$$

These two equations can be combined into a single Poisson equation for the pressure, which is solved by the incomplete Cholesky conjugate gradient solution technique[113]:

$$\nabla \cdot \left(\frac{1}{\rho^{n}} \nabla p^{n+1} \right) = \frac{\nabla \cdot \tilde{\boldsymbol{V}}}{\delta \boldsymbol{t}} \qquad [2.75]$$

In the solution of the free-surface equation [2.15], $\tilde{F}$ was defined via

$$\tilde{F} = F^{n} - \delta t \nabla \cdot (\boldsymbol{V} F^{n}) \qquad [2.76]$$

Then, the new fluid domain F^{n+1} can be calculated via a divergence correction to obtain

$$F^{n+1} = \tilde{F} + \delta t(\nabla \cdot V)F^{n} \qquad [2.77]$$

The energy equation [2.9] and/or species equation [2.10] are solved explicitly to determine the distributions of temperature and concentration respectively. Following the marker-and-cell scheme[114], the x, y and z velocity components are located at cell face centers on lines of constants x, y and z respectively, while the pressure, VOF function, temperature and scalar electric potential are located at cell centers.

In the solution of the electric potential equation [2.29], the governing differential equation is discretized in the following form:

$$\alpha_P\phi_P = \alpha_E\phi_E + \alpha_W\phi_W + \alpha_T\phi_T + \alpha_B\phi_B + S \qquad [2.78]$$

where the values of a_P, a_E, etc., are the coefficients resulting from Eqn [2.29] and S is the source term. The discretized equations were solved iteratively using the tridiagonal matrix algorithm[115] and the grid system is shown in Fig. 2.32.

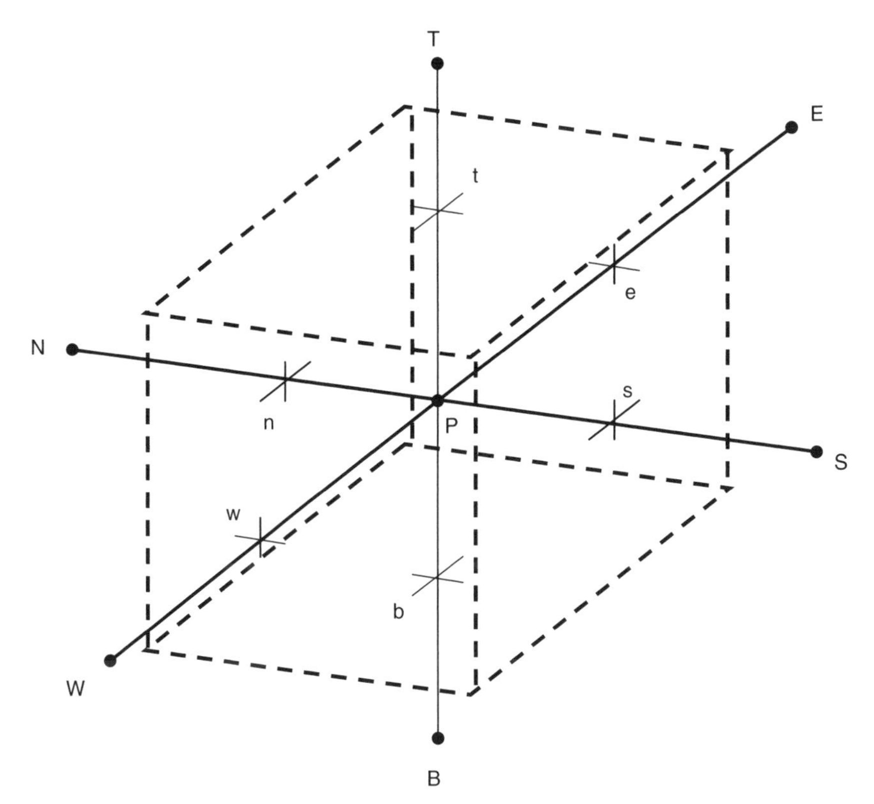

2.32 A cell in three dimensions and the neighboring nodes used in the calculations.

Convergence is declared when the following condition is satisfied:

$$\frac{\sum_{i=1}^{M}\sum_{k=1}^{N}|\phi_{i,k}^{m}-\phi_{i,k}^{m-1}|}{\phi_{\max}} < \varepsilon_{\phi}, \quad \varepsilon_{\phi} = 0.01 \qquad [2.79]$$

where M and N are the total numbers of nodes in the r and z directions respectively and m is the number of iterations.

As the weld pool and keyhole move in the welding direction, an adaptive grid system is employed, having finer grids in the weld pool and keyhole plasma zone. The finer grids concentrating on and around the weld pool and keyhole move with the weld pool and the keyhole as the welding proceeds. In the calculation, numerical stability is an important factor to be considered. The time step δt must be smaller than a certain critical value to prevent the unbounded growth of parasitic solutions of the difference equations. Once the mesh has been chosen, several restrictions are placed on δt to insure that it is below the critical value.

The first restriction is that material cannot move more than one cell width per time step. Therefore,

$$\delta t < \min\left\{\frac{\delta x_i}{|u_{i,j,k}|}, \frac{\delta y_j}{|v_{i,j,k}|}, \frac{\delta z_k}{|w_{i,j,k}|}\right\} \qquad [2.80]$$

where the minimum is to take over every cell in the mesh. Typically δt is chosen to be some fairly small fraction, say 0.25, of the minimum.

The explicit differencing of the viscous terms also limits the time step for the metal domain. It requires that

$$\upsilon \delta t < 0.5(\delta x_i^{-2} + \delta y_j^{-2} + \delta z_k^{-2})^{-1} \qquad [2.81]$$

where υ is the kinetic viscosity of the liquid metal.

Finally, the explicit treatment of the surface tension force requires that capillary waves do not travel more than one cell width in one time step. A rough estimate for this limit is

$$\gamma \delta t < \frac{\rho \delta x_{\mathrm{m}}^{3}}{4(1+\xi)} \qquad [2.82]$$

where δx_{m} is the minimum cell width in any direction anywhere in the mesh, γ is the surface tension coefficient and ξ is a factor that is zero in Cartesian coordinates, and unity in cylindrical coordinates.

2.6 Future trends

In general, welding is a very complicated technology involving many complicated physical processes such as arc–laser-induced plasma formation and recoil pressure formation, which are still not fully understood. All the results discussed in this chapter are based on the current existing published papers and some of these still need more investigation. With the advances of modern physics, the research on arc plasma formation, computation theories and the fast development of computer technology, people may have a better understanding of plasma formation, laser–plasma interaction, microstructure evolution, the deformed-surface-handling algorithm, etc. In that way, mathematical models can be employed to predict precisely the heat transfer phenomena in the welding process and to optimize the industrial welding processes to obtain quality welds with good mechanical and chemical properties.

2.7 Further reading

In this section, some common publications in welding fields are listed in the following.

1. Lancaster, J.F., *The Physics of Welding*, 2nd edition, Pergamon, Oxford, 1986.
2. Easterling, K., *Introduction to the Physical Metallurgy of Welding*, 2nd edition, Butterworth–Heinemann, Oxford, 1992.
3. Jenney, C.L., and O'Brien, A., *Welding Handbook*, 9th edition, American Welding Society, Miami, Florida, 2001.
4. Nipples, E.F., *Metals Handbook*, 9th edition, American Society For Metals, Metals Park, Ohio, 1983.
5. Kou, S., *Welding Metallurgy*, 2nd edition, John Wiley, New York, 2003.
6. Duley, W.W., *Laser Welding*, John Wiley, New York, 1999.
7. Dawes, C., *Laser Welding*, McGraw-Hill, New York, 1992.
8. Pionka, T.S., Voller, V., and Katgerman, L. (eds), *Modeling of Casting, Welding and Advanced Solidification Processes VI*, The Minerals, Metals and Materials Society, Warrendale, Pennsylvania, 1993.
9. Zumbrunnen, D.A. (ed.), *Heat and Mass Transport in Materials Processing and Manufacturing*, HTD-Vol. 261, American Society of Mechanical Engineers, New York, 1993.
10. *Journal of Applied Physics*, American Institute of Physics, College Park, Maryland.
11. *Journal of Physics D: Applied Physics*, Institute of Physics, London.
12. *Journal of Minerals, Metals and Materials*, The Minerals, Metals and Materials Society, Warrendale.
13. *Metallurgical and Materials Transactions*, ASM International, Materials Park, Ohio; The Minerals, Metals and Materials Society, Warrendale, Pennsylvania.
14. *Materials Science and Engineering*, Elsevier Science, Oxford.
15. *Welding in the World*, International Institute of Welding, Pergamon, Oxford.
16. *Acta Metallurgia et Materialia*, Elsevier Science, Tarrytown, New York.
17. *Proceedings of the International Congress on the Applications of Lasers and Electro-optics*, Laser Institute of America, Orlando, Florida.
18. *International Journal of Heat and Mass Transfer*, Science, Inc., Tarrytown, New York.

19. *Quarterly Journal of the Japan Welding Society*, Japan Welding Society, Tokyo.

2.8 References

1. Connor, L.P. (ed.), *Welding Handbook*, 8th edition, American Welding Society, Miami, Florida, 1987.
2. Rosenthal, D., 'Mathematical theory of heat distribution during welding and cutting', *Weld. J.*, **20**, 220-s, 1941.
3. Easterling, K., *Introduction to the Physical Metallurgy of Welding*, Butterworth–Heinemann, Oxford, 1992.
4. Kou, S., *Welding Metallurgy*, John Wiley, New York, 2003.
5. Jenney, C.L., and O'Brien, A., *Welding Handbook,* 9th edition, American Welding Society, Miami, Florida, 2001.
6. Kou, S., *Transport Phenomena and Materials Processing*, John Wiley, New York, 1996.
7. Oreper, G.M., Eagar, T.W., and Szekely, J., 'Convection in arc weld pools', *Weld. J.*, **62**, 307-s–312-s, 1983.
8. Lei, Y.P., and Shi, Y.W., 'Numerical treatment of the boundary conditions and source terms on a spot welding process with combining buoyancy-Marangoni-driven flow', *Numer. Heat Transfer*, **26B**, 455–471, 1994.
9. Matunawa, A., Yokoya, S., and Asako, Y., 'Convection in weld pool and its effect on penetration shape in stationary arc welds', *Q. J. Japan Weld. Soc.*, **6**, 3–10, 1987.
10. Zacharia, T., David, S.A., Vitek, J.M., and Debroy, T., 'Weld pool development during GTA and laser beam welding of Type 304 stainless steel, part 1 – theoretical analysis', *Weld. J.*, **68**, 499-s–509-s, 1989.
11. Choo, R.T.C., Szekley, J., and David, S.A., 'On the calculation of the free surface temperature of gas–tungsten arc weld pools from first principles: part 2. Modeling the weld pool and comparison with experiments', *Metall. Trans. B*, **23B**, 371–384, 1992.
12. Zacharia, T., Eraslan, A.H., and Aidun, K.K., 'Modeling of autogenous welding', *Weld. J.*, **67**, 53-s–62-s, 1988.
13. Zacharia, T., Eraslan, A.H., and Aidun, K.K., 'Modeling of nonautogenous welding', *Weld. J.*, **67**, 18-s–27-s, 1988.
14. Choo, R.T.C., Szekley, J., and Westhoff, R.C., 'Modeling of high current arcs with emphasis on free surface phenomena in the weld pool', *Weld. J.*, **69**, 346-s–361-s, 1990.
15. Tsai, M.C., and Kou, S., 'Electromagnetic-force-induced convection in weld pools with a free surface', *Weld. J.*, **69**, 241-s–246-s, 1990.
16. Kim, S.D., and Na, S.J., 'Effect of weld pool deformation on weld penetration in stationary gas tungsten arc welding', *Weld. J.*, **71**, 179-s–193-s, 1992.
17. Fan, H.G., Tsai, H.L., and Na, S.J., 'Heat transfer and fluid flow in a partially or fully penetrated weld pool in gas tungsten arc welding', *Int. J. Heat Mass Transfer*, **44**, 417–428, 2001.
18. Tsao, K.C., and Wu, C.S., 'Fluid flow and heat transfer in GMA weld pools', *Weld. J.*, **67**, 70-s–75-s, 1988.

19. Wang, Y., and Tsai, H.L., 'Impingement of filler droplets and weld pool dynamic during gas metal arc welding process', *Int. J. Heat Mass Transfer*, **44**, 2067–2080, 2001.
20. Wang, Y., and Tsai, H.L., 'Effects of surface active elements on weld pool fluid flow and weld penetration in GMAW', *Metall. Trans. B*, **32B**, 501–515, 2001.
21. Wang, Y., and Tsai, H.L., 'Effects of droplet size and drop frequency on flow patterns and weld penetration in GMAW', 2005 (to be published).
22. Kim, J.W., and Na, S.J., 'A study on the three-dimensional analysis of heat and fluid flow in gas metal arc welding using boundary-fitted coordinates', *Trans. ASME, J. Engng Industry*, **116**, 78–85, 1994.
23. Ushio, M., and Wu, C.S., 'Mathematical modeling of three-dimensional heat and fluid flow in a moving gas metal arc weld pool', *Metall. Trans.*, **28B**, 509–516, 1997.
24. Kothe, D.B., and R. C. Mjolsness, R.C., 'Ripple: a new model for incompressible flows with free surfaces', Report LA-UR-91–2818, Los Alamos National Laboratory, Los Alamos, New Mexico, USA, 1991.
25. Diao, Q.Z., and Tsai, H.L., 'Modeling of solute redistribution in the mushy zone during solidification of aluminum-copper alloys', *Metall. Trans.*, **24A**, 963–973, 1993.
26. Wang, Y., 'Modeling of three-dimensional gas metal arc welding process', PhD dissertation, University of Missouri, Rolla, Missouri, 1999.
27. Carman, P.C., 'Fluid flow through granular beds', *Trans. Inst. Chem. Engrs*, **15**, 150–166, 1937.
28. Kubo, K., and Pehlke, R.D., 'Mathematical modeling of porosity formation in solidification', *Metall. Trans.*, **16A**, 823–829, 1985.
29. Beavers, G.S., and Sparrow, E.M., 'Non-Darcy flow through fibrous porous media', *J. Appl. Mech.*, **36**, 711–714, 1969.
30. Sahoo, P., DeBroy, T., and McNallan, M.J., 'Surface tension of binary metal-surface active solute systems under conditions relevant to welding metallurgy', *Metall. Trans.*, **19B**, 483–491, 1988.
31. Zacharia, T., David, S.A., and Vitek, J.M., 'Effect of evaporation and temperature-dependent material properties on weld pool development', *Metall. Trans.*, **22B**, 233–241, 1992.
32. Allum, C.J., 'Metal transfer in arc welding as a varicose instability: I. varicose instabilities in a current-carrying liquid cylinder with surface charge', *J. Phys. D: Appl. Phys.*, **18**, 1431–1446, 1985.
33. Waszink, J.H., and Graat, L.H.J., 'Experimental investigation of the forces acting on a drop of weld metal', *Weld. J.*, **62**, 109-s–116-s, 1983.
34. Lancaster, J.F., *The Physics of Welding*, 2nd edition, Pergamon, Oxford, 1986.
35. Allum, C.J., 'Metal transfer in arc welding as a varicose instability: II. Development of model for arc welding', *J. Phys. D: Appl. Phys.*, **18**, 1447–1468, 1985.
36. Nemchinsky, V.A., 'Size and shape of the liquid droplet at the molten tip of an arc electrode', *J. Phys. D: Appl. Phys.*, **17**, 1433–1442, 1984.
37. Simpson, S.W., and Zhu, P.Y., 'Formation of molten droplets at a consumable anode in an electric welding arc', *J. Phys D: Appl. Phys.*, **28**, 1594–1600, 1995.
38. Haidar, J., 'A theoretical model for gas metal arc welding and gas tungsten arc welding', *J. Appl. Phys.*, **84**(7), 3518–3540, 1998.

39. Haidar, J., and Lowke, J.J., 'Predictions of metal droplet formation in arc welding', *J. Appl. Phys. D: Appl. Phys.*, **29**, 2951–2960, 1996.
40. Fan, H.G., and Kovacevic, R., 'Droplet formation, detachment, and impingement on the molten pool in gas metal arc welding', *Metall. Mater. Trans.*, **30B**, 791–801, 1999.
41. Zhu, F.L., 'A comprehensive dynamic model of the gas metal arc welding process', PhD Dissertation, University of Missouri, Rolla, Missouri, 2003.
42. Schlichting, H., *Boundary-layer Theory*, 6th edition, McGraw-Hill, New York, 1968.
43. Dinulescu, H.A., and Pfender, E., 'Analysis of the anode boundary layer of high intensity arcs', *J. Appl. Phys.*, **51**(6), 3149–3157, 1980.
44. Lowke, J.J, Kovitya, P., and Schmidt, H.P., 'Theory of free-burning arc columns including the influence of the cathode', *J. Phys D: Appl. Phys.*, **25**, 1600–1606, 1992.
45. Zhu, P.Y, Lowke, J.J., and Morrow, R., 'A unified theory of free-burning arcs, cathode sheaths and cathodes', *J. Phys D: Appl. Phys.*, **25**, 1221–1230, 1992.
46. Finkelnburg, W., and Segal, S.M., 'The potential field in and around a gas discharge, and its influence on the discharge mechanism', *Phys. Rev.*, **83**(3), 582–585, 1951.
47. Duley, W.W., *Laser Welding*, John Wiley, New York, 1999.
48. Ashby, M.F., and Easterling, K.E., 'The transformation hardening of steel surfaces by laser beams – I Hypo-eutectoid steels, *Acta Metall.*, **32**(11), 1935–1937, 1984.
49. Semak, V.V., Hopkins, J.A., McCay, M.H., and McCay, T.D., 'A concept for a hydrodynamic model of keyhole formation and support during laser welding', *Proc. 13th Int. Congr. on the Applications of Lasers and Electro-optics*, 1994, Laser Institute of America, Orlando, Florida, pp. 641–650, 1994.
50. Kroos, J., Gratzke, U., and Simon, G., 'Toward a self-consistent model of the keyhole in penetration laser beam welding', *J. Phys. D: Appl. Phys.*, **26**, 474–480, 1993.
51. Dowden, J., Postacioglu, N., Davis, M., and Kapadia, P., 'A keyhole model in penetration welding with a laser', *J. Phys. D: Appl. Phys.*, **20**, 36–44, 1987.
52. Andrews, J.G., and Atthey, D.R., 'Hydrodynamic limit to penetration of a material by a high-power beam', *J. Phys. D: Appl. Phys.*, **9**, 2181–2194, 1976.
53. Semak, V.V., Hopkins, J.A., McCay, M.H., and McCay, T.D., 'Dynamics of penetration depth during laser welding', *Proc. 13th Int. Congr. on the Applications of Lasers and Electro-optics*, 1994, Laser Institute of America, Orlando, Florida, pp. 17–20, 1994.
54. Hopkins, J.A., McCay, T.D., McCay, M.H., and Eraslan, A., 'Transient predictions of CO_2 laser spot welds in inconel 718', *Proc. 12th Int. Congr. on the Applications of Lasers and Electro-optics*, 1993, Laser Institute of America, Orlando, Florida, pp. 24–28, 1993.
55. Klemens, P.G., 'Heat balance and flow conditions for electron beam and laser welding', *J. Appl. Phys.*, **47**, 2165–2174, 1976.
56. Allmen, M., and Blatter, A., *Laser–Beam Interaction with Material*, 2nd edition, Springer, Berlin, 1995.
57. Solana, P., Kapadia, P., and Dowden, J., 'Surface depression and ablation for a weld pool in material processing: a mathematical model', *Proc. 17th Int. Congr.*

on the Applications of Lasers and Electro-optics, 1998, Laser Institute of America, Orlando, Florida, Section F, pp. 142–147, 1998.

58. Clucas, A., Ducharme, R., Kapadia, P, Dowden, J., and Steen, W., 'A mathematical model of laser keyhole welding using a pressure and energy balance at the keyhole walls', *Proc. 17th Int. Congr. on the Applications of Lasers and Electro-optics*, 1998, Laser Institute of America, Orlando, Florida, Section F, pp. 123–131, 1998.
59. Ducharme, R., Williams, K., Kapadia, P., Dowden, J., Steen, B., and Glowacki, M., 'The laser welding of thin metal sheets: an integrated keyhole and weld pool model with supporting experiments', *J. Phys. D: Appl. Phys.*, **27**, 1619–1627, 1994.
60. Sudnik, W., Rada, D., Breitschwerdt, S., and Erofeew, W., 'Numerical simulation of weld pool geometry in laser beam welding', *J. Phys. D: Appl. Phys.*, **33**, 662–671, 2000.
61. Zhou, J., 'Modeling the transport phenomena during laser welding process', PhD dissertation, University of Missouri, Rolla, Missouri, 2003.
62. Kapadia, P., Dowden, J., and Ducharme, R., 'A mathematical model of ablation in the keyhole and droplet formation in the plume in deep penetration laser welding', *Proc. 15th Int. Congr. on the Applications of Lasers and Electro-optics*, 1996, Laser Institute of America, Orlando, Florida, Section B, pp. 106–114, 1996.
63. Farson, D.F., and Kim, K.R., 'Simulation of laser evaporation and plume', *Proc. 17th Int. Congr. on the Applications of Lasers and Electro-optics*, 1998, Laser Institute of America, Orlando, Florida, Section F, pp. 197–206, 1998.
64. Miyamoto, I., Ohmura, E., and Maede, T., 'Dynamic behavior of plume and keyhole in CO_2 laser welding', *Proc. 16th Int. Congr. on the Applications of Lasers and Electro-optics*, 1997, Laser Institute of America, Orlando, Florida, Section G, pp. 210–218, 1997.
65. Siegel, R., and Howell, J.R., *Thermal Radiation Heat Transfer*, 3rd edition, Hemisphere, New York, 1992.
66. Ho, J.R., Grigoropoulos, C.P., and Humphrey, J.A.C., 'Gas dynamics and radiation heat transfer in the vapor plume produced by pulsed laser irradiation of aluminum', *J. Appl. Phys.*, **79**, 7205–7215, 1996.
67. Knight, C.J., 'Theoretical modeling of rapid surface vaporization with back pressure', *AIAA J.*, **17**(5), 519–523, 1979.
68. Anisimov, S.I., 'Vaporization of metal absorbing laser radiation', *Soviet Physics JETP*, **27**(1), 182–183, 1968.
69. Semak, V., and Matsunawa, A., 'The role of recoil pressure in energy balance during laser materials processing', *J. Phys. D: Appl. Phys.*, **30**, 2541–2552, 1997.
70. Kaplan, A., 'A model of deep penetration laser welding based in calculation of the keyhole profile', *J. Phys. D: Appl. Phys.*, **27**, 1805–1814, 1994.
71. Ishide, T., Tsubota, S., Nayama, M., Shimokusu, Y., Nagashima, T., and Okimura, K., '10 kW class YAG laser application for heavy components', *Proc. SPIE*, **3888**, 543–550, 1999.
72. Katayama, S., Seto, N., Kim, J., and Matsunawa, A., 'Formation mechanism and reduction method of porosity in laser welding of stainless steel', *Proc. 16th Int. Congr. on the Applications of Lasers and Electro-optics*, 1997, Laser Institute of America, Orlando, Florida, Section G, pp. 83–92, 1997.

73. Katayama, S., and Matsunawa, A., 'Formation mechanism and prevention of defects in laser welding of aluminum alloys', *Proc. 6th Int. Congr. on Welding and Melting by Election and Laser Beams*, Vol. 1, Toulon, France, 1998, Institute de Soudure, Paris, pp. 215–222, 1998.
74. Katayama, S., Seto, N., Kim, J., and Matsunawa, A., 'Formation mechanism and suppression procedure of porosity in high power laser welding of aluminum alloys', *Proc. 17th Int. Congr. on the Applications of Lasers and Electro-optics*, 1998, Laser Institute of America, Orlando, Florida, Section C, pp. 24–33, 1998.
75. Katayama, S., Seto, N., Mizutani, M., and Matsunawa, A., 'Formation mechanism of porosity in high power YAG laser welding', *Proc. 19th Int. Congr. on the Applications of Lasers and Electro-optics*, 2000, Laser Institute of America, Orlando, Florida, Section C, pp. 16–25, 2000.
76. Seto, N., Katayama, S., and Matsunawa, A., 'A high-speed simultaneous observation of plasma and keyhole behavior during high power CO_2 laser welding', *Proc. 18th Int. Congr. on the Applications of Lasers and Electro-optics*, 1999, Laser Institute of America, Orlando, Florida, Section E, pp. 19–27, 1999.
77. Katayama, S., Kohsaka, S., Mizutani, M., Nishizawa, K., and Matsunawa, A., 'Pulse shape optimization for defect prevention in pulsed laser welding of stainless steels', *Proc. 12th Int. Congr. on the Applications of Lasers and Electro-optics*, 1993, Laser Institute of America, Orlando, Florida, pp. 487–497, 1993.
78. Zhou, J., Tsai, H.L., and Wang, P.C., 'Investigations of mixing phenomena in hybrid laser–MIG welding', *Proc. 23rd Int. Congr. on the Applications of Lasers and Electro-optics*, San Francisco, California, USA, 4–7 October 2004, Laser Institute of America, Orlando, Florida, Paper 806, 2004.
79. Zhou, J., Tsai, H.L., Wang, P.C., and Menassa, R.J., 'Metal flow and porosity formation in pulsed laser keyhole welding', *Proc. 2004 ASME Heat Transfer–Fluids Engineering Conf.*, Charlotte, North Carolina, USA, 11–15 July 2004, American Society of Mechanical Engineers, New York, 2004.
80. Zhou, J., Tsai, H.L., and Wang, P.C., 'Effects of electromagnetic force on melt flow and weld shape in laser keyhole welding', *Proc. 23rd Int. Congr. on the Applications of Lasers and Electro-optics*, San Francisco, California, USA, 4–7 October 2004, Laser Institute of America, Orlando, Florida, Paper 807, 2004.
81. Summerviller, T., 'In the heat of the weld – latest advances in hybrid laser welding', http://www.cmit.csiro.au/innovation/2002–12/hybrid.cfm, 2002.
82. Steen, M., and Eboo, M., 'Arc augmented laser welding', *Metal Constr.*, **11**(7), 332, 1979.
83. Diebld, T.P., and Albright, C.E., 'Laser–GTA welding of aluminum alloy 5052', *Weld. J.*, **63**, 18–24, 1984.
84. Hamazaki, M., 'Effect of TIG or MIG augmented laser welding of thick mild steel plate', *Joining Mater.*, **7**, 31–34, 1988.
85. Dilthey, U., 'Laser arc hybrid welding – an overview', IIW Document XII-1710-02, International Institute of Welding, Roissy, 2002.
86. Staufer, H., Ruhrnobl, M., and Miessbacher, G., 'Hybrid welding for the automotive industry', *Ind. Laser Solutions*, **18**(2), 7–10, 2003.

87. Graf, T., and Staufer, H., 'Laser hybrid process at Volkswagen', IIW Document XII-1730-02, International Institute of Welding, Roissy, 2002.
88. Minami, K., Asai, S., Makino, Y., Shiihara, K., and Kanehara, T., 'Laser–MIG hybrid welding process for stainless steel vessels', IIW Document XII-1704-02, International Institute of Welding, Roissy, 2002.
89. Nielsen, S.E., Anderson, M.M., Kristensen, J.K., and Jensen, T.A., 'Hybrid welding of thick section C/Mn steel and aluminum', IIW Document XII-1731-02, International Institute of Welding, Roissy, 2002.
90. Engstrom, H., Nilsson, K., Flinkfeldt, J., Nilson, T., Skirfors, A., and Gustavsson, B., 'Laser hybrid welding of high strength steels', *Proc. 20th Int. Congr. on the Applications of Lasers and Electro-optics*, Jacksonville, USA, 15–18 October 2001, Laser Institute of America, Orlando, Florida, Paper 303, 2001.
91. Jokinen, T., Karhu, M., and Kujanpaa, V., 'Welding of thick austenitic stainless steel using Nd:YAG laser with filler wire and hybrid process', *Proc. 21st Int. Congr. on the Applications of Lasers and Electro-optics*, 2002, Scottsdale, Arizona, USA, Laser Institute of America, Orlando, Florida, Section A, 2002.
92. Nilsson, K., Engstrom, H., and Kaplan, A., 'Influence of butt- and T-joint preparation in laser arc hybrid welding', IIW Document XII-1732-02, International Institute of Welding, Roissy, 2002.
93. Ishide, T., Tsubota, S., Watanabe, M., and Ueshiro, K., 'Development of YAG laser and arc hybrid welding method', IIW Document XII-1705-02, International Institute of Welding, Roissy, 2002.
94. Ono, M., Shinbo, Y., Yoshitake, A., and Ohmura, M., 'Development of laser-arc hybrid welding', *NKK Tech. Rev.*, **86**, 2002.
95. Naito, Y., Katayama, S., and Matsunawa, A., 'Keyhole behavior and liquid flow in molten pool during laser-arc hybrid welding', *Proc. SPIE*, **4831**, 357–362, 2002.
96. Kutsuna, M., and Chen, L., 'Interaction of plasma in both CO_2 laser – MAG hybrid welding of carbon steel', IIW-Document XII-1708-02, International Institute of Welding, Roissy, 2002.
97. Zhou, J., Tsai, H.L, Wang, P.C., and Manessa, R.J., 'Modeling of hybrid laser-MIG keyhole welding process', *Proc. 22nd Int. Congr. on the Applications of Lasers and Electro-optics*, 2003, Laser Institute of America, Orlando, Florida, 2003.
98. Zhou, J., Zhang, W.H, Tsai, H.L, Marin, S.P., Wang, P.C., and Manessa, R.J., 'Modeling the transport phenomena during hybrid laser–MIG welding process', *Proc. ASME Heat Transfer Divison*, HTD-Vol. 374, No. 3, American Society of Mechanical Engineers, New York, pp. 161–168, 2003.
99. Xie, J., 'Dual beam laser welding', *Weld. J.*, **81**, 223-s–230-s, 2002.
100. Xie, J., 'Weld morphology and thermal modeling in dual-beam laser welding', *Weld. J.*, **81**, 283-s–290-s, 2002.
101. Gref, W., Hohenberger, B., Dausinger, F., and Hügel, H., 'Energy coupling and process efficiency in double-focus welding with Nd:YAG', *Proc. 21st Int. Congr. on the Applications of Lasers and Electro-optics*, 2002, Laser Institute of America, Orlando, Florida, 2002.
102. Grupp, M., Seefeld, T., and Sepold, G., 'Laser beam welding of aluminum alloys with diode pumped Nd:YAG lasers', *Proc. 20th Int. Congr. on the*

Applications of Lasers and Electro-optics, Jacksonville, Florida, USA, 15–18 October 2001, Laser Institute of America, Orlando, Florida, Section G, Paper 1607, 2001.

103. Iwase, T., Shibata, K., Sakamoto, H., Dausinger, F., Hohenberger, B., Muller, M., Matsunawa, A., and Seto, N., 'Real time X-ray observation of dual focus beam welding of aluminum alloys', *Proc. 19th Int. Congr. on the Applications of Lasers and Electro-optics*, 2000, Laser Institute of America, Orlando, Florida, Section C, pp. 26–34, 2000.
104. Fabbro, R., 'Basic processes in deep penetration laser welding', *Int. Congr. on the Applications of Lasers and Electro-optics*, 2002, Laser Institute of America, Orlando, Florida, 2002.
105. Coste, F., Janin, F., Jones, L., and Fabbro, R., 'Laser welding using Nd:Yag lasers up to 12 kW application to high thickness welding', *Int. Congr. on the Applications of Lasers and Electro-optics*, 2002, Laser Institute of America, Orlando, Florida, 2002.
106. Coste, F., Janin, F., Hamadou, M., and Fabbro, R., 'Deep penetration laser welding with Nd:Yag lasers combination up to 11kw laser power', *Proc. Int. Congr. on Laser Advanced Materials Processing*, Osaka, Japan 2002, SPIE, Bellingham, Washington, 2002.
107. Chen, T.C., and Kannatey-Asibu, E., Jr, 'Convection pattern and weld pool shape during conduction-mode dual beam laser welding', *Proc. North American Manufacturing Research Conf.*, Ann Arbor, Michigan, USA, 24 May 1996, Society of Manufacturing Engineers, Dearborn, Michigan, pp. 259–264.
108. Hou, C.A., 'Finite element simulation of dual-beam laser welding using ANSYS', *Proc. 4th Int. Cont. of the International Association for the Study of Traditional Environments*, May 1994, Swanson Analysis Systems, Pittsburgh, Pennsylvania, pp. 430–433, 1994.
109. Capello, E., Chiarello, P., Piccione, E., and Previtali, B., 'Analysis of high power CO_2 dual beam laser welding', *Advanced Manufacturing Systems and Technology, Courses and Lectures*, Publication 437, International Center for Mechanical Sciences, Udine, 2002.
110. Hu, J.L., and Tsai, H.L., 'Fluid flow and weld pool dynamics in dual-beam laser keyhole welding', *Proc. ASME Heat Transfer Division*, HTD-Vol. 374, No. 3, American Society of Mechanical Engineers, New York, pp. 151–159.
111. Zhou, J., Tsai, H.L., and Wang, P.C., 'Modeling the transport phenomena in moving 3-D dual-beam laser welding', *Proc. ASME Heat Transfer Conf.*, San Francisco, California, USA, 17–22 July 2005, American Society of Mechanical Engineers, New York, 2005.
112. Patankar, S.V., *Numerical Heat Transfer and Fluid Flow*, Hemisphere, New York, 1980.
113. Kerhaw, D.S., 'The incomplete Cholesky-conjugate gradient method for the interactive solution of systems of linear equations', *J. Comput. Phys.*, **26**, 43–65, 1978.
114. Welch, J.E., Harlow, F.H., Shannon, J.P., and Daly, B.J., 'The MAC method: A computing technique for solving viscous, incompressible, transient fluid-flow problems involving free surfaces', Report LA-3425, Los Alamos Scientific Laboratory, Los Alamos, New Mexico, 1966.
115. Versteeg, H.K., and Malalasekera, W., *An Introduction to Computational Fluid Dynamics, The Finite Volume Method*, Longman, London, 1999.

3
Thermal–metallurgical–mechanical interactions during welding

T. INOUE, Fukuyama University, Japan

3.1 Introduction

The effect of coupling between metallic structures, including the molten state, temperature, and stress and/or strain occurring in processes accompanied by phase transformation, sometimes plays an important role in such industrial processes as quenching, welding and continuous casting. Figure 3.1 represents the schematic features of the effect of metallothermomechanical coupling with the induced phenomena[1–8].

When the temperature distribution in a material varies, thermal stress ① is caused in the body, and the induced phase transformation ② affects the structural distribution, which is known as melting or solidification in the solid–liquid transition and pearlite or martensite transformation from austenite in the solid phase. Local dilatation due to structural changes in the body brings out the transformation stress ③ and interrupts the stress or strain field in the body.

In contrast with these phenomena, which are well known in ordinal analysis, arrows in the opposite direction indicate coupling in the following manner. Part of the mechanical work done by the existing stress in the material is converted into heat ④, which may be predominant in the case of inelastic deformation, thus disturbing the temperature distribution. The acceleration of phase transformation by stress or strain, which is called stress- or strain-induced transformation ⑤, has been discussed by metallurgists as one of leading parameters of transformation kinetics. Arrow ⑥ corresponds to the latent heat due to phase transformation, which is essential in determining the temperature.

The purpose of this chapter is to present a method that simulates such processes involving phase transformation when considering the effect of the coupling just mentioned. First, the formulation of the fundamental equations for stress–strain relationships, heat conduction and transformation kinetics based on continuum thermodynamics will be carried out, and then

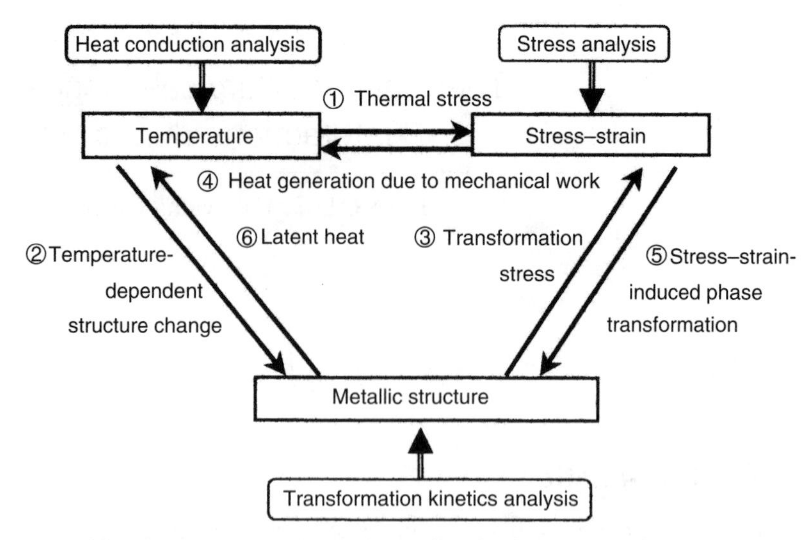

3.1 Metallo-thermo-mechanical coupling in the course of engineering processes incorporating phase transformation.

some examples of the numerical simulation of temperature, stress–strain and metallic structures in some welding processes will be presented.

3.2 Framework of governing equations

Consider a material undergoing structural change due to phase transformation as a mixture of N kinds of constituent[9]. Denoting the volume fraction of the Ith constituent as ξ_I (Fig. 3.2), the physical and mechanical properties x of the material are assumed to be a linear combination of the properties x_I of the constituent as

$$x = \sum_{I=1}^{N} x_I \xi_I \quad [3.1]$$

with

$$\sum_{I=1}^{N} \xi_I = 1, \quad [3.2]$$

where $\sum_{I=1}^{N}$ is the summation for subscript I from 1 to N. All material parameters appearing below are defined in the manner of Eqn [3.1].

The Gibbs free-energy density function G is defined as

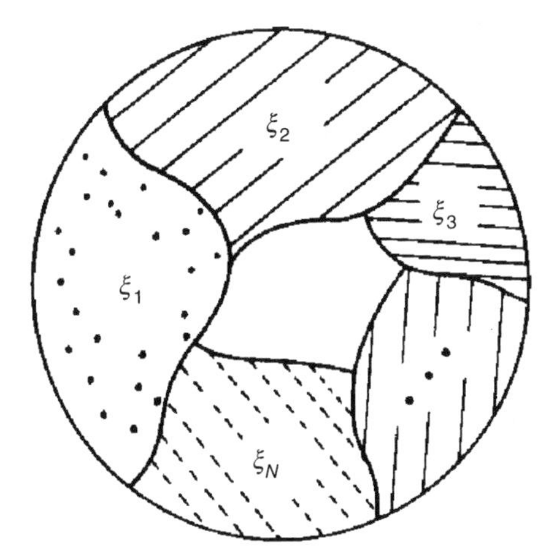

3.2 Illustration of a material point composed of *N* kinds of constituent.

$$G = U - T\eta - \frac{1}{\rho}\sigma_{ij}\varepsilon_{ij}^{e} \qquad [3.3]$$

where U, T, η and ρ are the internal energy density, temperature, entropy density and mass density respectively. The elastic strain rate $\dot{\varepsilon}_{ij}^{e}$ in Eqn [3.3] is defined as the subtraction of inelastic strain rate $\dot{\varepsilon}_{ij}^{i}$ from total strain rate $\dot{\varepsilon}_{ij}$, i.e.

$$\dot{\varepsilon}_{ij}^{e} = \dot{\varepsilon}_{ij} - \dot{\varepsilon}_{ij}^{i} \qquad [3.4]$$

The thermodynamic state of a material is assumed to be determined by stress σ_{ij}, temperature T, temperature gradient g_i (= grad T) and a set of internal variables of inelastic strain ε_{ij}^{i}, back stress α_{ij} and hardening parameter κ related to inelastic deformation, together with the volume fraction ξ_I of the constituents. Then, the general form of the constitutive equation can be expressed as

$$G = G(\sigma_{ij}, T, g_i, \varepsilon_{ij}^{i}, \alpha_{ij}, \kappa, \xi_I) \qquad [3.5]$$

$$\eta = \eta(\sigma_{ij}, T, g_i, \varepsilon_{ij}^{i}, \alpha_{ij}, \kappa, \xi_I) \qquad [3.6]$$

$$q_i = q_i(\sigma_{ij}, T, g_i, \varepsilon_{ij}^{i}, \alpha_{ij}, \kappa, \xi_I) \qquad [3.7]$$

$$\varepsilon_{ij}^{e} = \varepsilon_{ij}^{e}(\sigma_{ij}, T, g_i, \varepsilon_{ij}^{i}, \alpha_{ij}, \kappa, \xi_I) \qquad [3.8]$$

Here, q_i is the heat flux. The evolution equations for the internal variables are defined in the same form as Eqns [3.5]–[3.8], i.e.

$$\dot{\varepsilon}_{ij}^{i} = \dot{\varepsilon}_{ij}^{i}(\sigma_{ij}, T, g_i, \varepsilon_{ij}^{i}, \alpha_{ij}, \kappa, \xi_I) \qquad [3.9]$$

$$\dot{\alpha}_{ij} = \dot{\alpha}_{ij}(\sigma_{ij}, T, g_i, \varepsilon_{ij}^{i}, \alpha_{ij}, \kappa, \xi_I) \qquad [3.10]$$

$$\dot{\kappa} = \dot{\kappa}(\sigma_{ij}, T, g_i, \varepsilon^{\mathrm{i}}_{ij}, \alpha_{ij}, \kappa, \xi_I) \qquad [3.11]$$

$$\dot{\xi}_I = \dot{\xi}_I(\sigma_{ij}, T, g_i, \varepsilon^{\mathrm{i}}_{ij}, \alpha_{ij}, \kappa, \xi_I) \qquad [3.12]$$

Because of the restriction of the Clausius–Duhem inequality in strong form for every possible process

$$-\rho(\dot{G} + \eta\dot{T}) - \dot{\sigma}_{ij}\varepsilon^{\mathrm{e}}_{ij} + \sigma_{ij}\dot{\varepsilon}^{\mathrm{i}}_{ij} \geqslant 0 \qquad [3.13]$$

with

$$g_i q_i \leqslant 0 \qquad [3.14]$$

Eqns [3.5]–[3.12] are reduced to

$$G = G(\sigma_{ij}, T, \varepsilon^{\mathrm{i}}_{ij}, \alpha_{ij}, \dot{\kappa}, \xi_I) \qquad [3.15]$$

$$\varepsilon^{\mathrm{e}}_{ij} = \rho\frac{\partial}{\partial\sigma_{ij}}G(\sigma_{ij}, T, \varepsilon^{\mathrm{i}}_{ij}, \alpha_{ij}, \kappa, \xi_I) \qquad [3.16]$$

$$\eta = -\rho\frac{\partial}{\partial T}G(\sigma_{ij}, T, \varepsilon^{\mathrm{i}}_{ij}, \alpha_{ij}, \kappa, \xi_I) \qquad [3.17]$$

$$q_i = -k(\sigma_{ij}, T, \varepsilon^{\mathrm{i}}_{ij}, \alpha_{ij}, \kappa, \xi_I)g_i \qquad [3.18]$$

Here, Eqn [3.18] denotes the Fourier law with positive thermal conductivity k.

3.3 Stress–strain constitutive equation

To obtain an explicit expression for the elastic strain in Eqn [3.16], the Gibbs free energy G is assumed to be determined by that of constituent G_I in the form of Eqn [3.1] as

$$G(\sigma_{ij}, T, \varepsilon^{\mathrm{i}}_{ij}, \alpha_{ij}, \kappa, \xi_I) = \sum_{I=1}^{N}\xi_I G_I(\sigma_{ij}, T, \varepsilon^{\mathrm{i}}_{ij}, \alpha_{ij}, \kappa) \qquad [3.19]$$

When G_I is divided into the elastic and inelastic parts as

$$G_I(\sigma_{ij}, T, \varepsilon^{\mathrm{i}}_{ij}, \alpha_{ij}, \kappa, \xi_I) = G^{\mathrm{e}}_I(\sigma_{ij}, T) + G^{\mathrm{i}}_I(T, \varepsilon^{\mathrm{i}}_{ij}, \alpha_{ij}, \kappa) \qquad [3.20]$$

we can derive the elastic strain from Eqn [3.16] by expanding the elastic part G^{e}_I around the natural state, $\sigma_{ij} = 0$ and $T = T_0$, in terms of the representation theorem for an isotropic function:

$$\begin{aligned}\varepsilon^{\mathrm{e}}_{ij} = {} & \left(\sum_{I=1}^{N}\frac{1+\nu_I}{E_I}\xi_I\right)\sigma_{ij} - \left(\sum_{I=1}^{N}\frac{\nu_I}{E_I}\xi_I\right)\delta_{ij}\sigma_{kk} \\ & + \delta_{ij}\int_{T_0}^{T}\sum_{I=1}^{N}\alpha_I\xi_I \,\mathrm{d}T + \delta_{ij}\sum_{I=1}^{N}\beta_I(\xi_I - \xi_{I0})\end{aligned} \qquad [3.21]$$

Here, E_I, ν_I, α_I and β_I correspond to Young's modulus, Poisson's ratio, the thermal expansion coefficient and dilatation respectively of the Ith constituent.

3.3.1 Plastic strain rate

Assume that the evolution of back stress α_{ij} of the yield surface and hardening parameter κ can be determined by

$$\dot{\alpha}_{ij} = C(T, \kappa, \xi_I)\dot{\varepsilon}^{\mathrm{i}}_{ij} \tag{3.22}$$

$$\dot{\kappa} = \dot{\bar{\varepsilon}}^{\mathrm{i}} = \left(\frac{2}{3}\dot{\varepsilon}^{\mathrm{i}}_{ij}\dot{\varepsilon}^{\mathrm{i}}_{ij}\right)^{1/2} \tag{3.23}$$

where $\dot{\bar{\varepsilon}}^{\mathrm{i}}$ represents the equivalent inelastic strain rate. To take into account the effect of changing the structural fraction ξ_I, we take the form of the yield function as

$$F = F(\sigma_{ij}, T, \varepsilon^{\mathrm{i}}_{ij}, \xi_I) = \left[\frac{2}{3}(s_{ij} - \alpha_{ij})^2\right]^{1/2} - K(T, \kappa, \xi_I) \tag{3.24}$$

where s_{ij} $(= \sigma_{ij} - \frac{1}{3}\delta_{ij}\sigma_{kk})$ represents the deviatoric stress component. Employing the normality rule and the consistency relationship, the final form of the time-independent inelastic strain rate, or plastic strain rate, is

$$\dot{\varepsilon}^{\mathrm{i}}_{ij} = \Lambda\frac{\partial F}{\partial \sigma_{ij}} = \hat{G}\left(\frac{\partial F}{\partial \sigma_{kl}}\dot{\sigma}_{kl} + \frac{\partial F}{\partial T}\dot{T} + \sum_{I=1}^{N}\frac{\partial F}{\partial \xi_I}\dot{\xi}_I\right)\frac{\partial F}{\partial \sigma_{ij}} \tag{3.25}$$

with

$$\frac{1}{\hat{G}} = -\left[\frac{\partial F}{\partial \sigma_{mn}}\frac{\partial F}{\partial \varepsilon^{i}_{mn}} + \frac{\partial F}{\partial \sigma_{mn}}\frac{\partial F}{\partial \alpha_{mn}} + \left(\frac{2}{3}\frac{\partial F}{\partial \sigma_{mn}}\frac{\partial F}{\partial \sigma_{mn}}\right)^{1/2}\frac{\partial F}{\partial \kappa}\right] \tag{3.26}$$

Equation [3.25] means that plastic strain is induced not only by stress but also by the temperature and phase change.

3.3.2 Viscoplastic strain rate

The elastic–plastic constitutive relationship is suitable for describing the material behavior at relatively low temperatures. However, time dependence or viscosity might predominate at a higher temperature level, particularly when the material behaves like a viscous liquid beyond its melting point. In order to analyze such processes as welding and casting, in which melting and/or solidification of the metal are essential phenomena, adequate formulation of the viscoplastic constitutive model is needed.

Malvern[10] and Perzyna[11] proposed a viscoplastic constitutive equation for time-dependent inelastic strain rate $\dot{\varepsilon}_{ij}^{\mathrm{i}}$ in the form

$$\dot{\varepsilon}_{ij}^{\mathrm{i}} = \frac{1}{3\mu}\langle \psi(F) \rangle \frac{\partial F}{\partial \sigma_{ij}} \quad [3.27]$$

with the static yield function

$$F = \frac{f(\sigma_{ij}, T, \varepsilon_{ij}^{\mathrm{i}})}{K} - 1 \quad [3.28]$$

where μ and K denote the coefficient of viscosity and the static flow stress respectively, and

$$\langle \psi(F) \rangle = \begin{cases} 0 & \text{if } \psi(F) \leqslant 0 \\ \psi(F) & \text{if } \psi(F) > 0 \end{cases} \quad [3.29]$$

Equation [3.27] indicates that the inelastic strain rate is induced in an outer direction normal to the static yield surface F, and that the magnitude of the strain rate depends on the ratio of excess stress $\tilde{f} - K$ to the flow stress K. If we adopt the flow rule (Eqn [3.27]) to the liquid state, the flow stress tends to vanish ($K \to 0$) and the yield surface F expands infinitely ($F \to \infty$), which implies that the strain rate is infinite at low stresses. To compensate such an inconsistency occurring in a liquid, a modification to Eqn [3.28] is made such that

$$F = f(\sigma_{ij}, T, \varepsilon_{ij}^{\mathrm{i}}) - K(T, \kappa, \xi_I) \quad [3.30]$$

When we take the simple forms of functions ψ for F

$$\psi(F) = F \quad [3.31]$$

$$F = \left[\frac{3}{2}(s_{ij} - \alpha_{ij})(s_{ij} - \alpha_{ij}) \right]^{1/2} - K(T, \kappa, \xi_I) \quad [3.32]$$

Eqn [3.27] can be reduced to

$$\dot{\varepsilon}_{ij}^{\mathrm{i}} = \frac{1}{2\mu}\left\langle 1 - \frac{K(T, \kappa, \xi_I)}{[3(s_{kl} - \alpha_{kl})(s_{kl} - \alpha_{kl})]^{1/2}} \right\rangle (s_{ij} - \alpha_{ij}) \quad [3.33]$$

This constitutive relationship may be relevant to a liquid–solid transition region with high viscosity, as well as to a normal time-independent plastic body[6]. For instance, when flow stress K equals zero in Eqn [3.33], the total strain rate $\dot{\varepsilon}_{ij}$ ($= \dot{\varepsilon}_{ij}^{\mathrm{e}} + \dot{\varepsilon}_{ij}^{\mathrm{i}}$) is given by

$$\dot{\varepsilon}_{ij} = \frac{1+\nu}{E}\dot{\sigma}_{ij} - \frac{\nu}{E}\dot{\sigma}_{kk}\delta_{ij} + \frac{1}{2\mu}s_{ij} \quad [3.34]$$

when the effects of temperature and phase change are neglected for simplicity. This equation is equivalent to the Maxwell constitutive model for a viscoelastic body.

When the elastic component of shear deformation is sufficiently small compared with the viscoplastic component, as is usual for a viscous fluid, the Newtonian fluid model

$$\sigma_{ij} = 2\mu\dot{\varepsilon}_{ij} - \frac{2}{3}\mu\dot{\varepsilon}_{kk}\delta_{ij} - p\delta_{ij}, \quad p = -\frac{1}{3}\sigma_{kk} \qquad [3.35]$$

is obtainable from Eqn [3.34] by neglecting the elastic shear strain rate. Furthermore, when the elastic volume dilatation ε_{kk}^{e} is removed from Eqns [3.35], we have

$$\sigma_{ij} = 2\mu\dot{\varepsilon}_{ij} - p\delta_{ij} \qquad [3.36]$$

which represents the model for an incompressible Newtonian fluid. In the limiting case for an inviscid material ($\mu \to 0$), $\psi(F)$ in Eqn [3.27] tends to infinity and

$$\frac{1}{3\mu}\psi(F) \equiv \Lambda = \text{constant} \qquad [3.37]$$

should hold to give the form

$$\dot{\varepsilon}_{ij}^{i} = \Lambda\frac{\partial F}{\partial \sigma_{ij}} \qquad [3.38]$$

The parameter Λ can be easily determined by applying the consistency relationship, and thus we return again to the previous discussion for time-independent plastic strain. From the considerations just mentioned, the constitutive relationship developed in Eqn [3.33] seems to be useful for a wide range of metals from inelastic solids to viscous fluids.

3.4 Heat conduction equation

When we adopt Eqns [3.15]–[3.18] for the energy conservation law

$$\rho\dot{U} - \sigma_{ij}\dot{\varepsilon}_{ij} + \frac{\partial q_i}{\partial x_I} = 0 \qquad [3.39]$$

the equation of heat conduction

$$\rho c\dot{T} - k\frac{\partial^2 T}{\partial x_i \partial x_i} + \rho\sum_{I=1}^{N} l_I\dot{\xi}_I + T\frac{\partial \varepsilon_{ij}^{e}}{\partial T}\dot{\sigma}_{ij} + \left(\rho\frac{\partial H}{\partial \varepsilon_{ij}^{i}}\dot{\varepsilon}_{ij}^{i} + \rho\frac{\partial H}{\partial \alpha_{ij}}\dot{\alpha}_{ij} + \rho\frac{\partial H}{\partial \kappa}\dot{\kappa} - \sigma_{ij}\dot{\varepsilon}_{ij}^{i}\right) = \rho r \qquad [3.40]$$

holds with the enthalpy density $H(\sigma_{ij}, T, \xi_I)$ and latent heat l_I due to the increase in the Ith phase given by

$$l_I = \frac{\partial H}{\partial \xi_I} \tag{3.41}$$

The fifth term on the left-hand side of Eqn [3.40] denotes the heat generation by inelastic dissipation, which is significant when compared with the elastic work represented by the fourth term, and the third term arises from the latent heat through phase changes. Hence, it can be seen that Eqn [3.40] corresponds to the ordinal equation of heat conduction, provided that these terms are neglected.

3.5 Kinetics of phase transformation

During phase transformation, a given volume of material is assumed to be composed of several kinds of constituent ξ_I as expressed in Eqn [3.1]. We choose four kinds of volume fraction, namely liquid ξ_L, austenite ξ_A, pearlite ξ_P and martensite ξ_M, and other structures induced by precipitation by recovery effect, say, during the annealing process.

For the rate of solidification in steel, we employ the well-known lever rule (Fig. 3.3 is an example), and the volume fraction of austenite is

$$\xi_A = \frac{(T_L - T)/m_L}{(T - T_A)/m_A + (T_L - T)/m_L} \tag{3.42}$$

where T_L and T_A denote the liquidus and solidus temperatures respectively, and m_L and m_A are the gradients of the liquidus and solidus temperatures respectively with respect to the carbon content in the phase diagram.

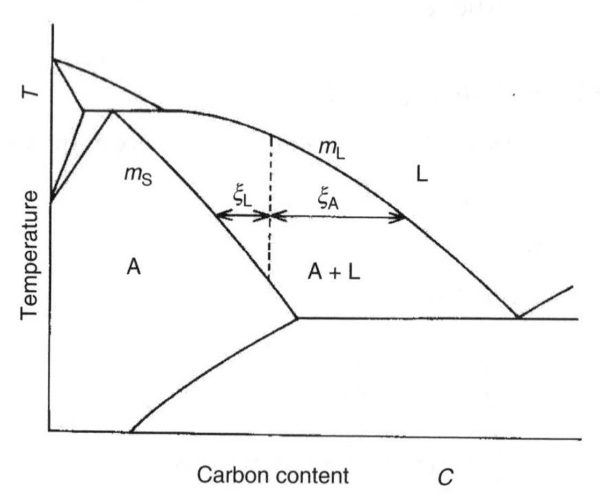

3.3 An example of a phase diagram.

When austenite is cooled in equilibrium, bainite, ferrite and carbite are produced in addition to pearlite, but for brevity all these structures resulting from a diffusion type of transformation are called pearlite. The nucleation and growth of pearlite in an austenitic structure are phenomenologically governed by the mechanism for a diffusion process, and Johnson and Mehl[12] proposed the following formula for the volume fraction ξ_P:

$$\xi_P = 1 - \exp(-V_e) \qquad [3.43]$$

where V_e is the extended volume of the pearlitic structure given by

$$V_e = \int_0^t \frac{4}{3}\pi R(t-\tau)^3 n \, d\tau \qquad [3.44]$$

Here, R is the moving rate of the surface of pearlite. Bearing in mind that the value of R is generally a function of stress as well as temperature, Eqn [3.44] may be reduced to

$$V_e = \int_0^t f(T, \sigma_{ij})(t-\tau)^3 \, d\tau \qquad [3.45]$$

The function $f(T, 0)$ can be determined by fitting the temperature–time–transformation diagram (TTT) or continuous-cooling transformation (CCT) diagram without stress, and $f(T, \sigma_{ij})$ may be given by the start-time or finish-time data for pearlite transformation with an applied stress, an example of which is shown in Fig. 3.4[13].

The empirical relationship for the austenite–martensite transformation is also obtainable by modifying the kinetic theory of Magee[14]. Assume that the growth of a martensite structure is a linear function of the increase in the difference ΔG in between the free energies of austenite and martensite as

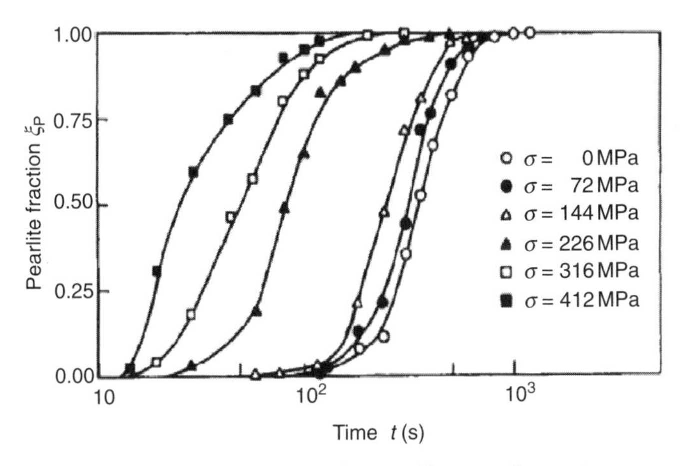

3.4 Dependence of stress on the pearlite transformation.

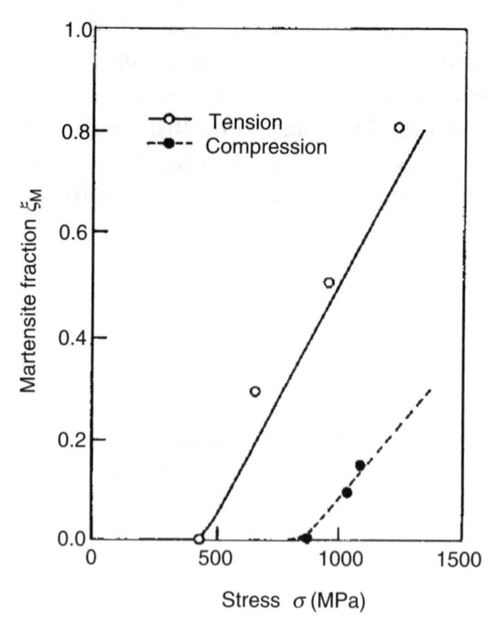

3.5 Dependence of stress on the induced martensite.

$$d\xi_M = -\bar{v}(1-\xi_M)\phi\, d(\Delta G) \qquad [3.46]$$

Regarding the free energy G as a function of temperature and stress, we can obtain the form of ξ_M by integrating Eqn [3.46] as

$$\xi_M = 1 - \exp[\phi_1(T - M_s) + \phi_2(\sigma_{ij})] \qquad [3.47]$$

The function ϕ_2 (σ_{ij}) is identified by data such as those shown in Fig. 3.5[15].

3.6 Simulating stresses in the welding process

3.6.1 Bead on the center of a circular plate

A bead is mounted in a center hole of a circular plate[16] of a low-carbon steel (SM41) for welding with a diameter of 300 mm and a thickness of 8 mm (Fig. 3.6) as the first simple example of axisymmetric welding simulation. Tension tests were carried out at temperature levels from ambient to 1100 °C under several strain rates in order to determine the material parameters of the viscoplastic constitutive relation [3.33]. A CCT diagram in Fig. 3.7 was used to determine the form of function F in Eqn [3.24], and the solidus temperature T_S and liquidus temperature T_L were 1480 °C and 1530 °C respectively.

In order to perform the calculation, the heat input from the electrode was assumed to be 299 J/s for the circular plates, and the heat transfer coef-

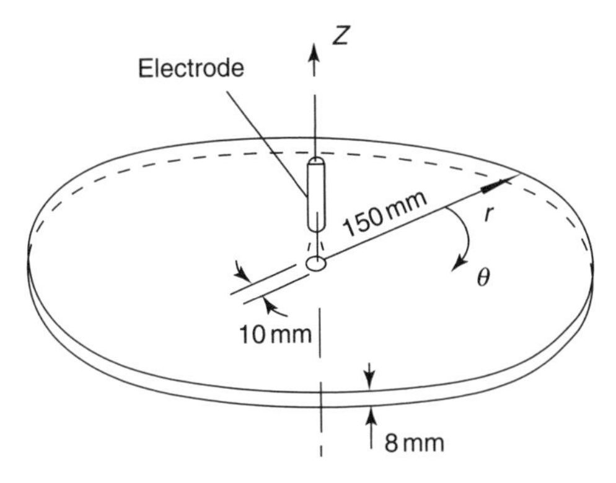

3.6 A circular plate with a hole.

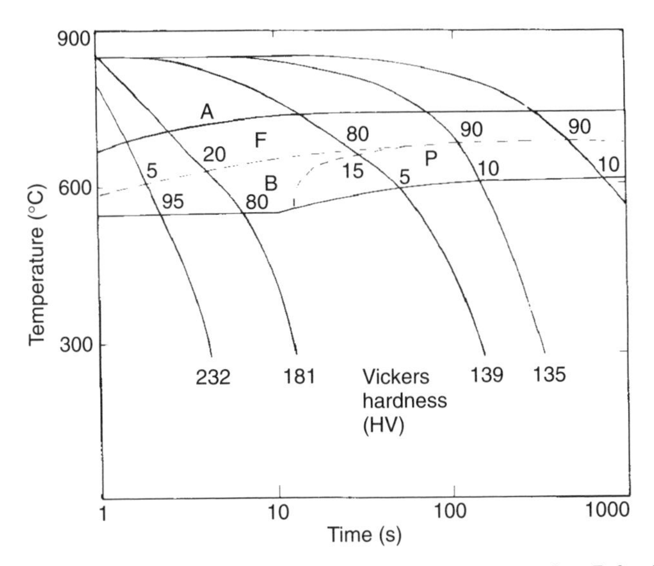

3.7 CCT diagram of the steel employed: A, austenite; F, ferrite; B, bainite; P, pearlite.

ficient on the surface of the plate during cooling was taken to be $2.31 \times 10^{-8}\,(T + T_a)(T^2 + T_a^2)\,\mathrm{W/m^2K^4}$. The effect of convection on the coefficient is not taken into account in the first approximation.

The finite-element method was used by dividing the plate into 29 elements along the radial direction, assuming for simplicity that there is no temperature or stress distribution in the z direction since the diameter is so much greater than the thickness.

The curves in Fig. 3.8 represent the temperature history at four characteristic points of r and agree fairly well with the experimental data

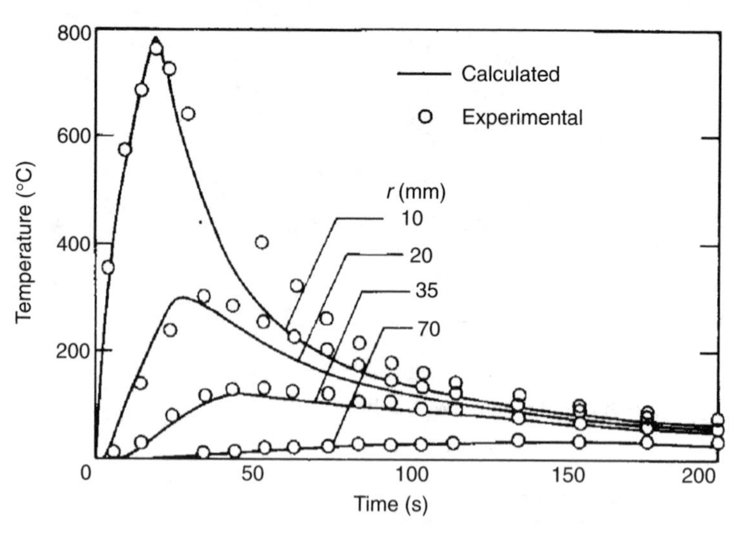

3.8 Temperature variation at characteristic points.

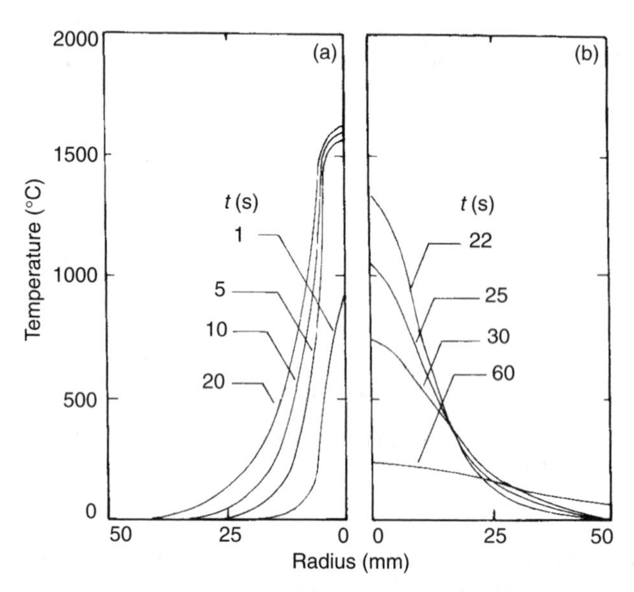

3.9 Changes in temperatures distributions near the hole in (a) the heating and (b) the cooling processes.

measured by thermocouples. The temperature distribution in Fig. 3.9 seems rather steep toward the welding point; so only the region of the plate near the hole is transformed into liquid. A similar situation is apparent from Fig. 3.10, representing the phase distribution; the region of liquid is

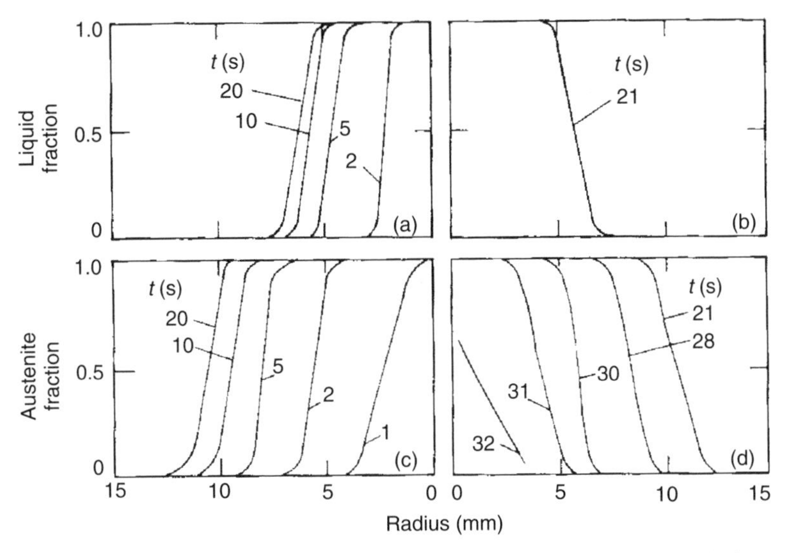

3.10 Changes in the volume fractions of liquid and solid in (a), (c) the heating and (b), (d) the cooling process.

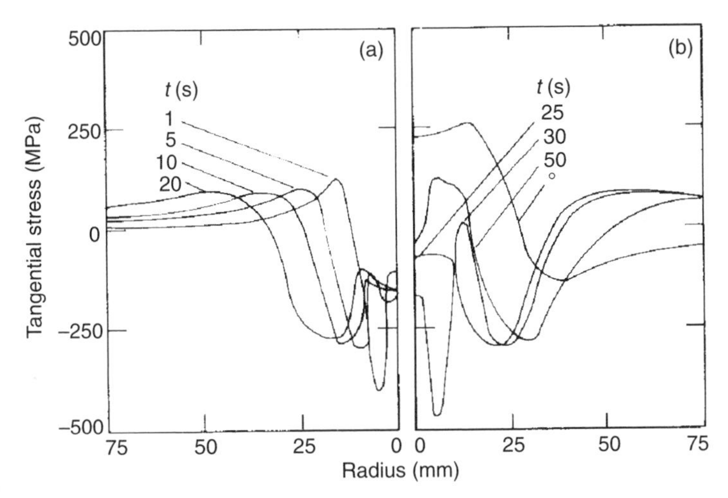

3.11 Changes in the tangential stress distributions in (a) the heating and (b) the cooling process.

restricted to $r < 6$ mm, and only the area for which $r < 10$ mm is heated beyond the A_{c3} temperature.

Figure 3.11 shows the calculated distribution of tangential stress. The complicated changes in the range $r = 5$–10 mm during heating, and at $r = 5$ mm for $t = 30$ s during cooling, affect the dependence of transformation stress on phase changes. The calculated residual stress distributions (curves

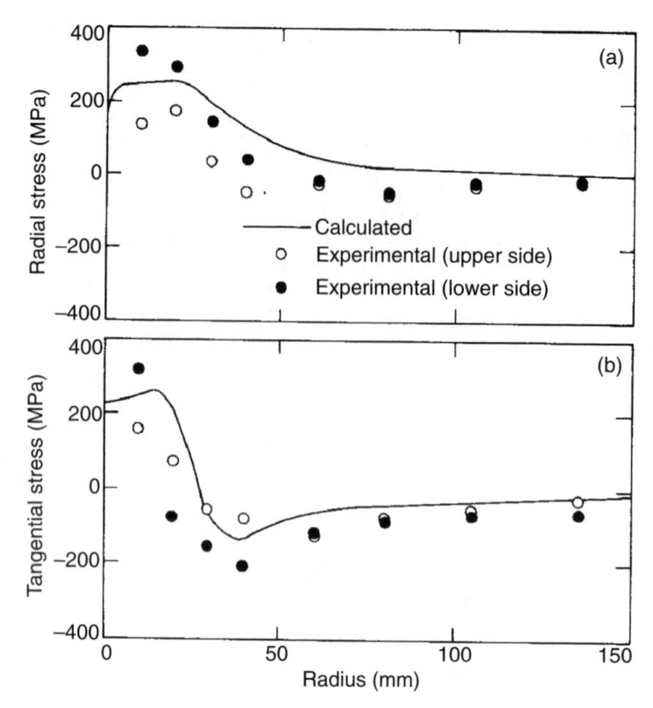

3.12 Distributions of (a) radial and (b) tangential residual stresses.

in Fig. 3.12) agree well with the stress measured by the X-ray diffraction technique which uses the premise that high stresses caused near the center decrease rapidly along the radial direction.

3.6.2 Butt welding of plates

Two rectangular plates of dimensions 500 mm × 145 mm × 8 mm of the same carbon steel (SM41) are butt welded[16] under conditions of 170 A and 28 V with a steady electrode travel speed of 4.7 mm/s along the longitudinal direction (Fig. 3.13). The calculated profiles of temperature are successively shown in the bird's-eye view of Fig. 3.14. The variation in the temperature at several points is given in Fig. 3.15 with data measured by thermocouples. Both the calculated and the experimental results show the temperature changes according to the travel of the electrode.

Since the thickness of the plate was sufficiently small compared with other dimensions, plane-stress conditions are assumed throughout the stress analysis. The calculated distribution of stress is partly shown in Fig. 3.16. It is shown that residual stress σ_x ($t = \infty$) at the periphery of the welding line is as high as the flow stress except at both edges, and the peaks in σ_y appear

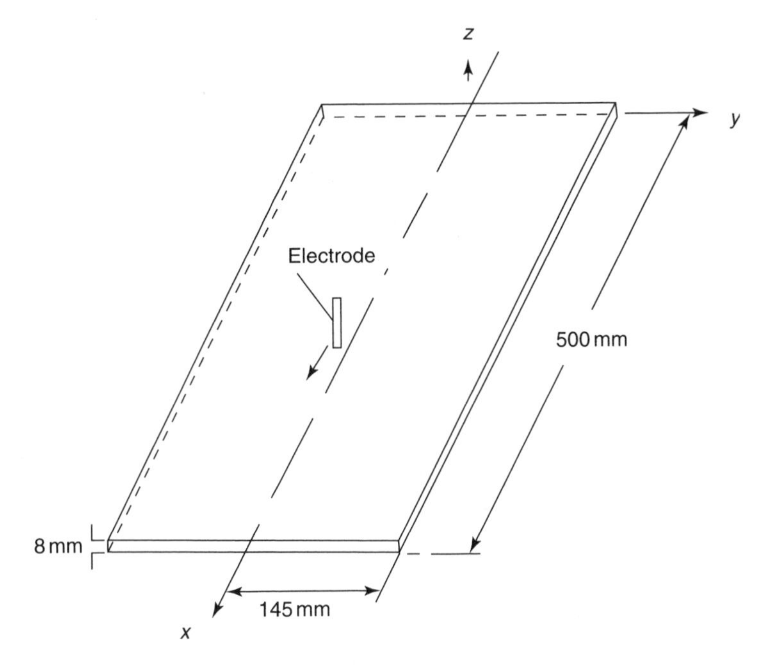

3.13 Butt-welded plates.

at about 100 mm inside the edge. The residual stresses at the midsection of the plate are plotted in Fig. 3.17 with the experimental data measured on the upper and lower sides of the plate by the X-ray diffraction technique.

3.6.3 Bead on plate

The equations developed in the previous section are now applied to simulation of the welding process of a bead on a plate[17] of type 304 stainless steel. The hatched cross-section in Fig. 3.18(a) is the region over which the calculation is executed and where the plane-strain condition is assumed. The electrode is assumed to travel along the x axis at a velocity of 5 mm/s with a bead width of 6 mm. The quantity of heat supplied is 4000 W, and the time required for the supply is 1.2 s. Figure 3.18(b) indicates the finite-element mesh division with 279 elements and 169 nodes. The displacement constraint condition is illustrated in Fig. 3.18(c), while no external force is applied.

Simulated variations in temperature, deformation, velocity and the x, y and z components of the stress fields with time are shown in Fig. 3.19, Fig. 3.20, Fig. 3.21, Fig. 3.22, Fig. 3.23 and Fig. 3.24 respectively. All figures show the results for a part of half of the cross-section within $y \leq 20$ mm, where the heat is supplied from the upper side. The change in distortion of

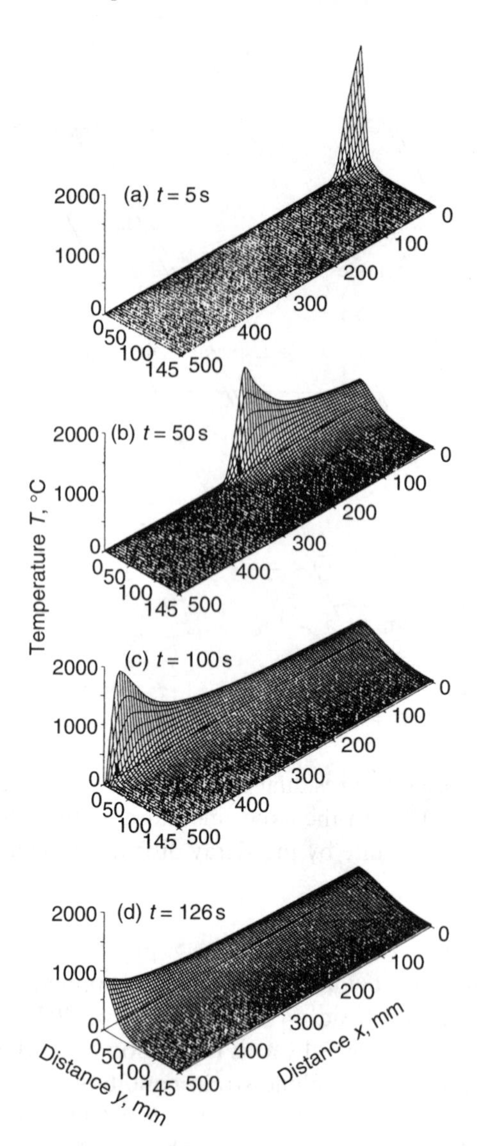

3.14 Bird's-eye-view of temperature distribution with successive movement of electrode.

the cross-section is represented as ten times the elongation throughout the figures.

It is known from Fig. 3.19 indicating the temperature distribution that the center is heated at the beginning of the welding operation, but that heat diffuses to the outer area at the final stage.

The distortion of finite elements is shown in Fig. 3.20. Lines near the left top indicate the volume fraction ξ_L of liquid phase, where $\xi_L = 1.0$ denotes

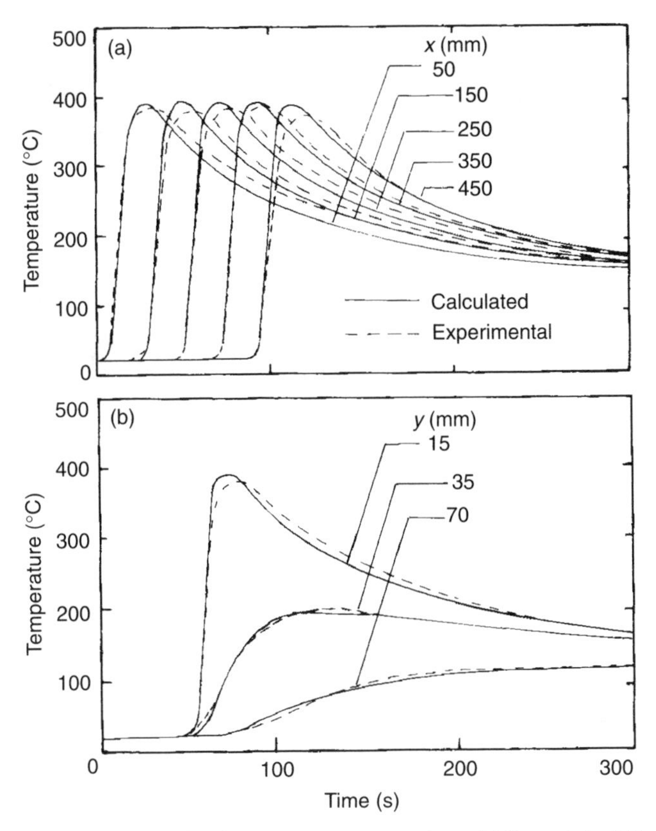

3.15 Variations in the temperature at the middle point of the plate.

the completely molten state. The supplied heat induces melting in the upper side of the section, and the deformation caused by thermal and transformation strain is observed in Fig. 3.20(a). The growth of the molten pool and the swelling of the surface become evident with the progress of the operation, as seen in Fig. 3.20(b) and Fig. 3.20(c); the induced deformations are depicted in the final stage of heating in Fig. 3.20(e). All this means that the plate deforms in the heating process to affect the final shape of the weldments.

Figure 3.21 indicates the change of velocity distribution. An upward flow can be observed at the beginning (Fig. 3.21(a)). The difference between the velocities of the liquid and solid phases is apparent in the mushy zone at $t = 0.6$ s in Fig. 3.21(b), and this occurs because the jump in velocity is caused by the difference between the densities of the solid and liquid phases on the interface in the melting process. In Fig. 3.21(a), Fig. 3.21(b) and Fig. 3.21(c) it can be observed that the jump affects the velocity distribution in the molten pool.

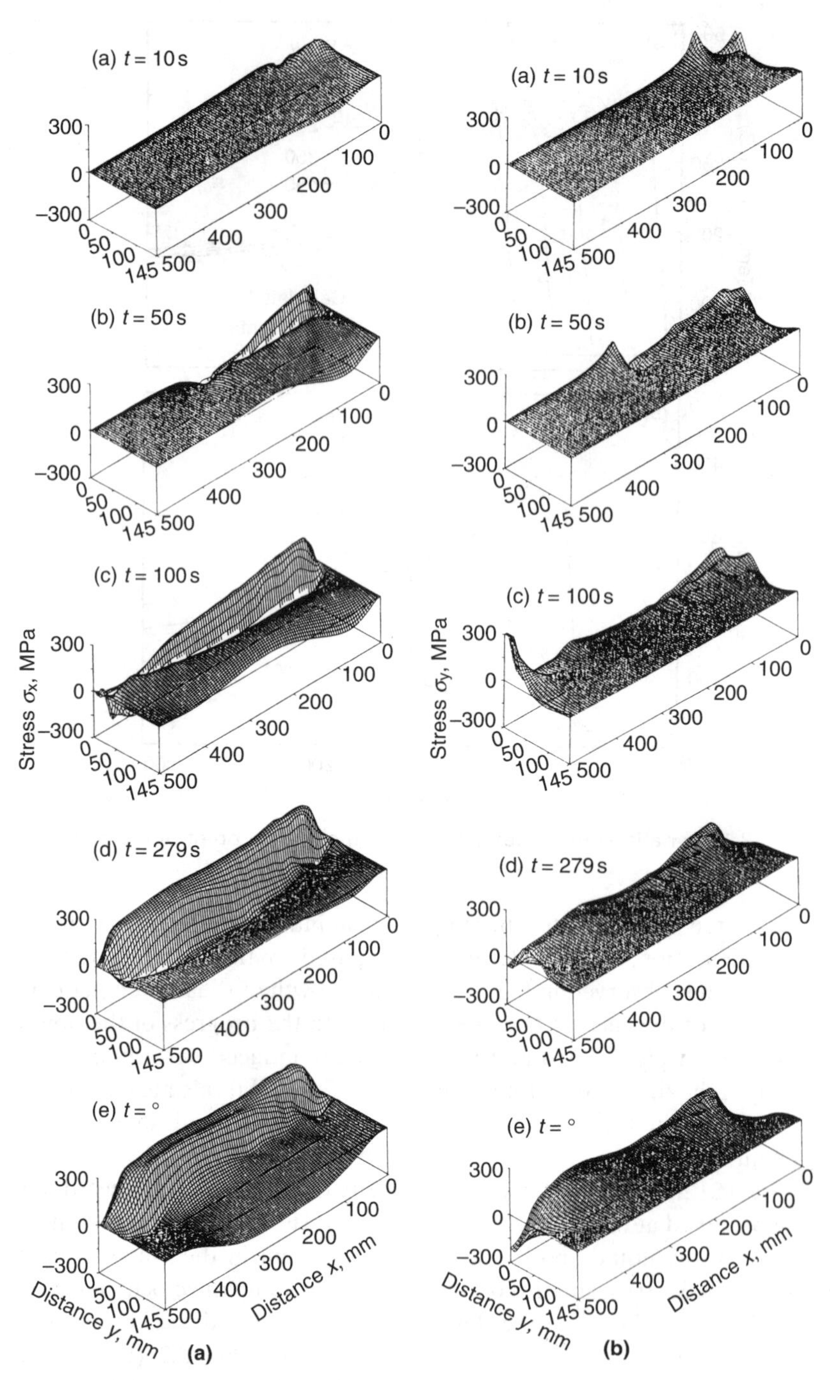

3.16 Bird's-eye-view of distribution of stresses: (a) x- and (b) y-direction.

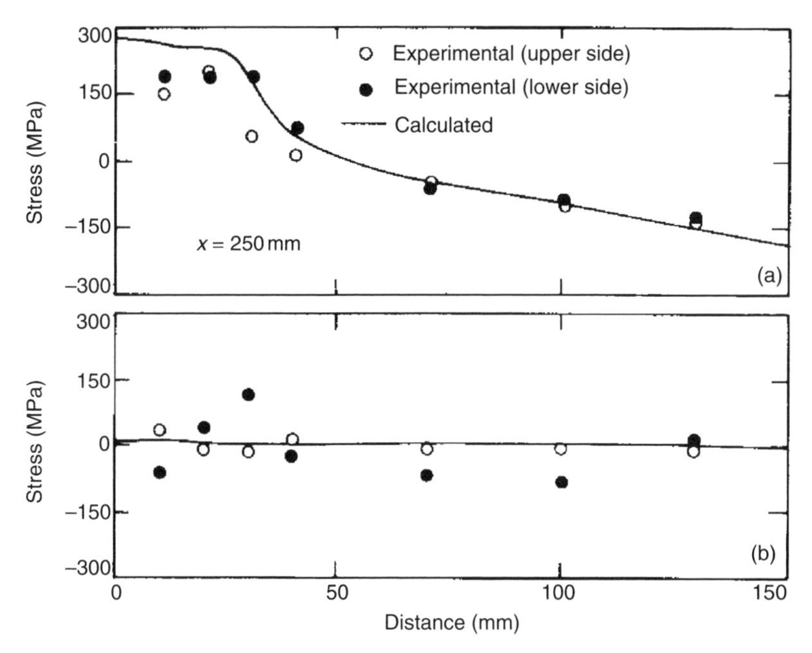

3.17 Residual stress distributions.

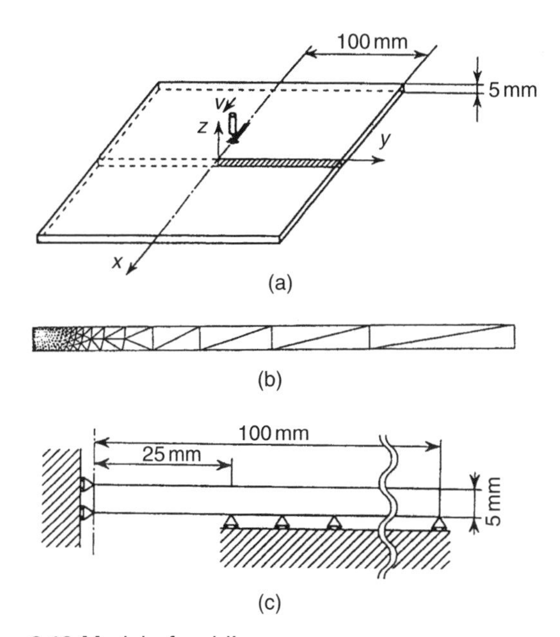

3.18 Model of welding.

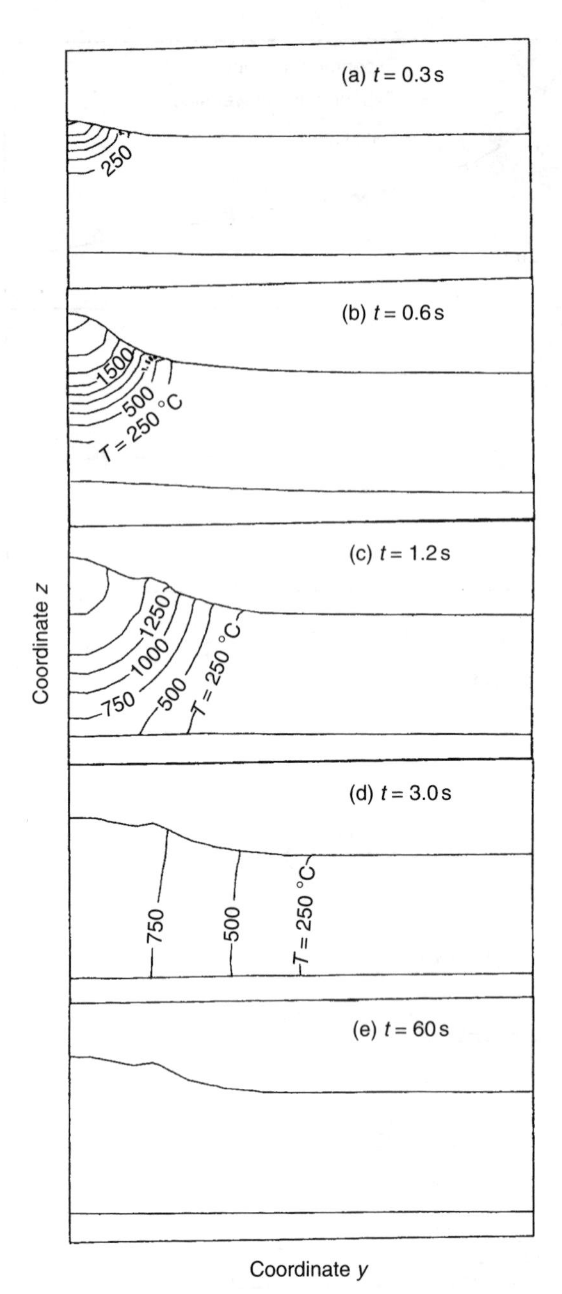

3.19 Temperature distributions.

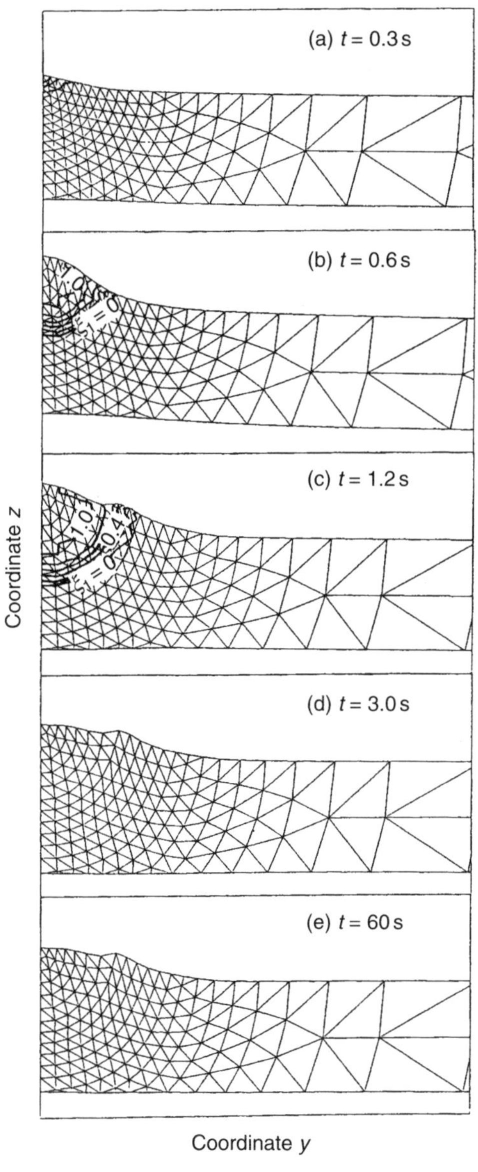

3.20 Progress of liquid phase with deformed finite-element mesh.

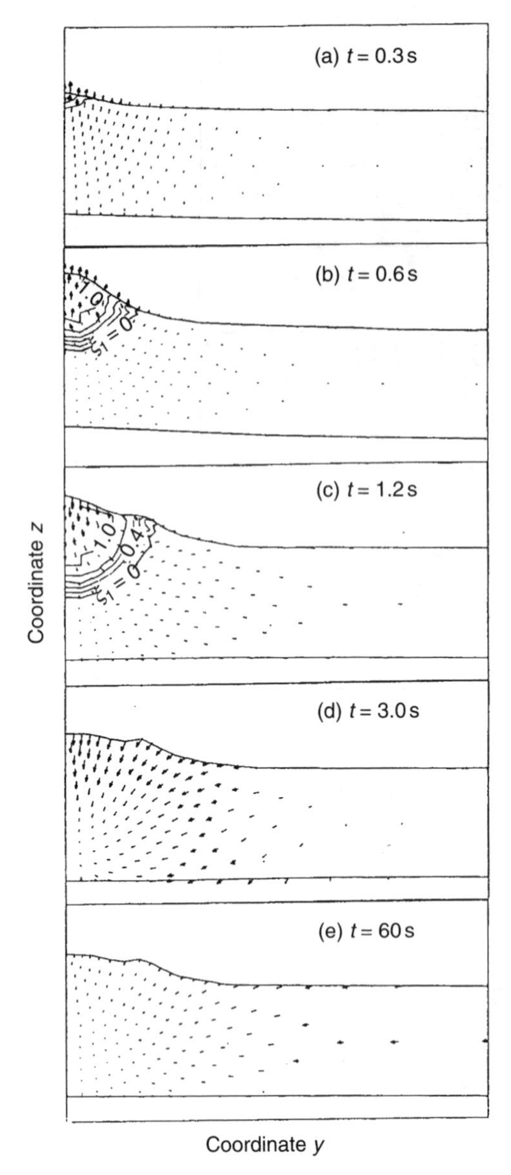

3.21 Mode of material flow.

The calculated results of the stress distributions are shown in Fig. 3.22, Fig. 3.23 and Fig. 3.24. In the distribution of longitudinal stress σ_x in Fig. 3.22, compressive stresses appear up to $t = 0.6$s in Fig. 3.22(b), but they become tension stresses in the final state (Fig. 3.22(e)) within and around the weld line. The location of maximum stress is around the molten pool in

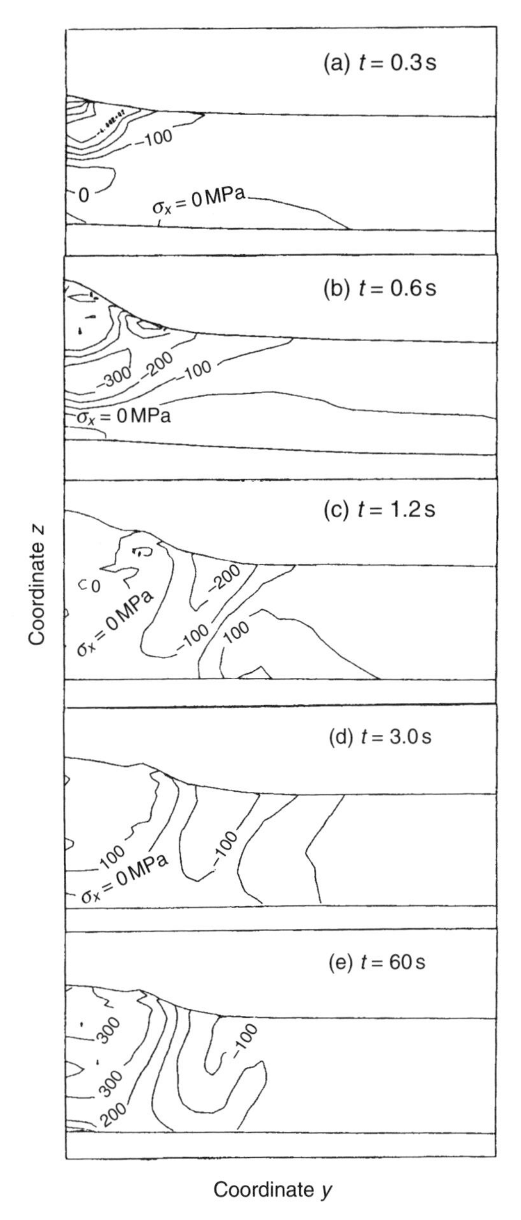

3.22 Distributions of stress in the x direction.

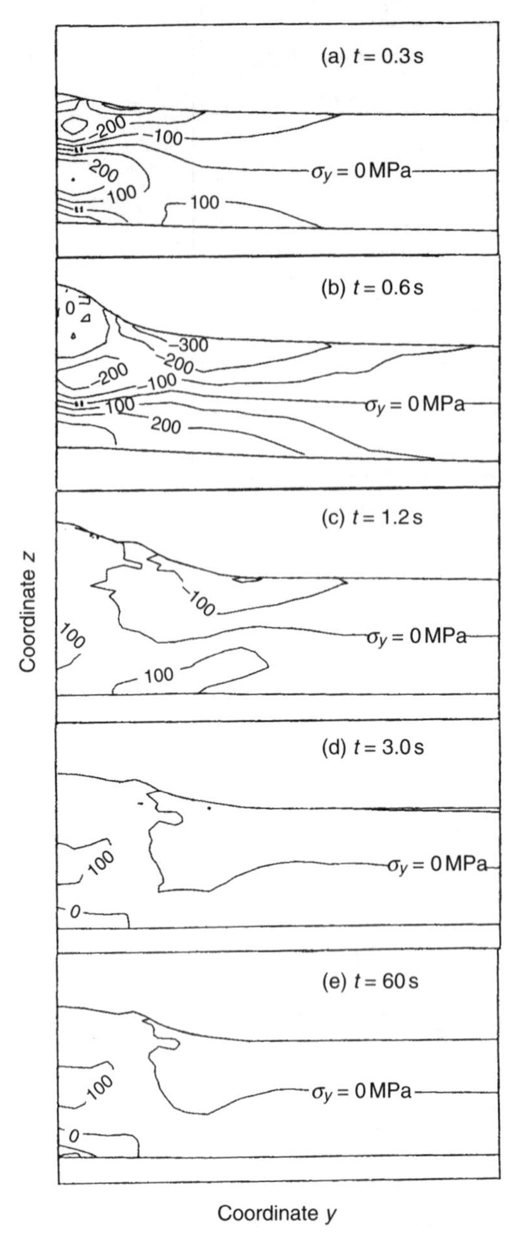

3.23 Distributions of stress in the *y* direction.

the final distribution, and this is because recrystallization occurs in the pool, which means that the method and procedure developed in this article can represent the phenomenon in contrast with the ordinal elastic–plastic analysis. The distribution of transverse stress σ_y involved is observed up to

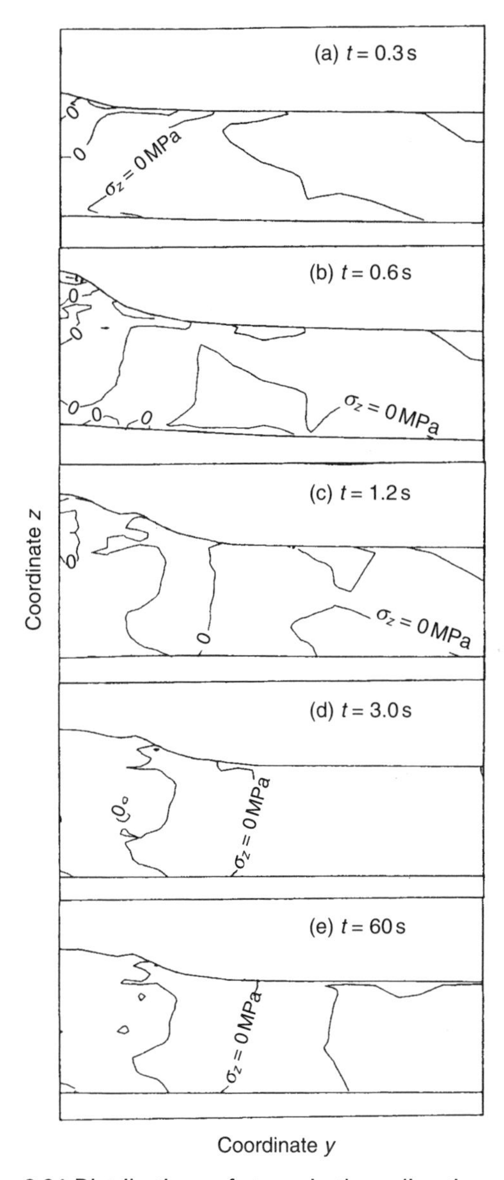

3.24 Distributions of stress in the *z* direction.

$t = 0.6$ s in Fig. 3.23, and this is considered to be caused by the temperature difference between the upper and lower sides of the plate. The distribution of stress σ_z is relatively small because of the thin thickness of the plate, as observed in Fig. 3.24.

3.7 Conclusions

Fundamental equations governing the temperature, stress and metallic structural distributions in steel undergoing phase transformation have been developed when the coupling effect between them is taken into account, and the procedure for solving the equations by the finite-element method has been presented. The theory was applied in order to simulate the welding process. The calculated results of the temperature, stress and structural changes were compared with the experimentally measured data to confirm the validity of the simulation.

3.8 References

1. Inoue, T., Nagaki, S., Kishino, T., and Monkawa, M., 'Description of transformation kinetics, heat conduction and elastic–plastic stresses in the course of quenching and tempering of some steels', *Ing.-Arch.*, **50**(5), 315–327, 1981.
2. Inoue, T., and Raniecki, B., 'Determination of thermal-hardening stresses in steels by use of thermoplasticity theory', *J. Mech. Physi. Solids*, **26**(3), 187–212, 1978.
3. Inoue, T., and Wang, Z., 'Coupling between stress, temperature and metallic structure during processes involving phase transformation', *Mater Sci. Technol.*, **1**(10), 845–851, 1985.
4. Inoue, T., 'Metallo-thermo-mechanical coupling – application to the analysis of quenching, welding and continuous casting processes', *Berg- und Huttenmannische Monatshefte*, **132**(3), 63–71, 1987.
5. Inoue, T., and Wang, Z.G., 'Metallo-thermo-mechanical simulation of some enginering processes incorporating phase transformation', (Computational Plasticity), *Current Japanese Materials Research*, Vol. 7 (eds T. Inoue, H. Kitagawa and S. Shima), Elsevier Applied Science, Barking, Essex, 1990, pp. 75–96.
6. Inoue, T., 'Inelastic constitutive relationships and applications to some thermomechanical processes involving phase transformation', *Thermal Stresses III* (ed. Richard, B. Hetnarski), North-Holland, Amsterdam, 1989, pp. 192–289.
7. Inoue, T., 'Coupling of stress–strain, thermal and metallurgical behaviors', *Handbook of Materials Behavior Models* (ed. J. Lemaitre), Academic Press, New York, 2002, pp. 875–886.
8. Inoue, T., 'Metallo-thermo-mechanics: application to quenching', *Handbook of Residual Stress and Deformation in Steel* (eds G. Totten, M. Howes and T. Inoue) ASM International, Materials Park, Ohio, 2002, pp. 296–311.
9. Bowen, B.M., *Theory of Mixture*, *Continuum Physics*, Vol. 6 (ed. A.C. Eringe), Academic Press, New York, 1976.
10. Malvern, L.E., *Introduction to the Mechanics of Continuous Medium*, Prentice-Hall, Englewood Cliffs, New Jersey, 1969.
11. Perzyna, P., 'Thermodynamic theory of viscoplasticity', *Adv. Appli. Mech.*, **9**, 315–354, 1971.
12. Johnson, A.W., and Mehl, R.F., 'Reaction kinetics in processes mucleation and growth', *Trans. AIME*, **135**, 416–456, 1939.

13. Bhattacharyya, S., and Kehl, G.L., 'Isothermal transformation of austenite under externally applied tensile stress', *Trans. ASM*, **47**, 351–379, 1955.
14. Magee, C.L., *The Nucleation of Mantensite*, American Society for Metals, Metals Park, Ohio, 1968.
15. Onodera, H., Gotoh, H., and Tamura, I., 'Effect of volume change in martensite transformation induced by tensile and compressive stresses in polycrystalline iron alloy, *Proc. 1st Japan Institute of Metals Int. Symp. on New Aspects of Martensitic Transformation*, Japan Institute of Metals, Sendai, 1976, pp. 327–332.
16. Wang, Z.G., and Inoue, T., 'Viscoplastic constitutive relation incorporating phase transformation: Application to welding, *Mater. Sci. Technol.*, **1**(10), 899–903, 1985.
17. Sakuma, A., and Inoue, T., 'Analysis of flow, deformation and stresses in melting/solidification process of welding by viscoplastic constitutive equation', *Proc. of 5th Int. Symp. on Plasticity and Current Applications*, Osaka, Japan, 1995, pp. 721–724.

4 Measuring temperature during welding

J. PAN, Tsinghua University, People's Republic of China

4.1 Introduction

The thermal process is central to the entire fusion-welding process. All the physical and chemical reactions in welding are related to the thermal process. It is closely related to the metallurgical reactions, crystallization and other phase transformations. The accurate measurement of the thermal process is a prerequisite for understanding and controlling metallurgical events, strain and stress during welding, residual stress and distortion after welding. Welding researchers have attached great importance to research on this topic and have performed many studies. There are two approaches; the first is the temperature-field calculation using heat transfer theory and the second is based on temperature-field measurement.

Basically there are two categories of temperature-measuring methods used in welding technology.

1. *Thermocouples*[1,2]. This is the conventional high-temperature measurement method. Thermocouple measurements are accurate and are widely used in research, e.g. for measuring the thermal cycle at a point. This method is suitable for measuring temperature at a point. Measuring a temperature field is difficult, particularly for welding processes, because many thermocouples are needed to map temperatures in three dimensions. Attaching thermocouples and measuring temperatures is laborious. Moreover, too many thermocouples change the temperature distribution and, owing to high temperatures, measuring the temperature near the arc is not possible. Therefore real-time control using thermocouples is not possible.
2. *Infrared radiation.* This is a noncontact measuring method that is superior to thermocouples because contacting the object is not necessary. The temperature field is not altered by the measuring method and waiting for temperature equilibration between the sensor and the object is not necessary. In addition, infrared radiation propagates at the speed of light; the measuring speed depends only on the response of

the measuring instruments. The particular advantage of this method is that measuring the temperature field and realizing real-time control are possible. The brightness method, radiation method and colorimetric method may refer to conventional temperature measurement methods using radiation[1,3].

Until now, there have been no reports about the application of the brightness method and colorimetric method for measuring temperature fields in welding. The colorimetric method has been used for detecting temperature at a point or the average temperature of a small area but not for temperature-field measurement. A direct measurement of a complete temperature field has been given by Chen *et al.*[4,5] in which an infrared camera was used. For the image taken by an infrared camera to describe the temperature field, its calibration is critical. Without calibration, the image is only a radiation intensity field and not the temperature field. Calibration is complicated because the gray levels of the image pixels depend on many factors such as the radiation wavelength, the radiation intensity, the conditions surrounding the detected point, the parameters of the atmosphere through which the radiation travels, the materials, the distance and the surface condition. Among these, the wavelength and radiation intensity are related to the temperature while the other factors are not related to temperature but affect the gray level of the pixels.

To overcome this problem, a colorimetric method using a charge-coupled device (CCD) camera was developed by the author with his colleagues[6,7]. The technology of this method has been described previously[8].

4.2 Principles

4.2.1 The radiation law for an absolute black body

A black body has the strongest emission power for a given temperature. An absolute black body is applied as a standard for evaluating emission power and is widely used for calibration of infrared instruments, for thermometers and for the measurement of radiation characteristics of various materials.

Planck radiation formula

The experimental data for black-body radiation are plotted as shown in Fig. 4.1.

In order to explain the experimental curve for black-body radiation, Planck hypothesized energy quantization. Planck regarded the radiation of a black-body cavity as the electromagnetic radiation of a cluster of oscillating dipoles (or simply oscillators). The energy of the oscillator cannot

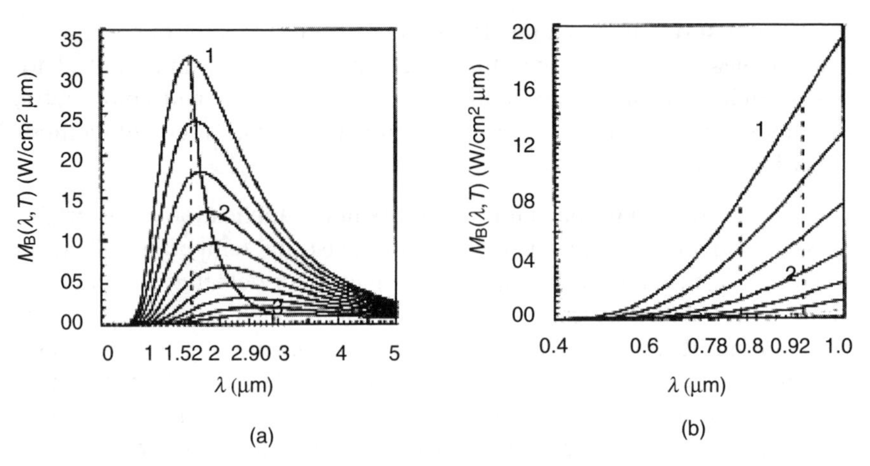

4.1 Experimental radiation curve of a black body: curves 1, T = 1900 K; curves 2, 1600 K; curve 3, 1000 K.

have any value; for a dipole of oscillating frequency γ, the minimum energy unit is $h\gamma$. A dipole in a certain state would have the energy $nh\gamma$ ($n = 0, 1, 2, \ldots$) where $h = 0.6262 \times 10^{-34}$ J s is the Planck constant and $h\gamma$ is called an energy quantum. Planck considered that the probability that an oscillator will have a certain energy level is proportional to $\exp(-nh\gamma/KT)$ where $K = 1.3806 \times 10^{-2}$ W s/K is the Boltzmann constant and T represents the absolute temperature. On the basis of these two hypotheses, the famous Planck radiation formula is written as follows:

$$M_\lambda = \frac{C_1}{\lambda^5}\left[\exp\left(\frac{C_2}{\lambda T}\right) - 1\right]^{-1} \text{ W/cm}^2\,\mu\text{m} \qquad [4.1]$$

where M_λ is the spectrum radiant exitance, $C_1 = 2\pi C^2 h = 3.7415 \times 10^{-16}$ W m² is the first radiation constant, $C = 2.9979 \times 10^8$ m/s is the velocity of light and $C_2 = 1.4388 \times 10^{-2}$ mK is the second radiation constant. hC/K

Stefan–Boltzmann law

Stefan observed that the total energy of a black body is proportional to the fourth order of the absolute temperature of the black body. He derived the formulas on the basis of the Maxwell theory. By integrating Eqn [4.1] over all wavelengths, the total power radiated by a black body over a half-sphere of space can be obtained as

$$M_B = \int_0^\infty M_\lambda \, d\lambda = \int_0^\infty \frac{C_1}{\lambda^5}\left[\exp\left(\frac{C_2}{\lambda T}\right) - 1\right]^{-1} d\lambda = \sigma T^4 \qquad [4.2]$$

This is the Stefan–Boltzmann law, where $\sigma = 5.670 \times 10^{-8}\,\text{W/m}^2\,\text{K}^4$ and is called the Stefan–Boltzmann constant.

It can be seen from Fig. 4.1 that the spectrum radiation power for an absolute black body has a maximum value at a certain wavelength, which depends on the absolute temperature of the body. The corresponding wavelength is designated as λ_m. The relationship between T and λ_m is given by

$$\lambda_m T = b \qquad [4.3]$$

This is the Wien displacement law, where $b = 2.897 \times 10^{-3}\,\text{m K}$.

For temperatures normally encountered, the wavelength corresponding to its maximum radiation power is in the infrared region.

4.2.2 Gray-body radiation

Most matter generally is not a black body; it has selective absorption characteristics. The absorptivity is large for certain wavelengths of electromagnetic radiation. In some categories of matter there is no obvious selective absorptivity; their emissivity is less than 1 but approximately a constant. This kind of matter is called a gray body.

For a gray body, the radiation law of a black body is applicable but emissivity should be taken into consideration. Thus the Planck formula becomes

$$M(\lambda, T) = \varepsilon(\lambda, T)\frac{C_1\lambda^{-5}}{\exp(C_2/\lambda T - 1)} \qquad [4.4]$$

and the Stefan–Boltzmann formula is

$$M_B = \varepsilon(T)\sigma T^4 \qquad [4.5]$$

4.2.3 An existing method for measuring the temperature field using thermal radiation

Because of the limitation of spectral response by the camera and the measured temperature range, radiation having a certain range of wavelength is used in existing measurement methods.

Assume that a radiation beam from the object is projected via an optical system onto the receiving element. Its output signal N can be expressed by the following equation:

$$\begin{aligned} N &= KF\eta(\lambda_m)\int_{\lambda_1}^{\lambda_2} \varepsilon(\lambda, T)M_\lambda\eta(\lambda)\gamma(\lambda)\varphi_1(\lambda)\varphi_2(\lambda)\Lambda\varphi_n(\lambda)\,d\lambda \\ &= KF\eta(\lambda_m)\int_{\lambda_1}^{\lambda_2} \varepsilon(\lambda, T)M_\lambda\eta(\lambda)\gamma(\lambda)\varphi(\lambda)\,d\lambda \end{aligned} \qquad [4.6]$$

where λ_1 and λ_2 are the upper and lower limits respectively of the received wavelength, K is the conversion coefficient of heat to electricity, F is the geometric factor of the receiving element and object, $\varepsilon(\lambda, T)$ is the emission coefficient, $\eta(\lambda_m)$ is the maximum spectral sensitivity of the receiving element, $\eta(\lambda)$ is the relative spectral sensitivity of the receiving element, $\gamma(\lambda)$ is the transmissivity of the filter, $\varphi_1(\lambda)$ is the transmissivity of the first optical element, $\varphi_2(\lambda)$ is the transmissivity of the second optical element, $\varphi_n(\lambda)$ is the transmissivity of the nth optical element, $\varphi(\lambda)$ is the product of transmissivity of all optical elements and M_λ is the radiance exitance of the object.

It is seen from Eqn [4.6] that the output signal depends on many factors including the geometric factors F of the receiving element and object and the emissivity $\varepsilon(\lambda, T)$ of the object. The other factors can be found for a given measurement system. Only these two factors F and $\varepsilon(\lambda, T)$ are variables that depend on the distance, material and surface conditions.

The effect of distance

Suppose that the other factors are kept constant, and only the distance is varied; the radiation intensity of the source S is I, and the angle between the normal line of the receiving surface and direction of the radiation is θ. Then the radiation power received can be written as (Fig. 4.2)

$$d\phi = I\,d\Omega = I\frac{dA\cos\theta}{d_0^2} \qquad [4.7]$$

In this equation, d_0 is the distance between the point source S and dA. Therefore the irradiance on dA by S is

$$E = \frac{d\phi}{dA} = I\frac{\cos\theta}{d_0^2} \qquad [4.8]$$

The output signal N of the receiver is proportional to the irradiance, i.e.

$$N \propto E, \quad N \propto \frac{\cos\theta}{d_0^2} \qquad [4.9]$$

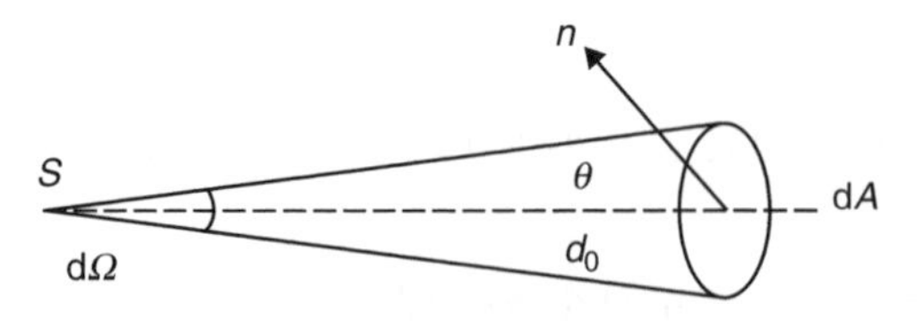

4.2 Radiance exitance on d*A* by point source *S*.

It can be concluded, therefore, that the output potential of the receiver is proportional to $\cos\theta$ and inversely proportional to the square of d_0.

It is seen that there is a strong influence of distance and θ on the measurement. Normally it is arranged so that the normal line of the receiving surface coincides with the radiation direction so that the influence of θ can be eliminated.

The influence of the material and the surface condition

It is seen that the output signal depends on the emissivity $\varepsilon(\lambda, T)$ of the object, which is, in turn, affected by many factors. It varies greatly. The amount of variation can be several orders of magnitude. This is the main factor that affects the accuracy of the measuring method.

4.2.4 Colorimetric image method

As described above, with measuring methods available currently, there is no corresponding relationship between the output potential and the temperature. The distance to the object and surface conditions affect it. In order to solve this problem, a colorimetric imaging method was proposed. It uses CCD camera to take two images at two different wavelengths, which are processed by computer to obtain the two-dimensional temperature field. The principle of the method is as follows.

From the Planck formula, one can obtain

$$M_{\lambda_1} = \varepsilon_{(\lambda_1,T)} \frac{C_1}{\lambda_1^5}\left[\exp\left(\frac{C_2}{\lambda_1 T}\right)-1\right]^{-1}, \quad M_{\lambda_2} = \varepsilon_{(\lambda_2,T)} \frac{C_1}{\lambda_2^5}\left[\exp\left(\frac{C_2}{\lambda_2 T}\right)-1\right]^{-1}$$

Dividing the first formula by the second gives

$$R = \frac{M_{\lambda_1}}{M_{\lambda_2}} = \frac{\varepsilon_{(\lambda_1,T)}}{\varepsilon_{(\lambda_2,T)}} \frac{(C_1/\lambda_1^5)[\exp(C_2/\lambda_1 T)-1]^{-1}}{(C_1/\lambda_2^5)[\exp(C_2/\lambda_2 T)-1]^{-1}} = \frac{\varepsilon_{(\lambda_1,T)}\lambda_2^5[\exp(C_2/\lambda_2 T)-1]^{-1}}{\varepsilon_{(\lambda_2,T)}\lambda_1^5[\exp(C_2/\lambda_1 T)-1]^{-1}} \quad [4.10]$$

where R is called the colorimetric ratio, λ_1 and λ_2 are two wavelengths and T is the absolute temperature of the object.

From Eqn [4.10] it can be seen that describing the temperature using the colorimetric ratio is an improved way because two wavelengths can be chosen so that the emissivities of the matter at the two wavelengths are similar and an accurate relationship between R and T can be obtained.

From Eqn [4.6], the output potential of the receiver in the wavelength band $\Delta\lambda_1$ near λ_1 is

$$N_1 = KF\eta(\lambda_m)\int_{\lambda_1}^{\lambda_1+\Delta\lambda_1} \varepsilon_{(\lambda_1,T)} \frac{C_1}{\lambda_1^5}\left[\exp\left(\frac{C_2}{\lambda_1 T}\right)-1\right]^{-1} \times \eta(\lambda_1)\gamma(\lambda_1)\varphi_1(\lambda_1)\varphi_2(\lambda_1)\Lambda\varphi_n(\lambda_1)\,\mathrm{d}\lambda$$

The output potential for the wavelength band $\Delta\lambda_2$ at λ_2 is

$$N_2 = KF\eta(\lambda_m)\int_{\lambda_2}^{\lambda_2+\Delta\lambda_2} \varepsilon_{(\lambda_2,T)} \frac{C_1}{\lambda_2^5}\left[\exp\left(\frac{C_2}{\lambda_2 T}\right)-1\right]^{-1} \times \eta(\lambda_2)\gamma(\lambda_2)\varphi_1(\lambda_2)\varphi_2(\lambda_2)\Lambda\varphi_n(\lambda_2)\,\mathrm{d}\lambda$$

Dividing the first formula by the second gives the colorimetric ratio

$$R = \frac{N_1}{N_2} = \frac{\varepsilon_{(\lambda_1,T)}}{\varepsilon_{(\lambda_2,T)}}\frac{\lambda_2^5\,\Delta\lambda_1}{\lambda_1^5\,\Delta\lambda_2}\frac{[\exp(C_2/\lambda_2 T)-1]\eta(\lambda_1)\gamma(\lambda_1)\varphi_1(\lambda_1)\varphi_2(\lambda_1)\Lambda\varphi_n(\lambda_1)}{[\exp(C_2/\lambda_1 T)-1]\eta(\lambda_2)\gamma(\lambda_2)\varphi_1(\lambda_2)\varphi_2(\lambda_2)\Lambda\varphi_n(\lambda_2)} \quad [4.11]$$

The following can be obtained for metals, because $\varepsilon_{(\lambda_1,T)} = \varepsilon_{(\lambda_2,T)}$:

$$R = \frac{\lambda_2^5\,\Delta\lambda_1}{\lambda_1^5\,\Delta\lambda_2}\frac{[\exp(C_2/\lambda_2 T)-1]\eta(\lambda_1)\gamma(\lambda_1)\varphi_1(\lambda_1)\varphi_2(\lambda_1)\Lambda\varphi_n(\lambda_1)}{[\exp(C_2/\lambda_1 T)-1]\eta(\lambda_2)\gamma(\lambda_2)\varphi_1(\lambda_2)\varphi_2(\lambda_2)\Lambda\varphi_n(\lambda_2)} \quad [4.12]$$

Equation [4.12] means that the potential response of the receiving element for a gray body is related to temperature and λ_1, λ_2, $\Delta\lambda_1$, $\Delta\lambda_2$, $\eta(\lambda_1)$, $\eta(\lambda_2)$, $\gamma(\lambda_1)$, $\gamma(\lambda_2)$, $\varphi_1(\lambda_1)$, $\varphi_2(\lambda_1)$, Λ, $\varphi_n(\lambda_1)$, $\varphi_1(\lambda_2)$, $\varphi_2(\lambda_2)$ and $\varphi_n(\lambda_2)$. Among these, all the factors except temperature are known and fixed for a given measuring system. Therefore, it can be concluded that temperature has a relationship with only the colorimetric ratio and has no relationship with material, distance, surface conditions, etc. Equation [4.12] is the general formula that describes the colorimetric ratio–temperature relationship for the colorimetric method.

4.3 Wavelength of the filter

The colorimetric ratio depends mainly on the radiation properties of the object, the transmissivity of the filter and the spectral response of the image-intensified charge-coupled device (ICCD); if canceling all common factors, Eqn [4.11] may be rewritten in a more exact form as follows:

$$R_{12}(T) = \frac{\int_{\lambda_1-\delta\lambda_1}^{\lambda_1+\delta\lambda_1} \gamma(\lambda)\eta(\lambda)\varepsilon(\lambda,T)M_B(\lambda,T)\,\mathrm{d}\lambda}{\int_{\lambda_2-\delta\lambda_2}^{\lambda_2+\delta\lambda_2} \gamma(\lambda)\eta(\lambda)\varepsilon(\lambda,T)M_B(\lambda,T)\,\mathrm{d}\lambda} \quad [4.13]$$

For the purpose of the following discussion, it was assumed that the wavelength was in a narrow band, and $\varepsilon(\lambda, T)$, $\eta(\lambda)$ and $M_B(\lambda, T)$ are slow functions. The formula can be written as

$$R_{12}(T) = \frac{\varepsilon(\lambda_1, T)\gamma(\lambda_1)\eta(\lambda_1)\lambda_2^5\ \delta\lambda_1}{\varepsilon(\lambda_2, T)\gamma(\lambda_2)\eta(\lambda_2)\lambda_1^5\ \delta\lambda_2}\exp\left[-\frac{c_2}{T}\left(\frac{1}{\lambda_1}-\frac{1}{\lambda_2}\right)\right] \quad [4.14]$$

In the measuring system, the sensitivity or the ratio of $R(T)$ to T reflects more clearly the response characteristics of the system. If the sensitivity is designated as $S(T)$, then

$$S(T) = \frac{\partial R(T)}{\partial T} = \frac{\displaystyle\int_{\lambda_1-\delta\lambda_1/2}^{\lambda_1+\delta\lambda_1/2} (c_2/\lambda T^2)\exp(c_2/\lambda T)[\exp(c_2/\lambda T)-1]^{-1} \times \varepsilon(\lambda, T)M_B(\lambda, T)\gamma_1(\lambda)\eta(\lambda)\,\mathrm{d}\lambda}{P_{\lambda_2}(T)}$$
$$-\frac{P_{\lambda_1}(T)\displaystyle\int_{\lambda_2-\delta\lambda_2/2}^{\lambda_2+\delta\lambda_2/2} (c_2/\lambda T^2)\exp(c_2/\lambda T)[\exp(c_2/\lambda T)-1]^{-1} \times \varepsilon(\lambda, T)M_B(\lambda, T)\gamma_2(\lambda)\eta(\lambda)\,\mathrm{d}\lambda}{P_{\lambda_2}^2(T)} \quad [4.15]$$

where

$$P_{\lambda_i}(T) = \int_{\lambda_i-\delta\lambda_i}^{\lambda_i+\delta\lambda_i} \gamma_i(\lambda)\eta(\lambda)\varepsilon(\lambda, T)M_B(\lambda, T)\,\mathrm{d}\lambda \quad (i = 1, 2)$$

The requirement for the system's response and sensitivity are that $R(T)$ must be linear with a steep slope in the measuring-temperature range T_1–T_2 so that the temperature scale is reasonably even and sensitive. More explicitly, $S(T_1) = S(T_2)$ and a large value of $S(T)$ are required.

The wavelengths of the maximum radiant exitance of the object at temperatures of 1100–1800 K are 2.616–1.599 μm. Obviously, the closer the chosen wavelengths are to these, the larger and better would be the values of $R(T)$ and $S(T)$, but they are not in the range of the ICCD's dynamic response range. Figure 4.3 shows the spectral response of the ICCD which is available at the moment on the market. It can be seen that the optimum response wavelength of the ICCD is 0.5 μm, which also cannot be used because of serious interference from ambient light (daylight). Therefore the applicable wavelength for the system is between 0.77 and 1.0 μm. In order to find the optimum wavelength for a given temperature range, numerical simulation was carried out by the author considering the radiant exitance of the measured body, the dynamic response of the ICCD and the transmissivity of the filters. Figure 4.4 shows the transmissivity curve of the filter. For this purpose the dynamic response curve of the ICCD and the transmissivity curve were expressed by mathematical functions. For calculation

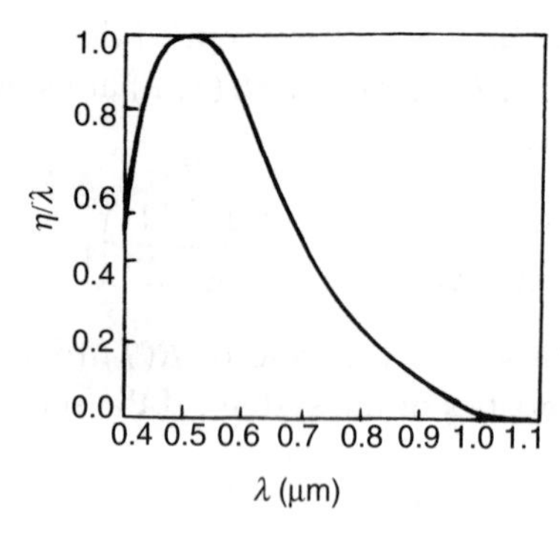

4.3 Spectral response of the ICCD.

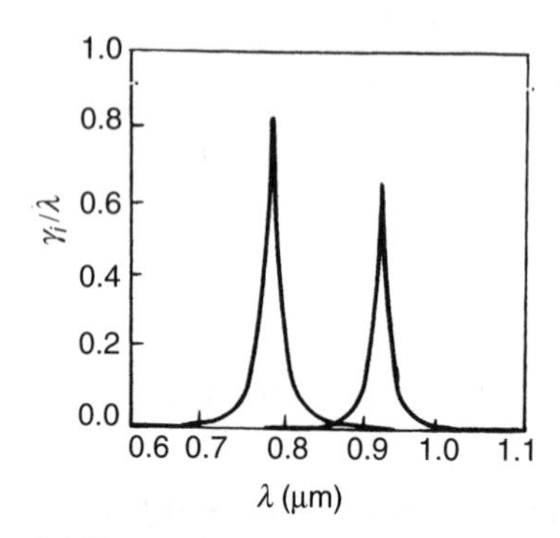

4.4 Transmissivity function of the filter.

of Eqn [4.13] the product of three functions was integrated using the approximation formula of a trapezoid.

Simulation results demonstrate that the wavelengths of the two colors that can meet the requirements of the system's response and sensitivity in the temperature range 1100–1800 K are shown in Fig. 4.5. Any two wavelengths in the figure could have a stable sensitivity of $S(T)$ in the assigned temperature range. Considering that the dynamic response of the ICCD should not be too small and avoiding the visible light range as far as possible, then it is reasonable that λ_2 could be in the range 0.90–0.948 μm and λ_1 in the range 0.77–0.80 μm. The possible combinations λ_1 and λ_2 are as follows: 0.77 and 0.905 μm; 0.780 and 0.920 μm; 0.790 and 0.934 μm; 0.800 and 0.948 μm. Figure 4.6 shows the relationship of $R(T)$ against T and Fig. 4.7 shows the relationship of $S(T)$ against T with different combination of two wavelengths obtained by numerical simulation. Detailed analysis shows that the best choice of the two wavelengths are $\lambda_1 = 0.780$ μm and $\lambda_2 = 0.920$ μm.

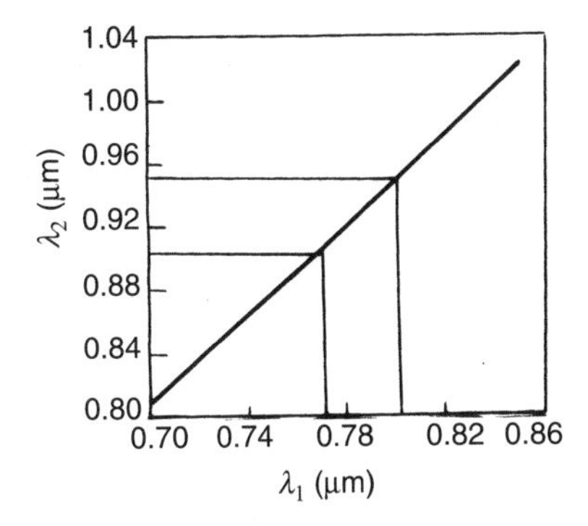

4.5 Curve for the selection of λ_1 and λ_2.

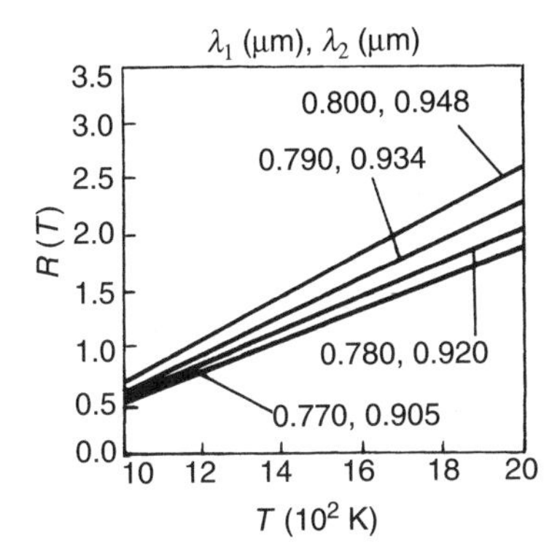

4.6 Relationship between $R(T)$ and T.

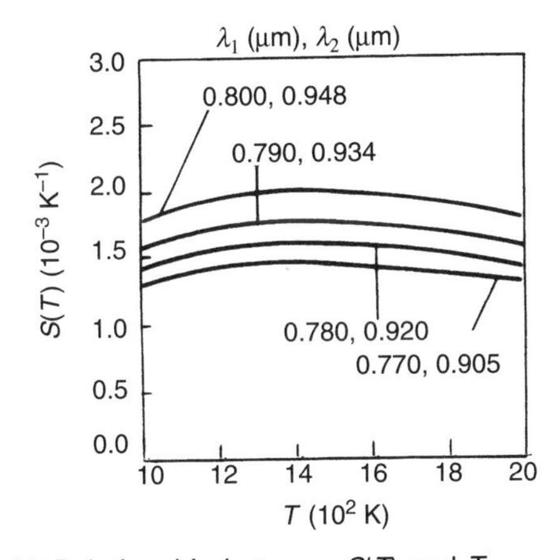

4.7 Relationship between $S(T)$ and T.

Simulation results shows that $\delta\lambda_1$ and $\delta\lambda_2$ exert an insignificant effect on the $R(T)$–T relationship. However, if $\delta\lambda$ is too large, the ICCD becomes saturated and, if $\delta\lambda$ is too small, a small signal is produced and, hence, a low signal-to-noise ratio. Therefore using 0.01–0.03 μm for $\delta\lambda_1$ and $\delta\lambda_2$ was recommended.

4.4 Division of the temperature field

The temperature field[9,10] in welding has a large range; the range of interest is at least 800–1400 °C. However, owing to the limited response capability of the CCD, it can detect only a small range of temperatures. Therefore it was proposed to divide the temperature field into several regions according to the temperature range. Using different exposure times for different regions, the radiation power of all regions could be limited to the response capability of the CCD and thus the temperature of the whole field could be obtained.

4.4.1 Dynamic response range of the image-intensified charge-coupled device

The most important characteristic of the ICCD is its dynamic response range. This is determined by the maximum electrical charge Q_{max} that can be stored in the potential pit and the minimum electrical charge Q_{min} due to the inevitable noise[11]. It can be seen that the charge in the pit depends not only on the irradiance at that point but also on the exposure time (the time integration of charge by the ICCD). The photoelectric characteristic of an ICCD is linear in the normal range as shown in Fig. 4.8. The electric charge signal that is converted by the ICCD may be written as

$$Q_{\lambda_0}(T) = K'\tau E(T, \lambda_0) \tag{4.16}$$

where K' is the coefficient decided by the optical imaging system. $Q' = \tau E(T, \lambda_0)$ is the exposure quantity; its units are lux seconds.

In cases when the exposure quantity of the electrical charge is larger than Q_{max} or smaller than Q_{min}, the signal induced by the irradiance would be saturated or submerged respectively in the noise.

The exposure quantity by which the quantity of electrical charge is determined can be regulated by the irradiance $E(T, \lambda_0)$ and the exposure time τ so that $Q_{min} < Q < Q_{max}$. The irradiance $E(T, \lambda_0)$ depends on the image distance l', aperture coefficient F, etc. For a case in which l' is unchanged and F is chosen, then the irradiance is a function of temperature only. Producing the quantity of electrical charge within the range from Q_{max} to Q_{min} for a different temperature range can be accomplished by using a different exposure time. This means the time should be taken as

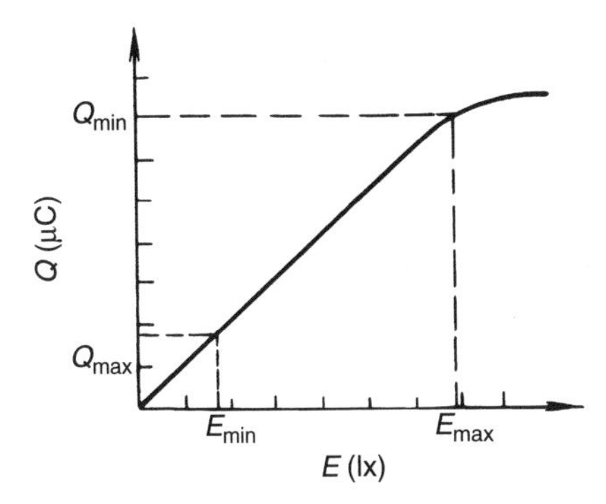

4.8 Photoelectric characteristics of the ICCD.

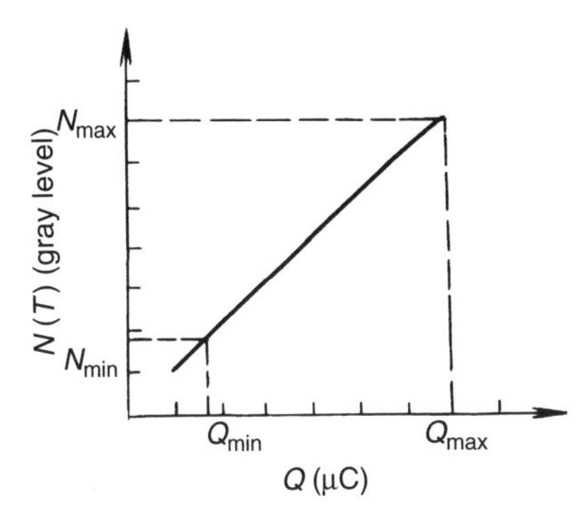

4.9 Analog-to-digital conversion characteristics of an image.

$$\frac{Q_{min}}{K'} < \tau E(T,\lambda) < \frac{Q_{max}}{K'} \qquad [4.17]$$

Obviously, when the welding temperature field has a large temperature range, satisfying this requirement using one exposure is difficult[12].

4.4.2 Method

Assuming that the ICCD has a minimum electrical charge quantity Q_{min} and a maximum Q_{max}, the corresponding gray level after 8 bit conversion is from N_{min} to N_{max}. Assuming also that the conversion is linear, as shown in Fig. 4.9, then the dynamic response range can be expressed as N_{max}/N_{min}. For the present system, $N_{min} \approx 50$ and $N_{max} \approx 240$. The ICCD has good linearity in this range.

Suppose that the temperature range is from T_1 to T_2 and the wavelength is λ; then the ratio of irradiance on the ICCD can be obtained from Eqn [4.1] as follows (the emissivity is not taken into consideration):

$$c(\lambda)=\frac{E(T_2,\lambda)}{E(T_1,\lambda)}=\frac{M_B(\lambda,T_2)}{M_B(\lambda,T_1)}=\exp\left[-\frac{c_2}{\lambda}\left(\frac{1}{T_2}-\frac{1}{T_1}\right)\right] \qquad [4.18]$$

For example, for the temperature range from $T_1 = 1073$ K to $T_2 = 1673$ K, $\lambda = 0.78\,\mu m$ and $c(\lambda) = 1200$, suppose that the irradiance due to T_1 makes up Q_{min} and N_{min}; then the Q produced by T_2 would be 1200 times that of Q_{min} and the gray level would be $1200N_{min} \gg 255$; obviously the ICCD is saturated. Therefore dividing the temperature field into several regions and using different exposure times is needed to keep the irradiance of the temperature range of all divisions within the dynamic response range of the ICCD[9,10].

Suppose that the exposure times are $\tau_1, \tau_2, \ldots, \tau_{m+1}$, and the temperature from T_1 to T_2 is divided into m segments (Fig. 4.10(a)).

In order to use the dynamic response range of the ICCD fully, the temperature range from T_1 to T_2 is divided into m sections T_1–T_1', T_1'–T_2', . . . , T_m'–T_2, according to geometric ratios and the corresponding exposure times are $\tau_1, \tau_2, \ldots, \tau_{m+1}$.

Then the following conditions must be satisfied:

$$\frac{M_B(\lambda,T_1')}{M_B(\lambda,T_1)}=\frac{M_B(\lambda,T_2')}{M_B(\lambda,T_1')}=\cdots=\frac{M_B(\lambda,T_2)}{M_B(\lambda,T_m')}=c'(\lambda)\leqslant\frac{N_{max}}{N_{min}} \qquad [4.19]$$

Transforming the formula shows that $T_1', T_2', \ldots, T_m'$ are only functions of T_1 and T_2 and are independent of λ:

$$\frac{1}{T_1'}-\frac{1}{T_1}=\frac{1}{T_2'}-\frac{1}{T_1'}=\cdots=\frac{1}{T_2}-\frac{1}{T_m'}=c_T \qquad [4.20]$$

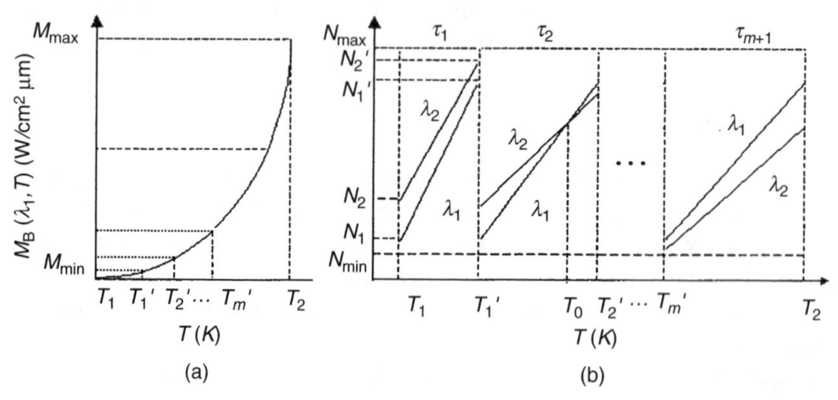

4.10 Division of the temperature field: (a) division of the temperature range from T_1 to T_2 by geometric ratio; (b) gray levels for each region.

Substituting Eqn [4.20] into Eqn [4.19] gives

$$c'(\lambda) = \exp\left(-\frac{c_2}{\lambda}c_T\right) \quad [4.21]$$

The exposure times $\tau_1, \tau_2, \ldots, \tau_{m+1}$, should be chosen so that the dynamic responses on the ICCD for T_1–T_1', T_1'–T_2', . . . , T_m'–T_2 are within the largest linear range. The corresponding gray levels are shown in Fig. 4.10(b).

For λ_1, the ideal case is

$$\tau_1 E(\lambda, T_1) = \tau_2 E(\lambda, T_1') = \Lambda \tau_{m+1} e(\lambda, T_m') \quad [4.22]$$

For λ_2, because $\lambda_2 > \lambda_1$, therefore $c'(\lambda_2) < c'(\lambda_1)$, which satisfies the condition for division given by Eqn [4.19].

Transforming Eqn [4.21] and combining it with Eqn [4.19] give

$$\frac{\tau_1}{\tau_2} = \frac{\tau_2}{\tau_3} = \cdots = \frac{\tau_m}{\tau_{m+1}} = c'(\lambda_1) \leqslant \frac{N_{\max}}{N_{\min}} \quad [4.23]$$

It is seen that, in order to satisfy the requirement of division of Eqn [4.19], the exposure time should be decided according to Eqn [4.23]. However, the exposure time of available ICCDs is discrete and the ratio of any neighboring two values may not be equal. However, it is advisable to satisfy Eqn [4.23] as closely as possible. For a two-color image, the irradiance of both wavelengths must be kept within the response range. Equation [4.14] shows that the related factors are the wavelengths λ_1 and λ_2, the bandwidths $\delta\lambda_2$ and $\delta\lambda_1$, the transmissivities I_1 and I_2 at the peak values of λ_1 and λ_2, and the response values $\eta(\lambda_1)$ and $\eta(\lambda_2)$ of the ICCD.

In order to use the dynamic range of the ICCD fully, locating the electrical charge quantity in the middle range of the temperature for when the two wavelengths are close to each other, i.e. $R(T_1)R(T_2) = 1$, was proposed so that the electrical charge quantity in the low-temperature and high-temperature ranges would not be too different for the two wavelengths, and a larger measurable temperature range could be obtained. In this case, the parameters of the filter could be found by transforming Eqn [4.14] as follows:

$$\left(\frac{I_1\,\delta\lambda_1\,\eta(\lambda_1)}{I_2\,\delta\lambda_2\,\eta(\lambda_2)}\right)^2 = \left(\frac{\lambda_1^5}{\lambda_2^5}\right)^2 \exp\left[C_2\left(\frac{1}{\lambda_1} - \frac{1}{\lambda_2}\right)\left(\frac{1}{T_1} + \frac{1}{T_2}\right)\right] \quad [4.24]$$

Thus there is a temperature T_0 in the range from T_1 to T_2 where $R(T_0) = 1$. From Eqn [4.14], T_0 can be determined by the following formula:

$$T_0 = c_2\left(\frac{1}{\lambda_2} - \frac{1}{\lambda_1}\right)\left\{\ln\left[\left(\frac{\lambda_1}{\lambda_2}\right)^5 \frac{\eta(\lambda_2)\,I_2\,\delta\lambda_2}{\eta(\lambda_1)\,I_1\,\delta\lambda_1}\right]\right\}^{-1} \quad [4.25]$$

The formula shows that T_0 can be determined by proper design of the peak transmissivity value of the two-color filter. In other words, the peak transmissivity value is an important factor for the design of the filter.

Combining Eqns [4.24] and [4.25], the relationship of T_0, T_1 and T_2 can be written as

$$\frac{2}{T_0}=\frac{1}{T_1}+\frac{1}{T_2} \qquad [4.26]$$

Substituting Eqn [4.25] into Eqn [4.14] gives

$$R(T)=\exp\left[-c_2\left(\frac{1}{T}-\frac{1}{T_0}\right)\left(\frac{1}{\lambda_2}-\frac{1}{\lambda_1}\right)\right] \qquad [4.27]$$

From this formula, $R(T) < 1$ when $T > T_0$ and $R(T) > 1$ when $T < T_0$.

4.4.3 Effect of the dynamic range

In the following paragraphs, the effects of the dynamic response range of the charge-coupled device of the ICCD on the measuring-temperature range, the division number, the choice of wavelength and the requirement of exposure time are discussed on the basis of Eqns [4.18], [4.19], [4.20] and [4.22].

1. *The requirement of exposure time.* The available exposure times for the ICCD applied in the present design were 16.7, 8, 4, 2, 1, 0.5, 0.25 and 0.1 ms. The exposure time should be as short as possible; therefore, $\tau_1 = 8$ ms, $\tau_2 = 2$ ms, $\tau_3 = 0.5$ ms and $\tau_4 = 0.1$ ms were chosen. Thus $\tau_1/\tau_2 = 4$, $\tau_2/\tau_3 = 4$ and $\tau_3/\tau_4 = 5$, which satisfy the requirement of Eqn [4.23]. Because of the exposure times available, dividing the temperature field into a maximum of four regions was allowed.
2. *The requirement on λ for a given range from T_1 to T_2 and the division number N.* Suppose that a temperature field from T_1 to T_2 is divided into n regions; c_T can be obtained from Eqn [4.20]. Substituting it into Eqn [4.21] and combining it with Eqns [4.19] and [4.18] gives

$$\lambda \geqslant -\frac{c_2}{\ln(N_{max}/N_{min})}c_T = -\frac{c_2}{n\ln(N_{max}/N_{min})}\left(\frac{1}{T_2}-\frac{1}{T_1}\right) \qquad [4.28]$$

   This equation shows that the smaller the dynamic range of the ICCD, the longer λ should be.
3. *The requirement of n for given values of the range from T_1 to T_2 and λ.* Equation [4.28] can be transformed to

$$n \geqslant -\frac{c_2}{\lambda\ln(N_{max}/N_{min})}\left(\frac{1}{T_2}-\frac{1}{T_1}\right) \qquad [4.29]$$

4. *The limitations on the lower temperature T_1 and the upper temperature T_2 when λ and n are given.* Similarly, Eqn [4.28] can be transformed to

$$\frac{1}{T_2} \geqslant \frac{1}{T_1} - \frac{\lambda n}{c_2} \ln\left(\frac{N_{\max}}{N_{\min}}\right) \quad [4.30]$$

Obviously, as the dynamic response range of the ICCD becomes larger, the measurable temperature range also can be larger.

4.5 Design of the sensor

The design of the two-color thermal-image sensor[12,13] determines the quality of the image. It should be handy and reliable and have a fast response. The main design problem is how to sample two-color images in as short a time as possible. The construction of the optical system, choice of devices, design of filters, image sampling, time-sequence control, etc., are described in the following sections.

4.5.1 Structure

There are two possible types of optical system for radiation-image formation as shown in Fig. 4.11: a reflection type (Fig. 4.11(a)) and a refraction type[14] (Fig. 14.11(b)). Because the system works at $\lambda_1 = 0.78\,\mu m$ and $\lambda_2 = 0.92\,\mu m$, all devices, including lenses, filters, etc., must have good transmissivity at these two wavelengths to ensure high image quality. Most existing infrared imaging systems are the reflection type because there are few

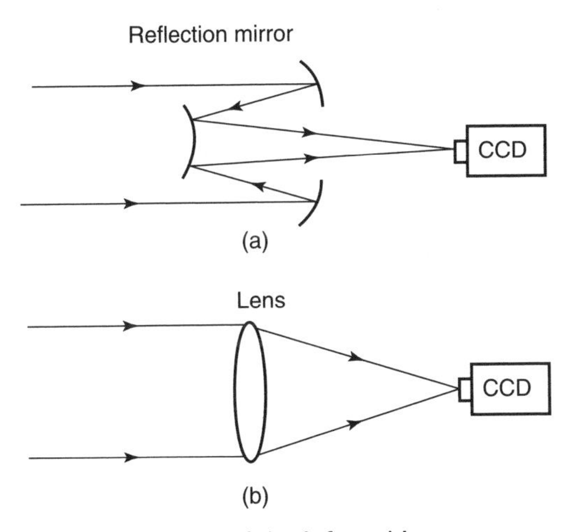

4.11 Optical path of the infrared image.

materials that can satisfy the physical, chemical and mechanical properties needed for an infrared refraction system. Designing an achromatic optical system is difficult. The reflection type has no chromatism and a wider applicable working wavelength range. The materials requirements are not rigorous. However, there are many shortcomings including a small field, bulkiness and eclipse in the major optical path. Therefore the refraction type was chosen for the present design.

For the system, the main aim was to obtain two thermal images. There were two possible choices for this, i.e. a single optical path (Fig. 4.12(a)) and a binary path[3] (Fig. 4.12(b)). In the binary-path design, a spectroscope was used that divided the radiation into two paths with different filters. Two ICCDs were used to form the image. In this way, two thermal images could be obtained simultaneously. However, very strict synchronization and accuracy are required for all optical devices in establishing these two optical paths. Practical experience demonstrated that this type of sensor was hard to fabricate and also was bulky. The single-path design was much simpler to construct and had greater system stability but it required a precise color modulator and two filters.

The optical system of the present design is shown in Fig. 4.13. On the color modulator there are two half-disk filters. The optical path and all the devices included were identical except for the wavelengths of the two filters. Therefore it was simple and stable. The focal length of the lens was f = 50 mm. In the case in which the object field was 50 mm × 40 mm and the

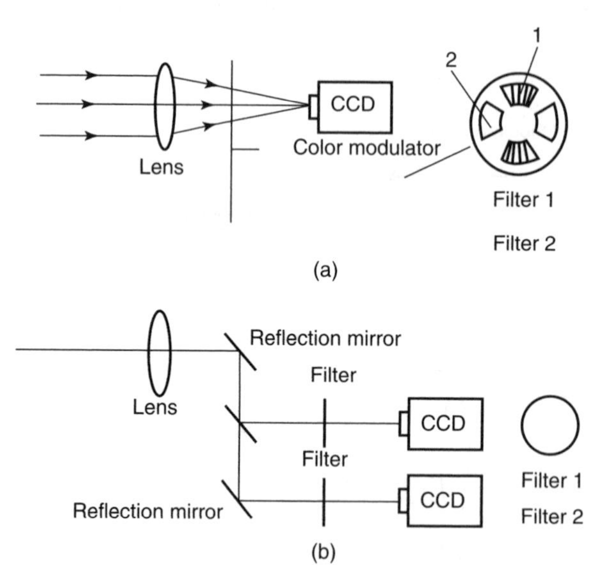

4.12 Optical paths for infrared imaging.

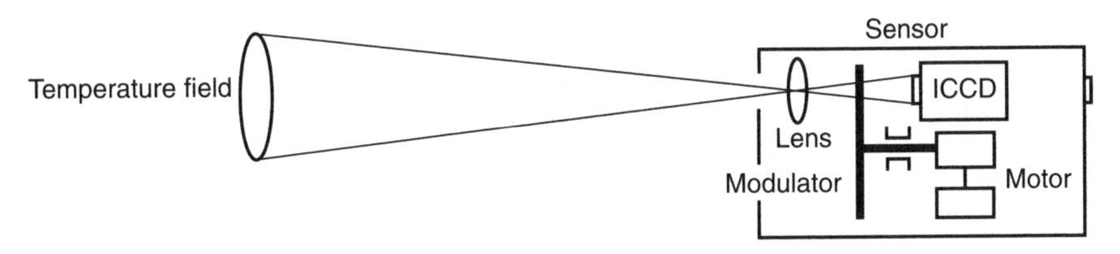

4.13 Single-path sensor.

optically sensitive surface area of the ICCD was 7.95 mm × 6.45 mm, the image distance l' and object distance l can be calculated as

$$l' = \frac{\alpha}{1+\alpha} f = 43.14\,\text{mm}$$
$$l = \alpha l' = 271.32\,\text{mm} \qquad [4.31]$$

where α is the magnification of the image. A different image size can be obtained by regulating the image distance.

4.5.2 Sensor components

The main components of the sensor are the lens and the ICCD; their characteristics determine the image quality.

Lens

In the choice of a lens, the following factors are considered.

1. The focal length is chosen according to the image and object distances, the image-field size and the space needed for mounting, the required temperature-field size, etc. For the present design, 50 mm was chosen.
2. The aperture coefficient F is chosen according to the minimum and maximum irradiances acceptable to the ICCD and should be coordinated with the exposure time. In the present design, F = 4.5, 5.6, 8, 11 and 16 were chosen. The irradiance of the ICCD decreases by one half when the F stop is increased by one step.
3. The vignetting coefficient H affects the evenness of the irradiance on the ICCD surface when a uniform source emits and projects radiation to the lens and is given by

$$H = \frac{D_\omega}{D} \qquad [4.32]$$

where D is the optical beam diameter projected by a point on the object located on the principal axis and D_ω is the actual beam width projected by a point located outside the principal axis. The angle between the major optical line of the point and the principal axis is ω. The beam width is measured on the intersecting plane formed by the object point and the principal axis. Because of the vignetting effect, the irradiance at the periphery of the image decreases; the larger the viewing field, the more serious this phenomenon becomes. Because the viewing field is small in the present case, H may be neglected.

4. Geometric parameters and lens quality.

Image-intensified charge-coupled device

An ICCD has many advantages including low weight, small volume, low-power requirement and robustness in operation. Moreover its resolving power, dynamic response, sensitivity, and real-time scanning and signal transmission are also superior. The main function of the ICCD is to accumulate and transmit electric charge. In the present system, a model MTV1881EX was chosen, the characteristics of which are shown in Table 4.1.

4.5.3 Two-color modulator and filter

The function of the color modulator is to alternate between the filters in order to take two images at different wavelengths. Because the temperature field changes rapidly, high-speed switching between the two filters is required to minimize the displacement of the temperature field between images. Figure 4.14(a) and Fig. 4.14(b) show the construction of the modulator. It consists of a disk with two half-circle holes as shown in Fig. 4.14(a). On the frame of the disk are two small holes that are called location holes and are used for a photoelectric element. When the hole passes through the photoelectric element, a signal is produced that instructs the computer to take an image. Figure 4.14(b) shows the two half-circle filters having wave-

Table 4.1 Characteristics of the ICCD

Parameter	Value
Image area	4.8 mm × 3.6 mm
Number of pixels	795 (horizontal) × 596 (vertical)
Horizontal frequency	15.625 kHz
Vertical frequency	50 Hz
Scanning system	CCIR Standard, 625 line, 50 field/s
Minimum irradiance	$0.01\lambda_x$ for output voltage –6 dB *at* f = 1.4 lens
Signal-to-noise ratio	>48 dB
Resolving power	Better than 600-line television
Shutter	Discrete; 8 grades

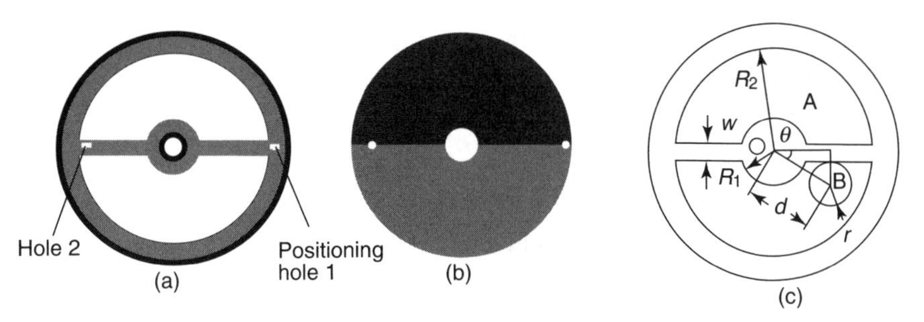

4.14 Construction of a two-color modulator.

lengths λ_1 and λ_2. Suppose that the center of the disk is O (see Fig. 4.14(c)), the outer diameters are R_1 and R_2, the width of the cross beam between the two half-circles is w, the center of the light beam is at B and its radius is r, the distance between the disk center and the center of the light beam is OB = d, the angle between OB and the cross beam is θ, and the angle used for taking the image is θ_0. Then the values of R_1, R_2 and θ should satisfy the following formula:

$$\theta_0 = \sin^{-1}\left(\frac{r+w/2}{d}\right), \quad \theta_0 < \theta < \pi - \theta_0, \quad R_1 < d - r, \quad R_2 > d + r \qquad [4.33]$$

It can be seen from these formulas that an appropriate design of the disk could increase the beam receiving angle and the switching speed of the filters. If $r + w/2 = d/2$, then $\theta_0 = 30°$ and the effective beam receiving angle would be $180° - 2 \times 30° = 120°$.

The maximum exposure time of the ICCD was 16.7 ms. In selecting the rotation speed of the color modulator, 40 ms was considered a satisfactory time for taking an image. If the switching time of the filters was 20 ms, then the total time for taking two images would be 2(40 + 20) = 120 ms. Therefore N = 8.33 rev/s was selected for the rotation speed of the modulator. A 6 V, 2400 rev/min direct-current miniature motor having a belt drive was chosen to rotate the modulator.

Because the images are taken during high-speed rotation of the filter, the filters should have uniform quality. Two specially made interference filters were designed according to the principle described in Section 4.4 and Eqn [4.24]. The filter parameters were $\lambda_1 = 0.78\,\mu m$, $\delta\lambda_1 = 0.0235\,\mu m$, $I_1 = 0.83$, $\lambda_2 = 0.92\,\mu m$, $\delta\lambda_2 = 0.0204\,\mu m$ and $I_2 = 0.662$.

4.5.4 Acquisition of the image signals and control

For dividing the temperature range, choosing several exposure times, which are controlled by the computer via an input–output interface, are necessary[15,16]. The switching of the imaging was controlled by a photoelectric signal produced using two small holes. A schematic diagram of the operation is shown in Fig. 4.15.

Exposure-time control

The exposure time was determined by a 4 bit logic circuit. The logic settings are shown in Table 4.2.

Acquisition of the image

Image acquisition was controlled by the signal generated from the holes on the modulator disk. When a hole reached a preset location, the photoelectric device transmitted a signal to the computer and started the sampling of the image. The time sequence is shown in Fig. 4.16.

The image signal-processing system

This system, which is shown in Fig. 4.17, consisted mainly of an image module and a computer. The image was sampled according to the location

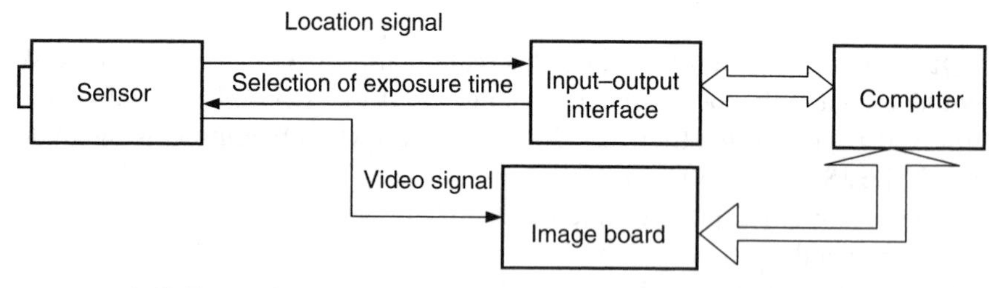

4.15 Two-color sensor and computer input–output interface.

Table 4.2 Exposure-time control (1, open circuit; o, short circuit)

	J_4	J_5	J_6	J_7
Close	0	×	×	×
1/60 ms	1	0	0	0
1/125 ms	1	0	0	1
1/250 ms	1	0	1	0
1/500 ms	1	0	1	1
1/1000 ms	1	1	0	0
1/2000 ms	1	1	0	1
1/4000 ms	1	1	1	0
1/10 000 ms	1	1	1	1

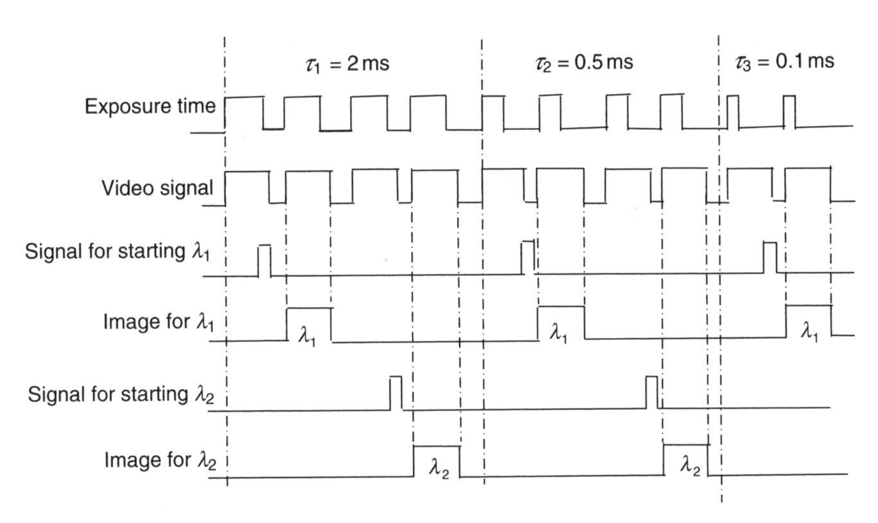

4.16 Time sequence for imaging.

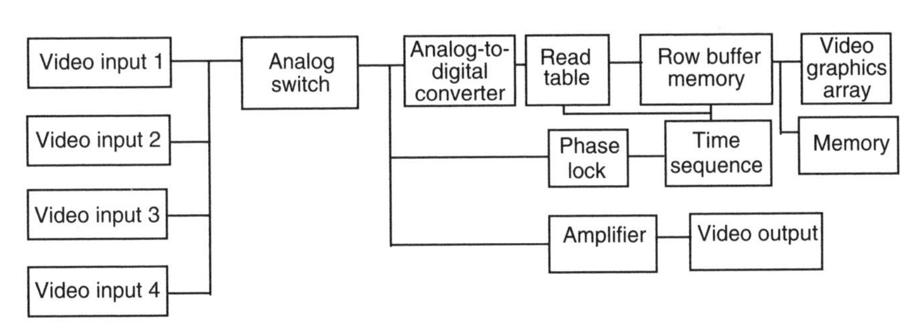

4.17 Schematic diagram of the image module.

pulse, which could be sent to the monitor for display or to the memory of the computer for processing. The software selected one of the four channels and the signals were sent to an analog-to-digital converter. After reading the table, the 8 bit image signals were sent to the buffer memory. At the same time as the signals were sent to the row buffer memory, they were taken by the computer via a VESA bus and sent to the computer's memory or the display module of the video graphics array for further processing.

4.6 Calibration of the system

After temperature-measuring instruments are built, they should be calibrated to establish the relationship between the output and temperature, which is called the calibration function or calibration curve[17,18]. For conventional instruments, a standard reference should be used, either near the object or inside the photoelectric scanning device. This requirement is hard to satisfy, particularly for temperature-field measurements.

The author proposed a special calibration method. First, off-line calibration was carried out with a known temperature reference so that the relationship between the gray ratio and temperature was obtained. In subsequent actual applications, the instrument was calibrated online to establish the relationship between gray level and temperature. Then the temperature-field data can be quickly acquired in actual measurements.

4.6.1 Offline calibration

From the principle of the measuring method, it is seen that a small distance variation within 40 mm does not greatly affect the gray ratio–temperature relationship. Proper selection of λ_1 and λ_2 confines the error due to surface conditions to a certain range. Therefore the factors that should be calibrated are the exposure of the ICCD and the aperture of the lens.

The conditions for calibration were as follows: Gleeble-1500 thermal-cycle simulation machine, a lens of aperture 8, ICCD exposures of 0.1, 0.5 and 2 ms, stainless steel material, specimen dimensions of 12 mm × 12 mm × 60 mm and a distance of 310 mm. The schematic diagram of the calibration system is shown in Fig. 4.18.

The procedure is as follows. Two thermal images were taken at the center of a uniform temperature field, the data were processed and the gray-level ratios were determined.

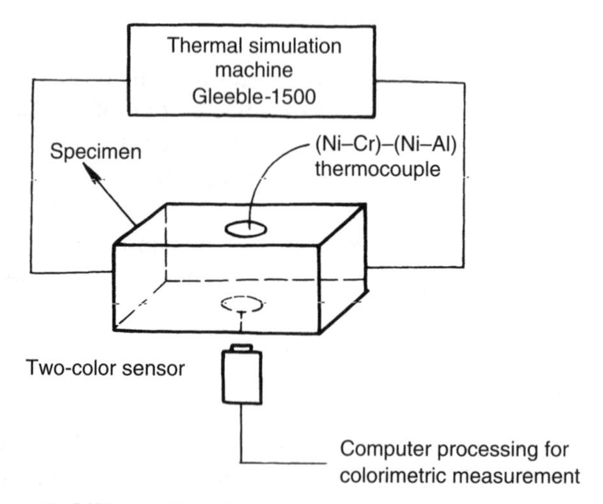

4.18 Offline calibration.

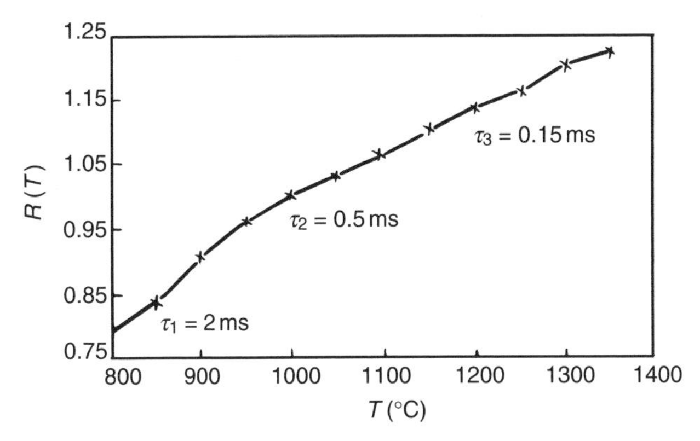

4.19 Relationship between the gray ratio and the temperature.

In order to avoid the thermocouple from influencing the temperature field, the thermocouple was mounted on the back of the specimen opposite the measured point. Measurements were then taken every 50 °C. The relation between the gray-level ratio and the temperature was recorded for each exposure. Three calibration curves were obtained by fitting the data, as shown in Fig. 4.19.

4.6.2 Online calibration

In the colorimetric method, much time is needed for data acquisition and processing for each temperature field. The first step is sampling of two thermal images, which requires 40 ms including the switching of the optical path. The second step is filtering of the image data, which also takes 40 ms if a (3 × 3) pixel average is used. This duration can be shortened if signals of only local parts of the temperature field are necessary. The first duration may be greatly shortened if a single image is used instead of two images. However, the gray level of a single image is sensitive to distance and cannot be used directly for assessment of the temperature.

It is known that the gray-level ratio has a definite relationship to temperature independent of the measurement distance. If the distance and material are kept constant, there is a definite relationship between the gray-level ratio and the gray level. In other words, there is a definite relationship between the gray level $N_{\lambda_1}(T)$ of the image wavelength λ_1 and the temperature. This relationship can be calibrated online for all three exposures. The method is shown in Fig. 4.20. Thus the temperature-field data can be established directly from its gray level from one image. The method is called a 'two-color calibration and one-color measuring method'. In this way, the process becomes simple and quick. Figure 4.21 shows an online calibration

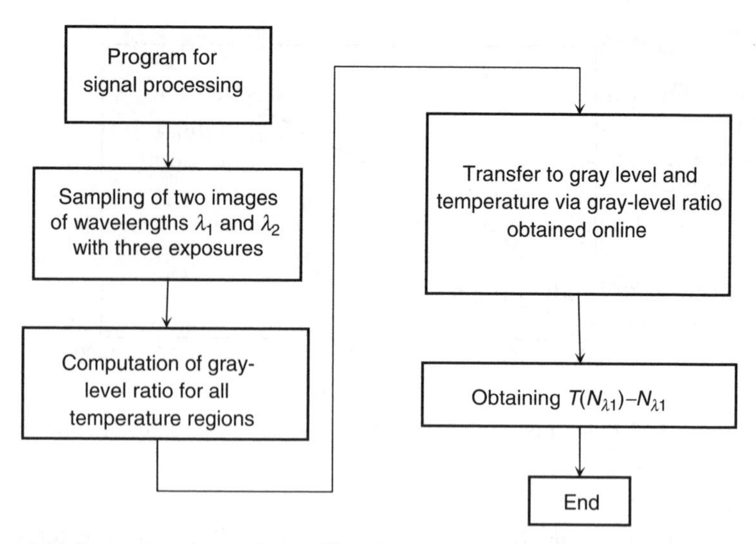

4.20 Procedure for online calibration.

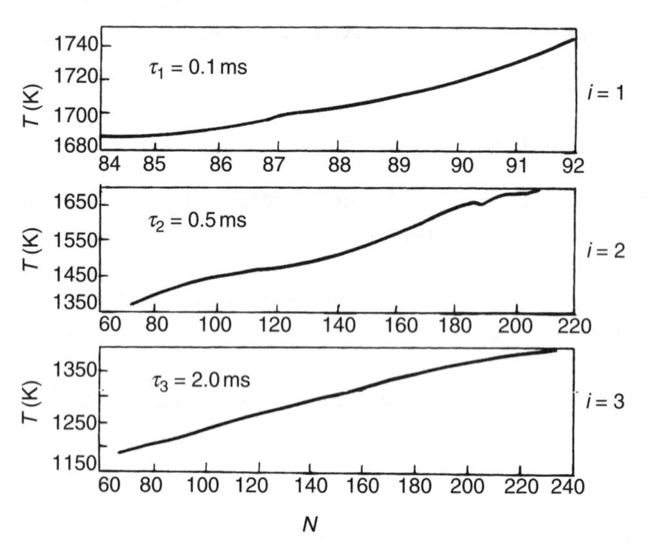

4.21 Online calibrated curves of temperature versus gray level N obtained from the CCD image taken with a filter of wavelength λ_1.

curve obtained under the following conditions: a distance of 310 mm, stainless steel, an aperture of 16 and exposure times of 0.1, 0.5 and 2 ms. By using these curves, the measuring time is shortened to one half of that for the two-color measurement method.

4.7 Measurement of the temperature field

The main problems for real-time measurement of a welding temperature field are the measuring speed and measurable temperature range. Applying combined offline and online calibration can shorten the measuring speed for one image to 40 ms.

Applying several exposures has extended the measurable range from 250 °C to 600 °C. Therefore adequate information about the welding temperature field can be obtained within 0.5 s, which makes real-time control possible and connection of the temperature regions to form a complete field, etc.

4.7.1 Acquisition of temperature-field signals

Measurements were conducted using tungsten–inert gas welding. The experimental conditions were flat welding without filler-wire addition, mild steel having dimensions of 60 mm × 150 mm × 2 mm, argon shielding gas at a flow rate of 0.5 m^3/h, a welding current of 128 A, an arc voltage of 14 V, a travel speed of 7.07 mm/s, a tungsten electrode of diameter 2.5 mm, a measuring distance of 310 mm, an aperture setting of 11 and exposure times of 0.1, 0.5 and 2 ms. Two images were taken through the filters designed for wavelengths λ_1 and λ_2 at an exposure time of 0.1 ms for the high-temperature region, at an exposure time of 0.5 ms for the medium-temperature region and at an exposure time of 2.0 ms for the low-temperature region. The time sequence for imaging is shown in Fig. 4.16 and the flow chart is shown in Fig. 4.22. Because the total imaging time was less than 60 ms, the influence of torch movement could be ignored. Using a two-color calibration method and a one-color measuring method, three images were obtained for λ_1 at three different exposure times.

4.7.2 Procedure for signal processing

Signal processing can be divided into three procedures: preprocessing of the thermal image, calibration of the gray level of each of the three regions and connection of the three temperature regions.

Preprocessing of the image

In practical measurements, there is contamination on the object's surface and noise produced by the optical system itself. Both of these cause error in the measurements; the gray-level ratio is particularly sensitive to noise.

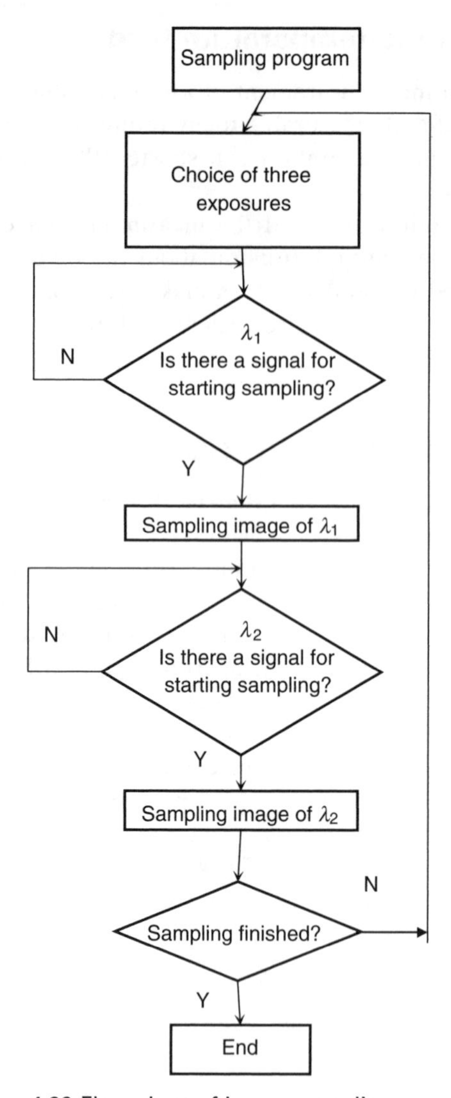

4.22 Flow chart of image sampling.

Therefore filtering the signals is necessary. The average gray level of 9 neighboring pixels normally is used, namely

$$\begin{aligned} N'(i, j) = (&N(i-1, j-1) + N(i-1, j) + N(i-1, j+1) \\ &+ N(i, j-1) + N(i, j) + N(i, j+1) \\ &+ N(i+1, j-1) + N(i+1, j) + N(i+1, j+1))/9 \end{aligned} \qquad [4.34]$$

where $N'(i, j)$ is the gray value of the row i, column j pixel.

Detection of the temperature in different regions

After filtering of the signals, two images in the high-temperature region, with gray levels $N(\lambda_1)$ and $N(\lambda_2)$, are selected. The pixels that have the same gray level N_1 in the image of wavelength λ_1 are selected and the pixels in the image of wavelength λ_2 that correspond to the pixels of gray level N_1 in the image of wavelength λ_1 are identified. The average gray value $N_2(N_1)$ of these pixels is calculated using the following formula:

$$R(N_1) = \frac{N_1}{N_2(N_1)} \qquad [4.35]$$

Taking advantage of the offline calibration curve (the R–T relationship curve), the temperature corresponding to N_1 can be found as $T(R(N_1))$.

In order to eliminate the misalignment of the images with wavelength λ_1 and λ_2 (in the present design, it would be no more than two pixels), the value of $T(N_1)$ is averaged with two neighboring ratio values:

$$T(N_1) = \frac{T(N_1 - 1) + T(N_1) + T(N_1 + 1)}{3} \qquad [4.36]$$

Similarly, from images in the medium-temperature region and the low-temperature region, $T_2(N_1)$ and $T_3(N_1)$ can be obtained in the same way as $T_1(N_1)$.

It is obvious now that $T_1(N_1)$, $T_2(N_1)$ and $T_3(N_1)$ were obtained on the basis of $R(N_1)$. However, on this basis the temperature values of the field can be evaluated directly from N_1 obtained from the three images of wavelength λ_1.

Connection of temperature regions

In order to obtain the complete temperature field, the three regions are connected together. To make this connection, the borders of each region have to be defined. The location where the gray level $N_1 < 50$ in the high-temperature image of wavelength λ_1 is defined as the border of the high-temperature region. Similarly, the location where $N_1 < 50$ in the medium-temperature image of wavelength λ_1 is defined as the border of the medium-temperature region. The location where $N_1 < 50$ in the low-temperature image of wavelength λ_1 is considered as a region that does not give reliable data. In order to put the temperature fields of three regions together and to display them as a single image, the temperature data are transformed into their gray levels. Supposing that the maximum and minimum temperatures obtained for the complete temperature field are T_{max} and T_{min}, transforming it into gray levels from 255 to 0 is required. The formula for transformation is

$$N[T_i(N_1)] = 255\frac{T_i(N_1) - T_{min}}{T_{max} - T_{min}} \quad [4.37]$$

where $i = 1, 2$ and 3 represent the high-temperature, medium-temperature and low-temperature regions respectively.

If the gray level N is known, the temperature can be found using the following formula:

$$T = T_{min} + (T_{max} - T_{min})\frac{N}{255} \quad [4.38]$$

The maximum temperature in the center is 1700 K and at the periphery is 1200 K. Each color band represents 50 K. A flow chart of the signal processing is shown in Fig. 4.23.

4.7.3 Connecting the temperature regions

Figure 4.24(a), Fig. 4.24 (b) and Fig. 4.24 (c) show three-dimensional diagrams of the temperature distribution in the low-, medium- and high-temperature regions respectively. Figure 4.24 (d) is the whole picture that

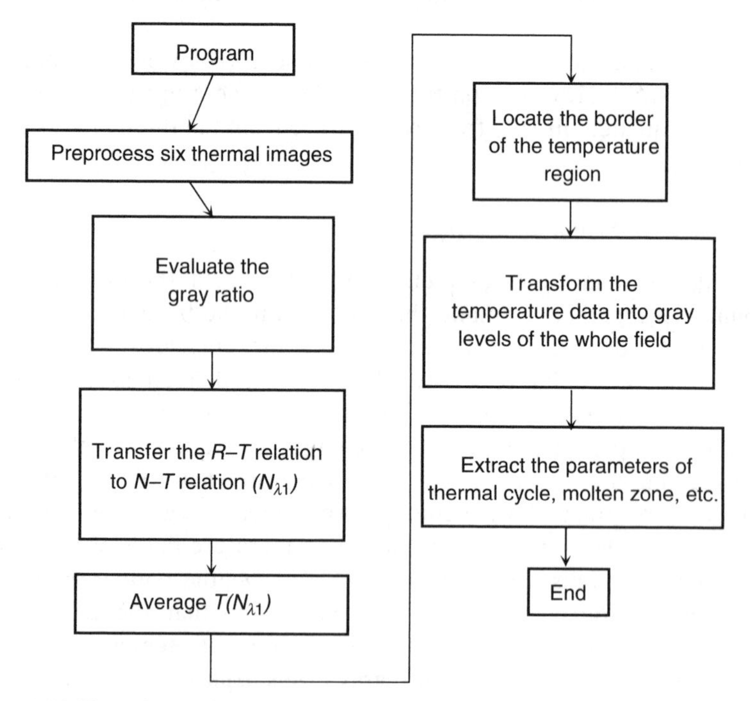

4.23 Flow chart of signal processing of the temperature field.

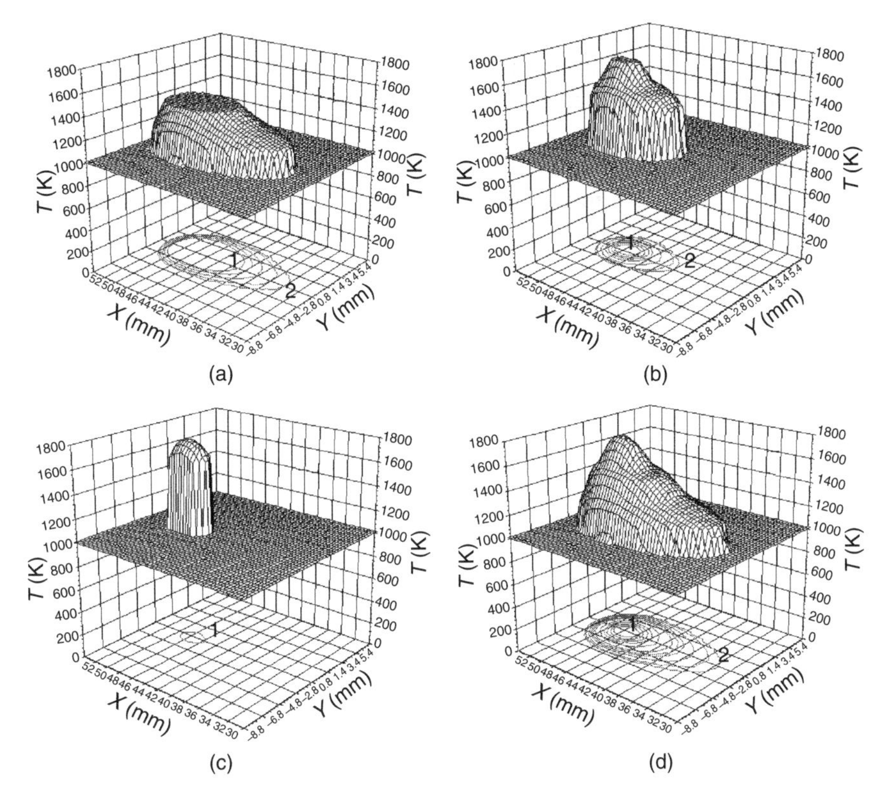

4.24 Temperature distributions of three regions and the complete field: (a) $\tau_1 = 2$ ms, low-temperature region (curve 1, 1400 K; curve 2, 1200 K); (b) $\tau_2 = 0.5$ ms, medium-temperature region (curve 1, 1700 K; curve 2, 1400 K); (c) $\tau_1 = 0.1$ ms, high-temperature region (curve 1, 1700 K); (d) complete temperature field (curve 1, 1700 K; curve 2, 1200 K).

incorporates the three connected regions where *XOY* is an isothermal plane. Each color grade is 50 K.

4.8 Applications

4.8.1 Verification of the mathematic model

The analysis of the temperature field is based mainly on three factors, namely the distribution of the heat source, the heat transfer laws and the boundary conditions of the workpiece. An analytical solution can be obtained only for the case of a workpiece half infinitely thick, half infinitely large or infinitely long. The heat source is assumed to have a point, line or surface shape. In fact, the heat source is not ideally distributed; therefore the temperatures cannot be analyzed even if the boundary conditions of the workpiece are ideal. Anyhow, an analytical function has been obtained when the heat source distribution is assumed to obey a Gaussian model, which approximates a real arc. Zhang[13] proposed a heat source following a double-elliptical Gaussian model that is closer to a real heat source than the simple Gaussian model. The analysis was performed using a finite workpiece thickness. The temperature distribution was obtained by numerical simulation and checked using the data obtained by actual measurements. It was proved that the proposed heat source model was closer to the actual conditions than the former model.

Bi-ellipse Gaussian model

The moving-line heat source and Gaussian-distributed heat source

In the welding of thin plates, the heat source is close to a line heat source. Assume that the effective power is q, the plate being welded is infinitely large, the heat source moves from the point O_0 with constant speed and after t s it reaches O, as shown in Fig. 4.25. The temperature distribution is[19]

$$T(x, y, t) = \frac{q}{4\pi\lambda h}\int_0^t \frac{\mathrm{d}t}{t' + \tau} \exp\left(-\frac{(x + vt')^2 + x^2}{4a(t' + \tau)} - b(t' + \tau) \right) \qquad [4.39]$$

where $r^2 = x^2 + y^2$, $x = x_0 - vt$, $y = y_0$, h is the plate thickness, b is the heat dissipation coefficient of the plate, v is the welding speed, q is the effective power of the welding power source, λ is the heat transmission coefficient, a is the heat diffusion coefficient, τ is the cooling time and t is the total welding time.

The period 0–t is divided into P subfields, the center coordinate of the subfields is t_i and its duration is $\Delta t = t/P$. The Gaussian integration is applied

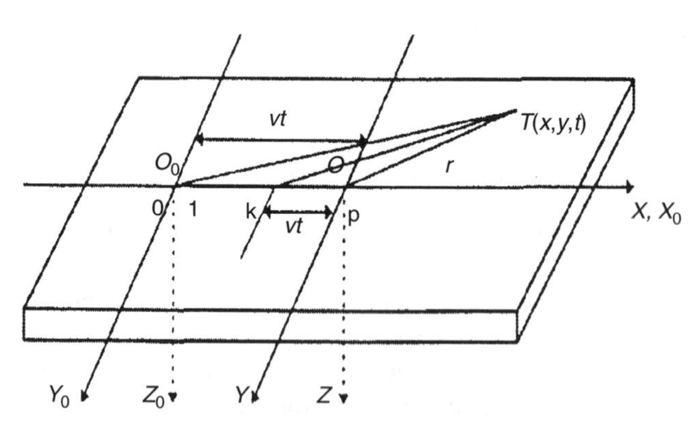

4.25 The coordinate system for calculation.

to each subfield. The numerical integration formula can be written as follows[19]:

$$T(x, y, t) = \sum_{i=1}^{P} \sum_{j=1}^{N} \frac{q}{4\pi\lambda h} \frac{H_j}{\sum N_i t_i' + \tau} \frac{\Delta t}{2} \times \exp\left[-\frac{(x + v\sum N_i t_i')^2 + x^2}{4a(\sum N_i t_i' + \tau)} - b(\sum N_i t_i' + \tau) \right] \quad [4.40]$$

where

$$\sum N_i t_i' = \frac{-(1-\xi_j)\xi_j}{2}\left(t_i' - \frac{\Delta t}{2}\right) + (1-\xi_j^2)t_i' + \frac{(1+\xi_j)\xi_j}{2}\left(t_i' + \frac{\Delta t}{2}\right) \quad [4.41]$$

where $j = 1, 2, 3$, $N = 4$, ξ_j is the coordinate of each Gaussian point and H_j is the integration coefficient.

For a half-infinite-thickness plate and a Gaussian heat source, the temperature-field distribution can be written as[20]

$$T(x, y, z, t) - T(x, y, z, 0) = \frac{q}{\rho c\pi} \int_0^t \frac{1}{[4\pi a(t-t')]^{1/2}[2a(t-t')+w^2]} \times \exp\left(-\frac{z^2}{4a(t-t')} - \frac{(x-vt')^2 + y^2}{4a(t-t')+2w^2} \right) dt' \quad [4.42]$$

where q is the effective power of the heat source, p is the material density, c is the specific heat capacity, w is the Gaussian distribution parameter, v is the welding speed, a is the thermal diffusion coefficient and t is the total welding time. Similarly the duration 0–t is divided into P subfields, and a Gaussian integration is applied to each subfield. The numerical integration formula can be written as follows:

$$T(x, y, z, t) - T(x, y, z, 0) = \sum_{l=-\infty}^{\infty} \sum_{i=1}^{P} \sum_{j=1}^{N} \frac{q}{\rho c\pi} \frac{1}{[4\pi a(t - \sum N_i t_i)]^{1/2}[2a(t - \sum N_i t_i) + w^2]} \times \exp\left(-\frac{z^2}{4a(t - \sum N_i t_i)} - \frac{(x - v\sum N_i t_i)^2 + y^2}{4a(t - \sum N_i t_i) + 2w^2} \right) \frac{H_j}{2} \Delta t \quad [4.43]$$

The bi-elliptical heat source distribution

The model of the heat source is shown in Fig. 4.26. The power density of the heat source is

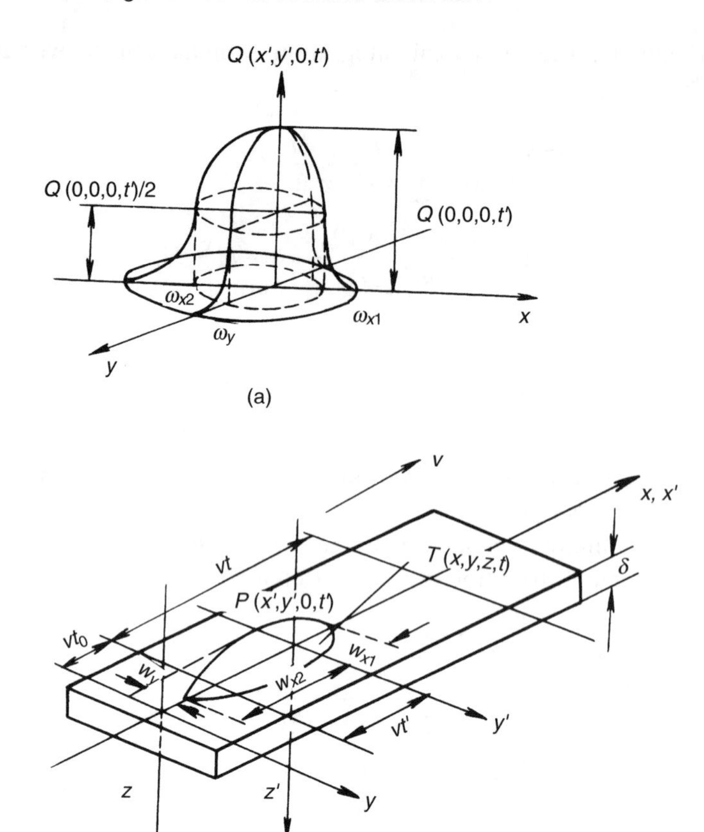

4.26 Bi-elliptical heat source distribution.

$$Q(x', y', 0, t') = \begin{cases} \dfrac{f_f q}{2\pi w_{x1} w_y} \exp\left(-\dfrac{x'^2}{2w_{x1}^2} - \dfrac{y'^2}{2w_y^2}\right), & x' > 0 \\ \dfrac{f_r q}{2\pi w_{x2} w_y} \exp\left(-\dfrac{x'^2}{2w_{x2}^2} - \dfrac{y'^2}{2w_y^2}\right), & x' < 0 \end{cases} \qquad [4.44]$$

where q is the effective power of the heat source, f_f and f_r are the coefficients of the power distribution in the front and rear parts respectively of the molten pool, with $f_f + f_r = 2$, v is the welding speed, w_{x1}, w_{x2} and w_y are the distributive coefficients of the bi-elliptical heat source and $x'y'z'$ is the coordinate system moving with the heat source.

The temperature distribution in a half-infinite-thickness body

Substituting the heat source formula into the heat flow formula, the analytical expression for the temperature distribution can be written as

$$dT(x, y, z, t) = \frac{2Q(x', y', 0, t')dx'dy'dt'}{\rho c[4\pi a(t - t' - \tau)]^{3/2}} \times \exp\left(-\frac{[x - x' - v(t' + t_0)]^2 + (y - y')^2 + z^2}{4a(t - t' - \tau)}\right) \quad [4.45]$$

where ρ is the material density, c is the specific heat capacity, a is the heat diffusion coefficient, vt_0 is the initial position of the heat source ($vt_0 = 0$) and τ is the cooling time ($\tau = 0$).

After integrating the heat source, the temperature distribution formula can be derived as follows:

$$dT(x, y, z, t)|_t = dt'\iint \frac{2Q(x', y', 0, t')}{\rho c[4\pi a(t - t')]^{3/2}} \times \exp\left(-\frac{[x - x' - vt']^2 + (y - y')^2 + z^2}{4a(t - t')}\right)dx'dy'$$
$$= \frac{dt'}{\rho c[4\pi a(t - t')]^{3/2}} \frac{q}{\pi w_y} \times \exp\left(-\frac{x^2 + y^2 + z^2}{4a(t - t')}\right)\left(\frac{f_f}{w_{x1}}\Delta_1 + \frac{f_r}{w_{x2}}\Delta_2\right)\Delta_3 \quad [4.46]$$

where Δ_1, Δ_2 and Δ_3 are as follows:

$$\Delta_1 = \left[\frac{\pi^{1/2}}{2}\left(\frac{2a(t - t') + w_{x1}^2}{4a(t - t')w_{x1}^2}\right)^{-1/2} + \int_0^{w_{x_1}^2 (x - vt')/2a(t - t') + w_{x_1}^2} \exp\left(-\frac{2a(t - t') + w_{x1}^2}{4a(t - t')w_{x1}^2} x''^2\right)dx''\right]$$
$$\times \exp\left[\frac{1}{4a(t - t')}\left(\frac{w_{x1}^2 (x - vt')^2}{2a(t - t')w_{x1}^2} + vt'(2x - vt')\right)\right] \quad [4.46a]$$

$$\Delta_2 = \left[\frac{\pi^{1/2}}{2}\left(\frac{2a(t - t') + w_{x2}^2}{4a(t - t')w_{x2}^2}\right)^{-1/2} - \int_0^{w_{x_2}^2 (x - vt')/2a(t - t') + w_{x_2}^2} \exp\left(-\frac{2a(t - t') + w_{x2}^2}{4a(t - t')w_{x2}^2} x''^2\right)dx''\right]$$
$$\times \exp\left[\frac{1}{4a(t - t')}\left(\frac{w_{x2}^2 (x - vt')^2}{2a(t - t') + w_{x2}^2} + vt'(2x - vt')\right)\right] \quad [4.46b]$$

$$\Delta_3 = \pi^{1/2}\left(\frac{2a(t-t')+w_y^2}{4a(t-t')w_y^2}\right)^{-1/2} \exp\left(\frac{w_y^2 y^2}{4a(t-t')[2a(t-t')+w_y^2]}\right) \quad [4.46c]$$

Integrating the temperature distribution over time, it can be found that

$$T(x, y, z, t) - T(x, y, z, 0) = \int_0^t \frac{1}{\rho c[4\pi a(t-t')]^{3/2}} \frac{q}{\pi w_y} \times \exp\left(-\frac{x^2+y^2+z^2}{4a(t-t')}\right)\left(\frac{f_f}{w_{x1}}\Delta_1 + \frac{f_r}{w_{x2}}\Delta_2\right)\Delta_3 \, dt' \quad [4.47]$$

The temperature-field distribution of a finite-thickness plate

For a plate thickness δ, the temperature distribution can be obtained from the formula above by the mirror-images method as follows.

$$T(x, y, z, t) - T(x, y, z, 0) = \sum_{l=-\infty}^{\infty} \int_0^t \frac{1}{\rho c[4\pi a(t-t')]^{3/2}} \frac{q}{\pi w_y} \times \exp\left(-\frac{x^2+y^2+(z-2l\delta)^2}{4a(t-t')}\right)\left(\frac{f_f}{w_{x1}}\Delta_1 + \frac{f_r}{w_{x2}}\Delta_2\right)\Delta_3 \, dt' \quad [4.48]$$

Dividing the time duration 0–t into p subfields, the numerical formula for calculation of the temperature distribution by Gaussian integration is

$$T(x, y, z, t) - T(x, y, z, 0) = \sum_{l=-\infty}^{\infty} \sum_{i=1}^{P} \sum_{i=1}^{N} \frac{1}{\rho c[4\pi a(t-\sum N_i t_i)]^{3/2}} \frac{q}{\pi w_y} \times \exp\left(-\frac{x^2+y^2+(z-2l\delta)^2}{4a(t-\sum N_i t_i)}\right)\left(\frac{f_f}{w_{x1}}\Delta_1 + \frac{f_r}{w_{x2}}\Delta_2\right)\Delta_3 \frac{\Delta t}{2} H_j \quad [4.49]$$

where Δ_1, Δ_2 and Δ_3 are as follows:

$$\Delta_1 = \left\{\frac{\pi^{1/2}}{2}\left(\frac{2a(t-\sum N_i t_i)+w_{x1}^2}{4a(t-\sum N_i t_i)w_{x1}^2}\right)^{-1/2} + \frac{w_{x1}^2(x-v\sum N_i t_i)}{M[2a(t-\sum N_i t_i)+w_{x1}^2]} \times \sum_{m=0}^{M} \exp\left[-\frac{2a(t-\sum N_i t_i)+w_{x1}^2}{4a(t-\sum N_i t_i)w_{x1}^2}\left(\frac{m w_{x1}^2(x-v\sum N_i t_i)}{M[2a(t-\sum N_i t_i)+w_{x1}^2]}\right)^2\right]\right\} \times \exp\left\{\frac{1}{4a(t-\sum N_i t_i)}\left[\frac{w_{x1}^2(x-v\sum N_i t_i)^2}{2a(t-\sum N_i t_i)+w_{x1}^2} + v\sum N_i t_i(2x - v\sum N_i t_i)\right]\right\} \quad [4.49a]$$

$$\Delta_2 = \left\{ \frac{\pi^{1/2}}{2} \left(\frac{2a(t - \sum N_i t_i) + w_{x2}^2}{4a(t - \sum N_i t_i) w_{x2}^2} \right)^{-1/2} - \frac{w_{x2}^2 (x - v \sum N_i t_i)}{M[2a(t - \sum N_i t_i) + w_{x2}^2]} \right.$$

$$\left. \times \sum_{m=0}^{M} \exp\left[-\frac{2a(t - \sum N_i t_i) + w_{x2}^2}{4a(t - \sum N_i t_i) w_{x2}^2} \left(\frac{m w_{x2}^2 (x - v \sum N_i t_i)}{M[2a(t - \sum N_i t_i) + w_{x2}^2]} \right)^2 \right] \right\}$$

$$\times \exp\left\{ \frac{1}{4a(t - \sum N_i t_i)} \left[\frac{w_{x2}^2 (x - v \sum N_i t_i)^2}{2a(t - \sum N_i t_i) w_{x2}^2} \right. \right.$$

$$\left. \left. + v \sum N_i t_i (2x - v \sum N_i t_i) \right] \right\} \qquad [4.49b]$$

$$\Delta_3 = \pi^{1/2} \left(\frac{2a(t - \sum N_i t_i) + w_y^2}{4a(t - \sum N_i t_i) w_y^2} \right)^{-1/2}$$

$$\times \exp\left(\frac{w_y^2 y^2}{4a(t - \sum N_i t_i)[2a(t - \sum N_i t_i) + w_y^2]} \right) \qquad [4.49c]$$

The numerical simulation formulas for calculating the temperature-field distribution under a bi-elliptical heat source have been defined. If the thickness is considered in Eqn [4.43], then Eqns [4.40], [4.43] and [4.49] can be used to calculate the temperature field for a plate with finite thickness but an infinite workpiece size. In the following section, practical measurements that were carried out are described and the data that were generated are used to check the theoretical results.

Comparison of calculated results with measured data

Numerical simulation and real-time measurement

The conditions for numerical simulation and measurement are shown in Table 4.3.

The experimental results are shown in Fig. 4.27 in which eight isothermal planes (*XOY*) are illustrated. The interval was 50 K, the inner isotherm was at 1450 K and the outer isotherm was at 1100 K.

Numerical simulation was carried out for a bi-elliptical Gaussian distribution, a Gaussian distribution and a line heat source using Eqns [4.49], [4.42] and [4.39]. The results are shown in Fig. 4.28, Fig. 4.29 and Fig. 4.30 respectively. The parameters for the bi-elliptical Gaussian heat source were $w_{x1} = 0.10$ cm, $w_{x2} = 0.30$ cm, $w_y = 0.14$ cm, $f_f = 0.6$ and $f_r = 1.4$. The parameter for the Gaussian heat source was $w = 0.14$ cm. Eight isothermal planes designated *XOY* are shown in all the figures. The intervals were 50 K, the inner isotherm was at 1450 K, and the outer isotherm was at 1100 K.

Table 4.3 Conditions for numerical simulation and experiment

Parameter	Value
Welding conditions	
Workpiece	Mild steel (60 mm × 150 mm × 2 mm)
Shielding gas	Argon (0.5 m^3/h)
Voltage	12 V
Current	60 A
Welding speed	5 mm/s
Electrode	Tungsten (diameter 2.5 mm)
Conditions for numerical simulation	
Material density	7.68 g/cm^3
Heat diffusion	0.08 cm^2/s
Heat dissipation coefficient	0
Specific heat capacity	0.679 J/g °C
Heat conduction coefficient	0.377 J/cm s °C
Heating efficiency	0.75

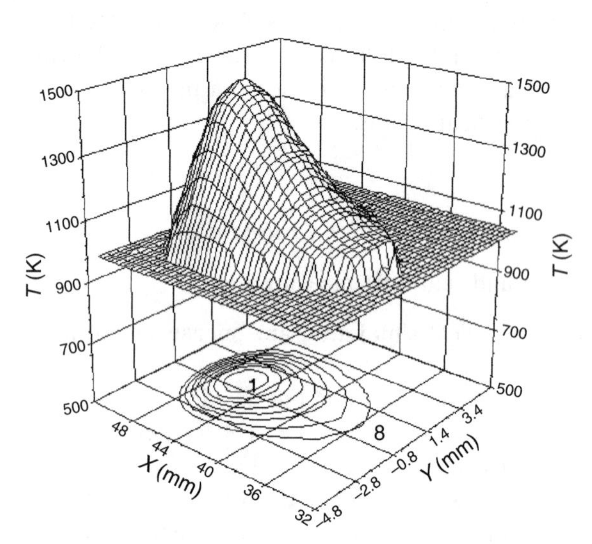

4.27 Experimental temperature distribution: curve 1, 1450 K; curve 8, 1100 K.

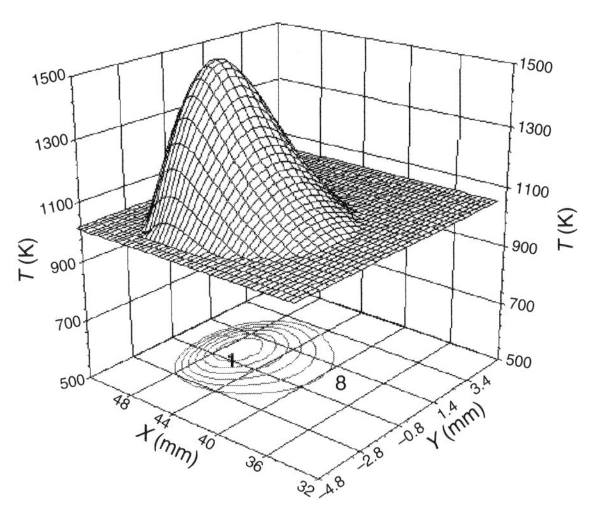

4.28 Numerical simulation results of a bi-elliptical Gauss heat source: curve 1, 1450 K; curve 8, 1100 K.

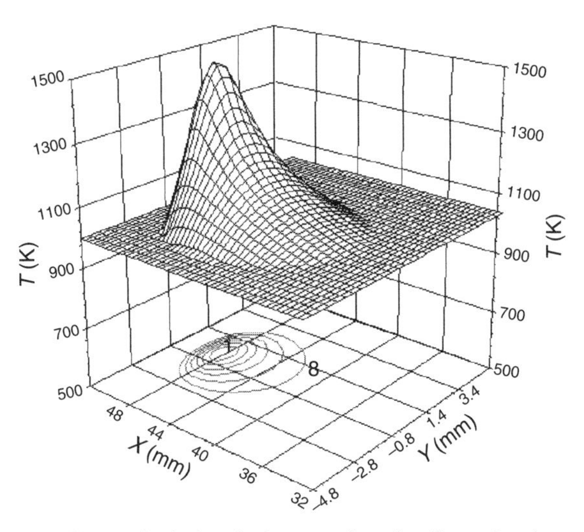

4.29 Numerical simulation results of a Gaussian heat source: curve 1, 1450 K; curve 8, 1110 K.

Comparison of results

Comparing Fig. 4.28, Fig. 4.29, and Fig. 4.30 with Fig. 4.27 shows that the bi-elliptical Gaussian model was the closest to the measured data. The difference between the results using the Gaussian model and the measured data was large. The difference between results for the line heat source and

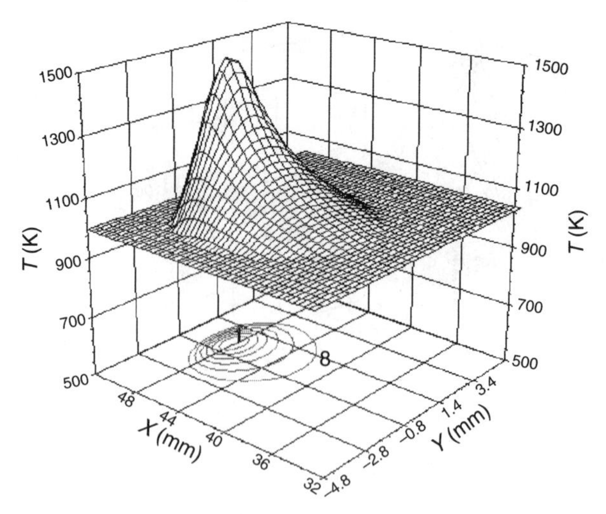

4.30 Numerical simulation results for a moving-line heat source: curve 1, 1450K; curve 8, 1100K.

Table. 4.4 Comparison of isotherms

Temperature (K)	Width (mm)		Error (%)	Length (mm)		Error (%)
	Actual	Bi-elliptical		Actual	Bi-elliptical	
1100	6.11	6.04	−1.15	13.48	12.35	−8.38
1150	5.54	5.56	0.36	11.52	10.61	−7.90
1200	4.80	4.92	2.50	9.37	9.26	−1.17
1250	4.32	4.42	2.31	8.27	8.00	−3.26
1300	3.75	3.89	3.73	6.74	6.90	−2.37
1350	3.18	3.22	1.26	5.39	5.60	3.89
1400	2.69	2.66	−1.12	4.31	4.39	1.86
1450	1.87	1.80	−3.74	2.94	3.10	5.44

the experimental data was larger than that for the Gaussian model. In order to compare them in more detail, the isotherms are given in Table 4.4. The temperature distribution curves in the longitudinal and transverse directions across the weld pool are plotted in Fig. 4.31.

The same conclusion can be drawn if the curves in Fig. 4.31 are compared. The Gaussian heat source showed a large difference from the actual temperature distribution; the line heat source showed an even larger difference than the former. Therefore, simulating the temperature distribution using a line heat source is not useful.

Comparing the numerical simulation results with the measured data is valuable. The measured results can be used to correct or modify the numer-

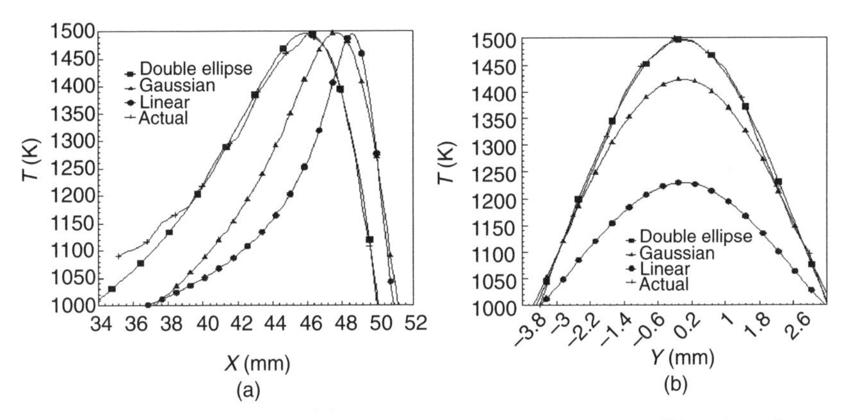

4.31 (a) Longitudinal ($Y = 0$ mm) and (b) transverse ($X = 46$ mm) temperature distributions for three different heat sources.

ical simulation model. Also, the numerical simulation model can be used to analyze the temperature-distribution phenomena for investigation of the metallurgical and physical processes in welding. For example, the numerical simulation can be used to find the three-dimensional temperature distribution or the temperature distribution at low temperatures, which is not possible using experimental measurements.

4.8.2 Extraction of the thermal cycle parameters

Thermal cycle parameters on the root side of the bead

Much information may be obtained from the temperature field on the root side of the bead, e.g. the isotherms and the temperature distribution in the longitudinal and transverse directions, the thermal cycle at any arbitrary point in the temperature field, the width of the pool and the characteristic parameters of the thermal cycle. In the following paragraphs, the isotherm parameters, longitudinal and transverse temperature distributions and thermal cycle parameters are extracted from the measured temperature field shown in Fig. 4.27.

Isotherm parameters

The isotherm parameters are the width and length of the isotherm that reflect the energy input of welding. The isotherm near the melting point represents the shape of molten pool. Table 4.5 lists the extracted isotherm parameters from Fig. 4.27.

Table 4.5 Isotherm parameters

T (K)	Width (mm)	Length (mm)
1050	6.76	15.52
1100	6.11	13.48
1150	5.54	11.52
1200	4.80	9.37
1250	4.32	8.27
1300	3.75	6.74
1350	3.18	5.39
1400	2.69	4.31
1450	1.87	2.94

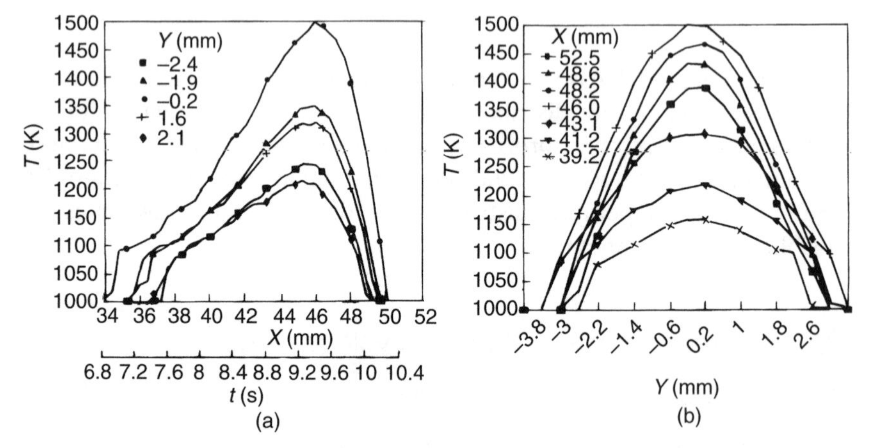

4.32 Temperature distributions of a weld in (a) the longitudinal and (b) the transverse directions.

Table 4.6 Thermal cycle parameters

Y (mm)	Heating rate ω_H (K/s)	Maximum temperature T_m (K)	High-temperature hold duration (s)	Cooling rate ω_c (K/s)
–2.4	427.35	1243.75	0.74	111.48
–1.9	732.60	1350.00	2.34	128.20
–0.2	1121.80	1500	2.72	106.80
1.6	512.82	1312.50	2.34	128.20
2.1	284.90	1218.75	0.74	111.48

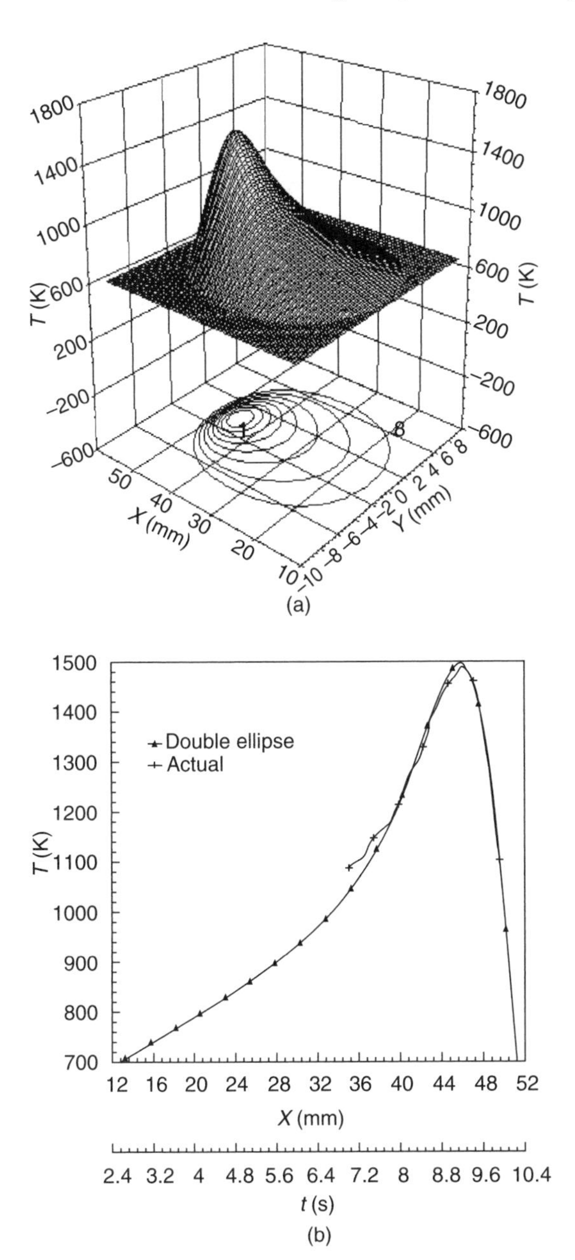

4.33 (a) The entire field (curve 1, 1400 K; curve 8, 700 K), (b) the longitudinal ($Y = 0$ mm) and (c) the transverse ($X = 46$ mm) longitudinal temperature distributions obtained by extrapolation.

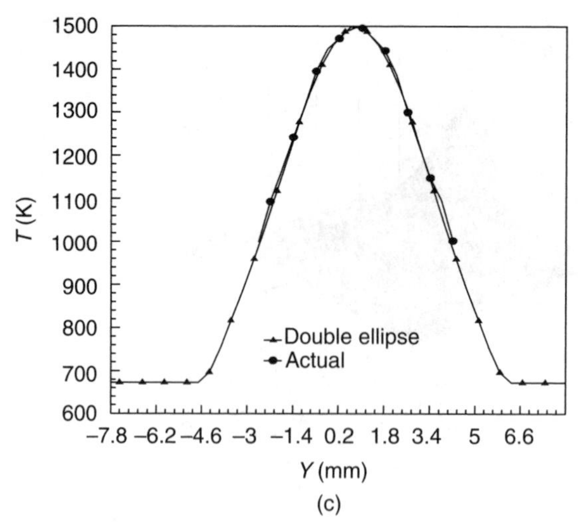

4.33 Continued

Longitudinal and transverse temperature distributions

The longitudinal and transverse temperature distributions at the weld pool can be identified easily from Fig. 4.27 and are illustrated in Fig. 4.32. In Fig. 4.32(a), the X abscissa designates the welding direction. The upper abscissa values represent the location on the workpiece and the lower values represent the travel time of the moving torch. The five curves are the temperature distributions of the point, the Y coordinates of which are –2.4, –1.9, –0.2, 1.6 and 2.1 mm. These show that the temperature gradient at the front is much larger than at the rear. In Fig. 4.32(b), Y represents the transverse position of the point. The seven curves are the temperature distributions at different cross lines, the X coordinates of which are 39.2, 41.2, 43.1, 46.0, 48.2, 48.6 and 52.5 mm. All these distributions are symmetrical.

Thermal cycle parameters

When the welding parameters are kept unchanged during the welding process, the temperature field is a quasistatic field. The temperature distribution along an arbitrary longitudinal line of the weld can be regarded as the thermal cycle of any point on the longitudinal line. Therefore the five curves in Fig. 4.32(a) are the thermal cycles of various points on these five lines. The thermal cycle parameters can thus be found easily from these curves (Table 4.6). The heating rates and cooling rates are the values when the temperature is at 1100 K. The high-temperature hold duration is the time above 1100 K. These parameters are valuable for the study of the microstructure, stress and strain of the weld.

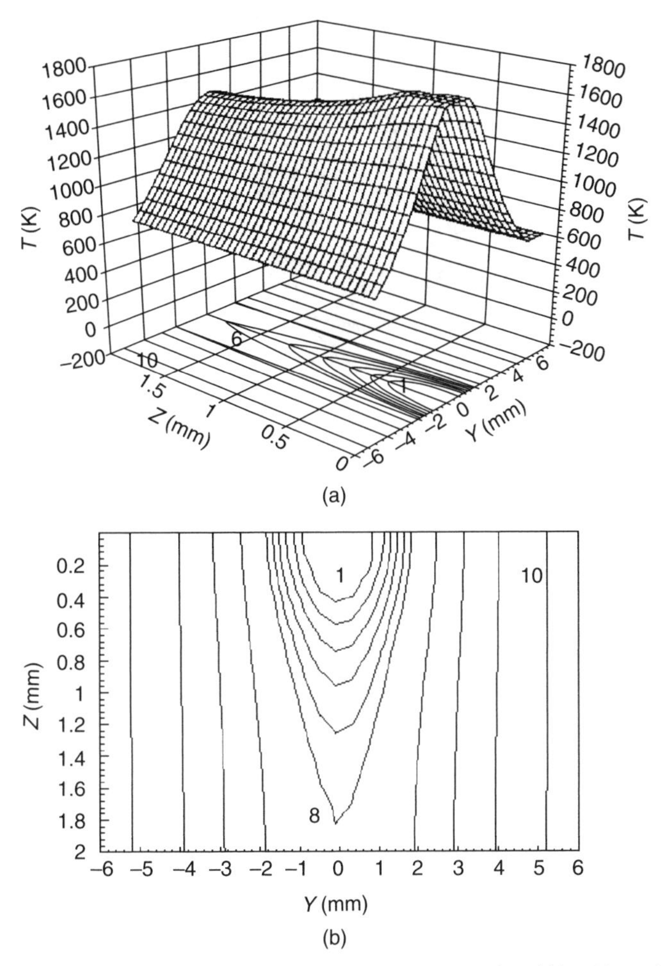

4.34 Temperature distribution on a cross-section ($X = 46$ mm): (a) three-dimensional distribution (curve 1, 1750 K; curve 6, 1500 K; curve 10, 700 K); (b) isotherms (curve 1, 1750 K; curve 6, 1500 K; curve 10, 700 K).

Extrapolation of the two-dimensional temperature field

Taking advantage of the numerical simulation method, the temperature distribution in the low-temperature range (lower than that the ICCD can measure by the colorimetric method) can be obtained by extrapolation. Figure 4.33(a) shows the results obtained by extrapolation using a measured temperature field. The extrapolation extends the temperature to 700 K. Figure 4.33(b) and Fig. 4.33 (c) show the longitudinal and transverse temperature distributions respectively obtained by measurement and extrapolation. Figure 4.33(b)

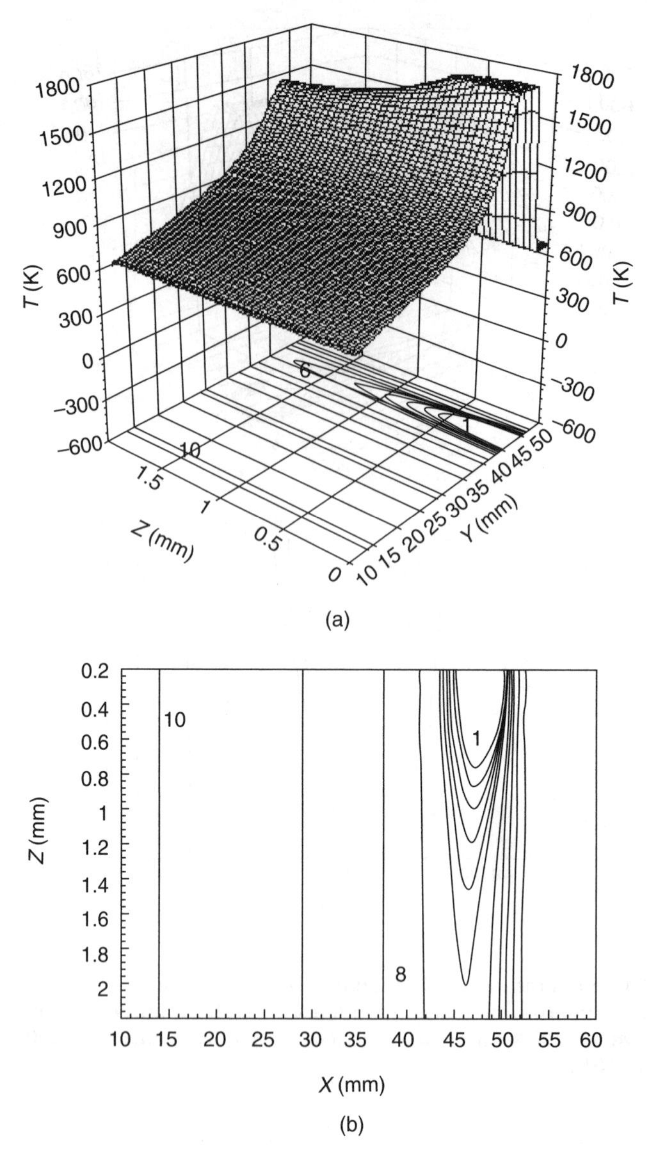

4.35 Temperature distribution on a longitudinal plane ($Y = 0$ mm): (a) three-dimensional distribution (curve 1, 1750 K; curve 6, 1500 K; curve 10, 700 K); (b) isotherms (curve 1, 1750 K; curve 8, 1500 K; curve 10, 700 K).

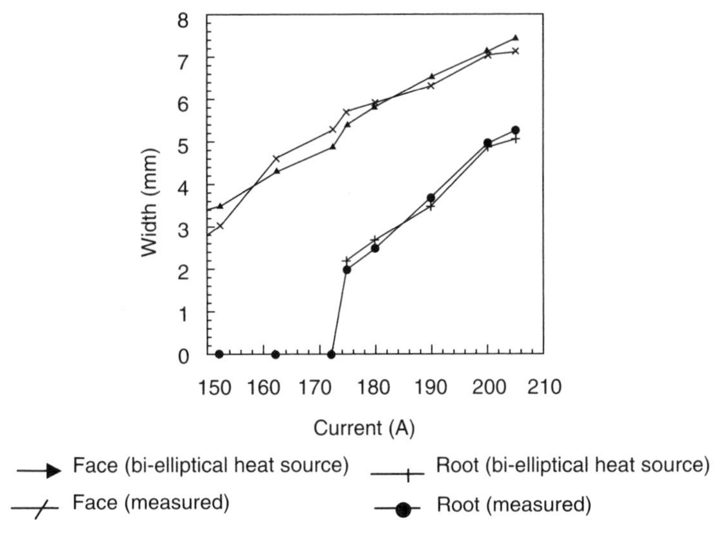

4.36 Calculated and measured widths of the weld root for different welding currents ($v = 7.07$ mm/s).

Table 4.7 Welding parameters

Experiment	Welding current (A)	Welding voltage (V)	Total power (W)
1	152.5	14.0	2135
2	162.5	15.2	2470
3	172.5	15.2	2622
4	175.0	16.0	2800
5	180	16.0	2880
6	190	16.0	3040
7	200	16.8	3360
8	205	16.8	3444

shows that the cooling time $t_{8/5}$ is 3.6 s, which is one of the most important parameters that determine the microstructure of the heat-affected zone.

4.8.3 Three-dimensional welding temperature field

Results of extrapolation

The analytical model for numerical simulation was established using measurements of the temperature field on the root side of the weld bead. The three-dimensional temperature field can be derived by numerical simulation using this model but this temperature field cannot be experimentally

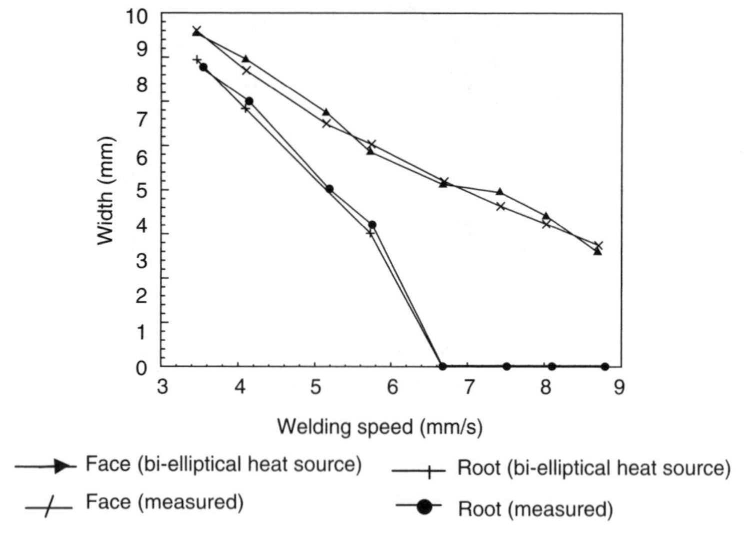

4.37 Calculated and measured widths of the weld pool for different welding speeds (current, 165 A).

measured. Figure 4.34 and Fig. 4.35 show the results obtained by the analytical model. They are based on the experimental results shown in Fig. 4.27. The welding and simulation parameters are listed in Table 4.3. The arc position was at $X = 50$ mm and $Y = 0$ mm. Figure 4.34(a) shows the temperature distribution on the transverse plane at $X = 46$ mm. Figure 4.34(b) shows six isotherms from 1500 to 1750 K at intervals of 50 K, and five isotherms from 700 to 1500 K at intervals of 200 K. Figure 4.35(a) shows the temperature distribution on the transverse plane at $Y = 0$ mm. Figure 4.35(b) shows six isotherms from 1500 to 1750 K at intervals of 50 K and five isotherms from 700 to 1500 K at intervals of 200 K.

Verification

It is difficult to compare the results obtained by the analytical method and the experimental method for the three-dimensional temperature field because there is no way to measure the three-dimensional temperature field. However, the results can be verified indirectly by comparing the two-dimensional temperature field obtained by numerical simulation. Figure 4.36 shows the isotherm width at the face and root of the weld obtained by both methods. The welding parameters are listed in Table 4.7 with the welding speed held constant. The conditions for the numerical simulation are a plate thickness of 2 mm, a bi-elliptical heat source, $w_{x1} = 0.15$ cm, $w_{x2} = 0.35$ cm, $w_y = 0.25$ cm and the conditions shown in Table 4.3.

The isotherm width was also measured and calculated for different welding speeds at constant welding current and power (165 A and 15.4 V giving a total power of 2541 W). The experimental and simulation conditions were the same as those for Fig. 4.36. The results for eight different speeds, namely 3.45, 4.08, 5.13, 5.17, 6.67, 7.41, 8.00 and 8.7 mm/s are shown in Fig. 4.37. The figure shows that the difference between the numerical values and the measured data at the face of the weld is greater than that at the back. But the error is within 8%. Therefore, the numerical simulation method can be used to define the width of the molten pool on the front side.

4.9 References

1. Zhu, L.Z., *Principle and Application of High-temperature Measurement* (in Chinese), Science Publishers, Beijing, 1991.
2. Cheng, G.S., *Detection Technique of the Welding Process* (in Chinese), North West Industries Press, Xian, 1988.
3. You, F.E., *Principle of Temperature-measuring Instrument by Radiation and Inspection* (in Chinese), China Metrology Press, Beijing, 1990.
4. Chen, D.H., Wu, L., Xu, Q.H., and Li Y., 'Study on measuring and controlling the welding process by the digital image technique' (in Chinese), *Trans. China Weld. Inst*, **6**(1), 23–30, 1985.
5. Chen, D.H., Wu, L., and Xu, Q.H., 'Study on temperature field measurement by microcomputer' (in Chinese), J. *Harbin Univ. Technol.*, (4), 1–11, 1981.
6. Pan J.L., He, F.D., and Su, Y., 'Sensor for measuring the welding temperature field' (in Chinese), China Patent ZL 93 2 13847.0, 1993.
7. Pan, J.L., Liao, B.J., and Zhang, H., 'Sensor for real-time measurement of the dynamic temperature field' (in Chinese), China Patent ZL 96 2 11738.2, 1996.
8. Pan, J.L., *Arc welding control*, Woodhead Publishing, Cambridge, 2003.
9. Zhang, H., Pan, J.L., and Liao B.J., 'Real-time measurement of the temperature field by ICCD, 1. Principle and method of temperature field division' (in Chinese), *Sci. China*, **27**(5), 418–423, 1997.
10. Zhang, H., Pan, J.L., and Liao, B.J., 'Real-time measurement of the temperature field with ICCD as a sensor, I. Principle and method for division of the temperature range', *Sci. China*, Ser. E, **41**(3), 295–301, 1998.
11. Wang, Q.Y., and Sun, X.Z., *Applied Technique of CCD* (in Chinese), Publishing House of Tianjin University, Tianjin, 1993.
12. Su, Y., 'Welding temperature field colorimetric measuring system' (in Chinese), Doctor's Dissertation, Tsinghua University, Beijing, 1993.
13. Zhang, H., 'Real-time measurement of the welding temperature field and its application' (in Chinese), Doctor's Dissertation, Tsinghua University, Beijing, 1997.
14. Sha, Z.X., *Characteristics and Selection of Camera Lens* (in Chinese), China Photography Press, Beijing, 1989.
15. Zhang, H., Pan, J.L., and Liao, B.J., 'Microcomputer processing technique of temperature field obtained by real-time colorimetric measurement' (in Chinese), *J. Tsinghua Univ.* (*Sci. Edn*), **37**(11), 30–34, 1997.

16. Zhang, H., and Pan, J.L., 'Application of a microcomputer in the temperature field: measurement by two-color image colorimetric method' (in Chinese), *Proc. Conf. on the Application of Microcomputers in Welding Technology*, Chinese Welding Society, Taiyuan, 1966, pp. 252–261.
17. Zhang, H., Pan, J.L., and Liao, B.J., 'Calibration of temperature-field-measuring system based on two-color ICCD images' (in Chinese), *Chin. J. Mech. Engng*, **9**(8), 94–97, 1998.
18. Zhang, H., Pan, J.L., and Wang, H.Q., 'A new method for measuring temperature field by two-color calibration and one-color measurement', *Proc. SPIE*, **3558**, 498–504, 1998.
19. Zhang, W.Y., *Heat Transfer in Welding* (in Chinese), China Machine Press, Beijing, 1989.
20. Eagar, T.W., and Tsai, N.S., 'Temperature fields produced by traveling distributed heat sources', Weld. J., **62**, 346-s–355-s, 1983.
21. Wu, C.S., *Numerical Analysis of Heat Transfer in Welding* (in Chinese), China Machine Press, Beijing, 1989.

5
Modeling the effects of welding

P. MICHALERIS, The Pennsylvania State University, USA

5.1 Introduction

Modeling of welding distortion and residual stress has been an active research area since the late 1970s. Some of the first publications in weld modeling were by Hibbit and Marcal[1], Argyris *et al.*[2] and Papazoglou and Masubuchi[3]. Significant research in the 1980s includes the development of the 'double-ellipsoid' heat input model by Goldak *et al.*[4], and the modeling of phase transformations[5–7]. Most of the weld modeling in the 1970s and 1980s involved two-dimensional (2D) models transverse to the welding direction using either plane-strain or generalized plane-strain conditions. These models demonstrated good correlations with experimental measurements for residual stress. However, these models were not capable of predicting angular[8] distortion and longitudinal buckling and bowing[9].

Significant developments in weld modeling in the 1990s included the use of three-dimensional (3D) moving-source models[10–12], the development sensitivity formulations[13,14], and the development of decoupled 2D weld process and 3D structural response models[9]. 3D moving-source models demonstrated the capability to model all distortion modes. However, they proved to be computationally costly for actual industrial applications. The decoupled 2D–3D approach by Michaleris and DeBiccari[9] mapped longitudinal plastic strain components only and was developed specifically for modeling welding-induced buckling in large ship panels. The approach did not account for angular distortion and sequencing effects.

More recent developments in the area of weld modeling include the development of Eulerian models for modeling long steady welds[15,16] and the development of a decoupled 3D weld modeling and 3D structural response approach[17]. Eulerian models offer an efficient approach to modeling very long uniform welds.

This chapter presents an overview of the solid mechanics issues for modeling welding processes. A review of continuum mechanics formulations is presented first, followed by a discussion of numerical formulations. The chapter concludes with a discussion on material properties.

5.2 Theoretical issues and background

This section presents a review of continuum mechanics related to weld process modeling. Small- and large-deformation kinematic formulations are presented first, followed by formulations of the equilibrium governing equation. The section concludes with a discussion of the material constitutive response. A more detailed presentation of non-linear finite-element formulations has been presented previously[18–22].

5.2.1 Kinematic formulations

When a body undergoes a large deformation, the deformed configuration may be significantly different than the undeformed. Measures of deformation, strain and stress can be defined with respect to either the undeformed configuration or the deformed configuration. Derivations with respect to the undeformed, or more generally the reference, configuration lead to total Lagrangian formulations. Derivations with respect to the deformed, or more generally the current, configuration lead to updated Lagrangian formulations.

In a stationary (Lagrangian) frame define the reference (undeformed) configuration of a body as V_X and current (deformed) configuration V_x (Fig. 5.1). A material point in the reference configuration and a material point in the deformed (current) configurations are denoted by $\boldsymbol{X}$ and $\boldsymbol{x}$ respectively. The deformation is denoted by $\boldsymbol{u}$ which is related to $\boldsymbol{X}$ and $\boldsymbol{x}$ by

$$\boldsymbol{x} = \boldsymbol{X} + \boldsymbol{u} \tag{5.1}$$

The deformation gradient $\mathbf{F}$ is defined as follows:

$$\mathbf{F} = \frac{\partial \boldsymbol{x}}{\partial \boldsymbol{X}} = \begin{bmatrix} \frac{\partial x_1}{\partial X_1} & \frac{\partial x_1}{\partial X_2} & \frac{\partial x_1}{\partial X_3} \\ \frac{\partial x_2}{\partial X_1} & \frac{\partial x_2}{\partial X_2} & \frac{\partial x_2}{\partial X_3} \\ \frac{\partial x_3}{\partial X_1} & \frac{\partial x_3}{\partial X_2} & \frac{\partial x_3}{\partial X_3} \end{bmatrix} \tag{5.2}$$

The determinant of $\mathbf{F}$ is denoted by J and it called the Jacobian of the deformation gradient

$$J = \det(\mathbf{F}) \tag{5.3}$$

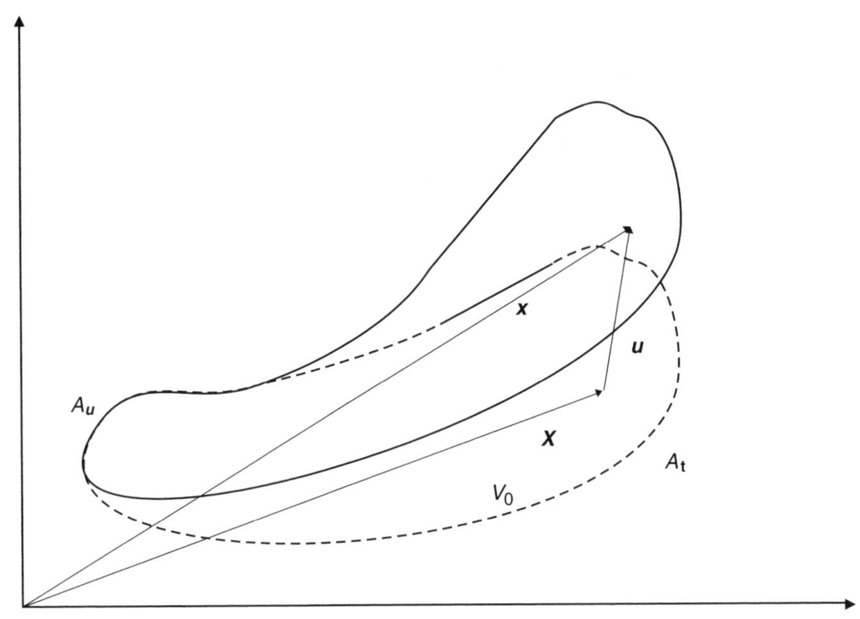

5.1 A body with volume *V*, prescribed displacement as surface A_u and prescribed surface flux on surface A_t.

An infinitesimal line segment in $\mathrm{d}\boldsymbol{X}$ in the reference configuration and the corresponding line segment in the current configuration $\mathrm{d}\boldsymbol{x}$ are related through

$$\mathrm{d}\boldsymbol{x} = \mathbf{F}\,\mathrm{d}\boldsymbol{X} \qquad [5.4]$$

The displacement gradient **D** is defined as follows:

$$\mathbf{D} = \frac{\partial \boldsymbol{u}}{\partial \boldsymbol{X}} = \begin{bmatrix} \frac{\partial u_1}{\partial X_1} & \frac{\partial u_1}{\partial X_2} & \frac{\partial u_1}{\partial X_3} \\ \frac{\partial u_2}{\partial X_1} & \frac{\partial u_2}{\partial X_2} & \frac{\partial u_2}{\partial X_3} \\ \frac{\partial u_3}{\partial X_1} & \frac{\partial u_3}{\partial X_2} & \frac{\partial u_3}{\partial X_3} \end{bmatrix} \qquad [5.5]$$

Substitution of Eqn [5.1] in Eqn [5.2] yields the following relation between the deformation gradient **F** and the displacement gradient **D**:

$$\mathbf{F} = \frac{\partial \boldsymbol{x}}{\partial \boldsymbol{X}} = \frac{\partial (\boldsymbol{X} + \boldsymbol{u})}{\partial \boldsymbol{X}} = \mathbf{I} + \mathbf{D} \qquad [5.6]$$

where **I** is the identity matrix.

Green strain

The Green strain $\boldsymbol{E}$ is a common strain measure with respect to the undeformed (reference) configuration. It is defined such that it measures the change in length of an infinitesimal segment in the current (deformed) ($\mathrm{d}s$) and reference (undeformed) ($\mathrm{d}S$) configurations:

$$\mathrm{d}s^2 - \mathrm{d}S^2 = \mathrm{d}\boldsymbol{x}\cdot\mathrm{d}\boldsymbol{x} - \mathrm{d}\boldsymbol{X}\cdot\mathrm{d}\boldsymbol{X} = 2\mathrm{d}\boldsymbol{X}\cdot\boldsymbol{E}\cdot\mathrm{d}\boldsymbol{X} \quad [5.7]$$

Using Eqn [5.4], $\mathrm{d}x \cdot \mathrm{d}x$ rewritten as

$$\begin{aligned}\mathrm{d}\boldsymbol{x}\cdot\mathrm{d}\boldsymbol{x} &= (\mathbf{F}\,\mathrm{d}\boldsymbol{X})\cdot(\mathbf{F}\,\mathrm{d}\boldsymbol{X}) = (\mathbf{F}\,\mathrm{d}\boldsymbol{X})^{\mathrm{T}}\cdot(\mathbf{F}\,\mathrm{d}\boldsymbol{X})\\ &= \mathrm{d}\boldsymbol{X}^{\mathrm{T}}\,\mathbf{F}^{\mathrm{T}}\cdot\mathbf{F}\,\mathrm{d}\boldsymbol{X} = \mathrm{d}\boldsymbol{X}\cdot(\mathbf{F}^{\mathrm{T}}\,\mathbf{F})\mathrm{d}\boldsymbol{X}\end{aligned} \quad [5.8]$$

Substituting Eqn [5.8] into Eqn [5.7] we obtain

$$\mathrm{d}\boldsymbol{X}(\mathbf{F}^{\mathrm{T}}\mathbf{F} - \mathbf{I})\cdot\mathrm{d}\boldsymbol{X} = 2\mathrm{d}\boldsymbol{X}\cdot\boldsymbol{E}\cdot\mathrm{d}\boldsymbol{X} \Rightarrow \boldsymbol{E} = \frac{1}{2}(\mathbf{F}^{\mathrm{T}}\mathbf{F} - \mathbf{I}) \quad [5.9]$$

The Green strain $\boldsymbol{E}$ is computed in terms of the displacement gradient $\mathbf{D}$ by substituting Eqn [5.6] into Eqn [5.9]:

$$\begin{aligned}\boldsymbol{E} &= \frac{1}{2}(\mathbf{F}^{\mathrm{T}}\mathbf{F} - \mathbf{I})\\ &= \frac{1}{2}\left[(\mathbf{I}+\mathbf{D})^{\mathrm{T}}(\mathbf{I}+\mathbf{D}) - \mathbf{I}\right]\\ &= \frac{1}{2}(\mathbf{D}+\mathbf{D}^{\mathrm{T}}) + \frac{1}{2}\mathbf{D}\mathbf{D}^{\mathrm{T}}\end{aligned} \quad [5.10]$$

From Eqn [5.10], the Green strain is written in engineering notation $\boldsymbol{\varepsilon}$ as follows:

$$\boldsymbol{\varepsilon} = \begin{bmatrix}\varepsilon_{11}\\ \varepsilon_{22}\\ \varepsilon_{33}\\ \gamma_{23}\\ \gamma_{31}\\ \gamma_{12}\end{bmatrix} = \begin{bmatrix}\dfrac{\partial u_1}{\partial X_1}\\ \dfrac{\partial u_2}{\partial X_2}\\ \dfrac{\partial u_3}{\partial X_3}\\ \dfrac{\partial u_2}{\partial X_3}+\dfrac{\partial u_3}{\partial X_2}\\ \dfrac{\partial u_1}{\partial X_3}+\dfrac{\partial u_3}{\partial X_1}\\ \dfrac{\partial u_2}{\partial X_1}+\dfrac{\partial u_1}{\partial X_2}\end{bmatrix} + \begin{bmatrix}\dfrac{1}{2}\left[\left(\dfrac{\partial u_1}{\partial X_1}\right)^2+\left(\dfrac{\partial u_2}{\partial X_1}\right)^2+\left(\dfrac{\partial u_3}{\partial X_1}\right)^2\right]\\ \dfrac{1}{2}\left[\left(\dfrac{\partial u_1}{\partial X_2}\right)^2+\left(\dfrac{\partial u_2}{\partial X_2}\right)^2+\left(\dfrac{\partial u_3}{\partial X_2}\right)^2\right]\\ \dfrac{1}{2}\left[\left(\dfrac{\partial u_1}{\partial X_3}\right)^2+\left(\dfrac{\partial u_2}{\partial X_3}\right)^2+\left(\dfrac{\partial u_3}{\partial X_3}\right)^2\right]\\ \dfrac{\partial u_1}{\partial X_2}\dfrac{\partial u_1}{\partial X_3}+\dfrac{\partial u_2}{\partial X_2}\dfrac{\partial u_2}{\partial X_3}+\dfrac{\partial u_3}{\partial X_2}\dfrac{\partial u_3}{\partial X_3}\\ \dfrac{\partial u_1}{\partial X_1}\dfrac{\partial u_1}{\partial X_3}+\dfrac{\partial u_2}{\partial X_2}\dfrac{\partial u_1}{\partial X_3}+\dfrac{\partial u_3}{\partial X_1}\dfrac{\partial u_3}{\partial X_3}\\ \dfrac{\partial u_1}{\partial X_2}\dfrac{\partial u_1}{\partial X_1}+\dfrac{\partial u_2}{\partial X_2}\dfrac{\partial u_2}{\partial X_1}+\dfrac{\partial u_3}{\partial X_2}\dfrac{\partial u_3}{\partial X_1}\end{bmatrix} \quad [5.11]$$

The Green strain is invariant for rigid-body rotation; i.e. ${}^{n}\mathbf{F} = \boldsymbol{R}^{n-1}\mathbf{F}$:

$$\begin{aligned} {}^{n}\boldsymbol{E} &= \frac{1}{2}({}^{n}\mathbf{F}^{\mathrm{T}}\,{}^{n}\mathbf{F} - \mathbf{I}) \\ &= \frac{1}{2}({}^{n-1}\mathbf{F}^{\mathrm{T}}\boldsymbol{R}^{\mathrm{T}}\boldsymbol{R}^{n-1}\mathbf{F} - \mathbf{I}) \\ &= \frac{1}{2}({}^{n-1}\mathbf{F}^{\mathrm{T}}\,{}^{n-1}\mathbf{F} - \mathbf{I}) \\ &= {}^{n-1}\boldsymbol{E} \end{aligned} \tag{5.12}$$

Almansi strain

The Almansi strain $\boldsymbol{A}$ (also called the Eulerian strain) is a common strain measure with respect to the deformed (current) configuration. It is defined as follows:

$$\mathrm{d}s^2 - \mathrm{d}S^2 = \mathrm{d}\boldsymbol{x}\cdot\mathrm{d}\boldsymbol{x} - \mathrm{d}\boldsymbol{X}\cdot\mathrm{d}\boldsymbol{X} = 2\,\mathrm{d}\boldsymbol{x}\cdot\boldsymbol{A}\cdot\mathrm{d}\boldsymbol{x} \tag{5.13}$$

Using Eqn [5.4] and Eqn [5.13] the Almansi strain is computed as

$$\mathbf{A} = \frac{1}{2}(\mathbf{I} - \mathbf{F}^{-\mathrm{T}}\mathbf{F}^{-1}) \tag{5.14}$$

Noting that

$$\mathbf{F}^{-1} = \frac{\partial \boldsymbol{X}}{\partial \boldsymbol{x}} = \frac{\partial(\boldsymbol{x} - \boldsymbol{u})}{\partial \boldsymbol{x}} = \mathbf{I} - \frac{\partial \boldsymbol{u}}{\partial \boldsymbol{x}} = \mathbf{I} - \mathbf{d} \tag{5.15}$$

where $\mathbf{d} = \partial\boldsymbol{u}/\partial\boldsymbol{x}$ is analogous to the displacement gradient $\mathbf{D} = \partial\boldsymbol{u}/\partial\boldsymbol{X}$. However, the gradient is the deformed configuration $\boldsymbol{x}$.

Substitution of Eqn [5.15] into Eqn [5.14] yields the following expression for the Almansi strain:

$$\boldsymbol{A} = \frac{1}{2}(\mathbf{d} + \mathbf{d}^{\mathrm{T}}) - \frac{1}{2}\mathbf{d}\mathbf{d}^{\mathrm{T}} \tag{5.16}$$

Using engineering notation, the Almansi strain is written as

$$\boldsymbol{\varepsilon}^{\mathrm{A}} = \begin{bmatrix} \varepsilon_{11}^{\mathrm{A}} \\ \varepsilon_{22}^{\mathrm{A}} \\ \varepsilon_{33}^{\mathrm{A}} \\ \gamma_{23}^{\mathrm{A}} \\ \gamma_{31}^{\mathrm{A}} \\ \gamma_{12}^{\mathrm{A}} \end{bmatrix} = \begin{bmatrix} \frac{\partial u_1}{\partial x_1} \\ \frac{\partial u_2}{\partial x_2} \\ \frac{\partial u_3}{\partial x_3} \\ \frac{\partial u_2}{\partial x_3} + \frac{\partial u_3}{\partial x_2} \\ \frac{\partial u_1}{\partial x_3} + \frac{\partial u_3}{\partial x_1} \\ \frac{\partial u_2}{\partial x_1} + \frac{\partial u_1}{\partial x_2} \end{bmatrix} - \begin{bmatrix} \frac{1}{2}\left[\left(\frac{\partial u_1}{\partial x_1}\right)^2 + \left(\frac{\partial u_2}{\partial x_1}\right)^2 + \left(\frac{\partial u_3}{\partial x_1}\right)^2\right] \\ \frac{1}{2}\left[\left(\frac{\partial u_1}{\partial x_2}\right)^2 + \left(\frac{\partial u_2}{\partial x_2}\right)^2 + \left(\frac{\partial u_3}{\partial x_2}\right)^2\right] \\ \frac{1}{2}\left[\left(\frac{\partial u_1}{\partial x_3}\right)^2 + \left(\frac{\partial u_2}{\partial x_3}\right)^2 + \left(\frac{\partial u_3}{\partial x_3}\right)^2\right] \\ \frac{\partial u_1}{\partial x_2}\frac{\partial u_1}{\partial x_3} + \frac{\partial u_2}{\partial x_2}\frac{\partial u_2}{\partial x_3} + \frac{\partial u_3}{\partial x_2}\frac{\partial u_3}{\partial x_3} \\ \frac{\partial u_1}{\partial x_1}\frac{\partial u_1}{\partial x_3} + \frac{\partial u_2}{\partial x_2}\frac{\partial u_1}{\partial x_3} + \frac{\partial u_3}{\partial x_1}\frac{\partial u_3}{\partial x_3} \\ \frac{\partial u_1}{\partial x_2}\frac{\partial u_1}{\partial x_1} + \frac{\partial u_2}{\partial x_2}\frac{\partial u_2}{\partial x_1} + \frac{\partial u_3}{\partial x_2}\frac{\partial u_3}{\partial x_1} \end{bmatrix} \tag{5.17}$$

Finally, from Eqn [5.14] the Almansi strain is related to the Green strain as

$$\boldsymbol{E} = \mathbf{F}^{\mathrm{T}}\boldsymbol{A}\mathbf{F} \quad [5.18]$$

$$\boldsymbol{A} = \mathbf{F}^{-\mathrm{T}} - \boldsymbol{E}\mathbf{F}^{-1} \quad [5.19]$$

Cauchy stress

The Cauchy stress $\boldsymbol{\sigma}$ is defined as the true stress at the current (deformed) configuration

$$\boldsymbol{t} = \boldsymbol{\sigma} \cdot \boldsymbol{n} \quad [5.20]$$

where $\boldsymbol{t}$ is the traction applied to a surface with outward normal $\boldsymbol{n}$.

For a rigid-body rotation ${}^{n}\mathbf{F} = \boldsymbol{R}^{n-1}\mathbf{F}$, the Cauchy stress transforms as

$${}^{n}\boldsymbol{\sigma} = \boldsymbol{R}^{n-1}\boldsymbol{\sigma}\boldsymbol{R}^{\mathrm{T}}$$

Second Piola–Kirchhoff stress

The counterpart of Eqn [5.20] in the reference configuration is

$$\boldsymbol{t}_X = \boldsymbol{P} \cdot \boldsymbol{n}_X \quad [5.21]$$

where $\boldsymbol{P}$ is the first Piola–Kirchhoff stress.

The second Piola–Kirchhoff stress $\boldsymbol{S}$ is defined as follows:

$$\mathbf{F}^{-1}\boldsymbol{t}_X = \boldsymbol{S} \cdot \boldsymbol{n}_X \quad [5.22]$$

From Eqns [5.22] and [5.21] the first and second Piola–Kirchhoff stresses are related through

$$\boldsymbol{P} = \mathbf{F}\boldsymbol{S} \quad [5.23]$$

$$\boldsymbol{S} = \mathbf{F}^{-1}\boldsymbol{P} \quad [5.24]$$

To derive the transformations between the different measures of strain, equate the force at a segment on the surface[19,20] A_t:

$$\boldsymbol{t}_X \, \mathrm{d}A_X = \boldsymbol{t}_X \, \mathrm{d}A_X \Rightarrow$$

$$\boldsymbol{P} \cdot \boldsymbol{n}_X \, \mathrm{d}A_X = \boldsymbol{\sigma} \cdot \boldsymbol{n}_X \, \mathrm{d}A_X \Rightarrow$$

$$\boldsymbol{P} \cdot \left(\frac{1}{J}\mathbf{F}^{\mathrm{T}}\right)\boldsymbol{n}_X \, \mathrm{d}A_X = \boldsymbol{\sigma} \cdot \boldsymbol{n}_X \, \mathrm{d}A_X \Rightarrow$$

$$\boldsymbol{P} = J\boldsymbol{\sigma}\mathbf{F}^{-\mathrm{T}} \Rightarrow \quad [5.25]$$

$$\boldsymbol{\sigma} = \frac{1}{J}\boldsymbol{P}\mathbf{F}^{\mathrm{T}} \quad [5.26]$$

Table 5.1 Comparison of stress measures

	Cauchy stress $\boldsymbol{\sigma}$	First Piola–Kirchhoff stress $\boldsymbol{P}$	Second Piola–Kirchhoff stress $\boldsymbol{S}$
$\boldsymbol{\sigma}$		$\frac{1}{J}\boldsymbol{P}\mathbf{F}^{\mathrm{T}}$	$\frac{1}{J}\mathbf{F}\boldsymbol{S}\boldsymbol{F}^{\mathrm{T}}$
$\boldsymbol{P}$	$J\sigma\mathbf{F}^{-\mathrm{T}}$		$\mathbf{F}\boldsymbol{S}$
$\boldsymbol{S}$	$J\mathbf{F}^{-\mathrm{T}}\sigma\mathbf{F}^{-\mathrm{T}}$	$\mathbf{F}^{-1}\boldsymbol{P}$	

From Eqns [5.25], [5.26], [5.23] and [5.24] the following are obtained:

$$\boldsymbol{\sigma} = \frac{1}{J}\mathbf{F}\boldsymbol{S}\mathbf{F}^{\mathrm{T}} \quad [5.27]$$

$$\boldsymbol{S} = J\mathbf{F}^{-1}\boldsymbol{s}\mathbf{F}^{-\mathrm{T}} \quad [5.28]$$

Table 5.1 is a comparison chart for the stress measures.

If the Cauchy stress $\boldsymbol{\sigma}$ is symmetric, then the second Piola–Kirchhoff stress $\boldsymbol{S}$ is also symmetric. However, the first Piola–Kirchhoff stress $\boldsymbol{P}$ is not symmetric.

The second Piola–Kirchhoff stress is invariant under a rigid-body rotation ${}^{n}\mathbf{F} = \boldsymbol{R}^{n-1}\mathbf{F}$

$$
\begin{aligned}
{}^{n}\boldsymbol{\sigma} &= \boldsymbol{R}\,{}^{n-1}\boldsymbol{\sigma}\boldsymbol{R}^{\mathrm{T}} \Rightarrow \\
\frac{1}{{}^{n}J}(\boldsymbol{R}\,{}^{n-1}\mathbf{F})\,{}^{n}\boldsymbol{S}\boldsymbol{R}\,{}^{n-1}\mathbf{F}^{\mathrm{T}} &= \frac{1}{{}^{n-1}J}\boldsymbol{R}\,{}^{n-1}\mathbf{F}^{n-1}\boldsymbol{S}^{n-1}\mathbf{F}^{\mathrm{T}}\boldsymbol{R}^{\mathrm{T}} \Rightarrow \\
\frac{1}{{}^{n-1}J}\boldsymbol{R}\cdot({}^{n-1}\mathbf{F}\,{}^{n}\boldsymbol{S}\,{}^{n-1}\mathbf{F}^{\mathrm{T}})\boldsymbol{R}^{\mathrm{T}} &= \frac{1}{{}^{n-1}J}\boldsymbol{R}\,{}^{n-1}\mathbf{F}\,{}^{n-1}\boldsymbol{S}\,{}^{n-1}\mathbf{F}^{\mathrm{T}}\boldsymbol{R}^{\mathrm{T}} \Rightarrow \\
{}^{n}\boldsymbol{S} &= {}^{n-1}\boldsymbol{S}
\end{aligned}
\quad [5.29]
$$

5.2.2 Equilibrium: momentum balance

Current configuration: updated Lagrangian formulation

Linear momentum balance of a differential volume in the current configuration yields

$$\int_V \boldsymbol{b}\,\mathrm{d}V + \int_A \boldsymbol{t}\,\mathrm{d}A = \int_V \boldsymbol{b}\,\mathrm{d}V + \int_A \boldsymbol{\sigma}\cdot\boldsymbol{n}\,\mathrm{d}A = \mathbf{0} \quad [5.30]$$

Application of divergence theorem yields

$$\int_V (\boldsymbol{b} + \nabla_x \cdot \boldsymbol{\sigma})\,\mathrm{d}V = \mathbf{0} \quad [5.31]$$

Enforcing that Eqn [5.31] holds for all V yields

$$\nabla_x \cdot \boldsymbol{\sigma} + \boldsymbol{b} = \mathbf{0} \qquad [5.32]$$

where the operator ∇_x denotes differentiations in the current (deformed configuration). In contrast the operator ∇_X denotes differentiations in the reference configuration. From Eqn [5.4], ∇_X and ∇_x are related through

$$\nabla_x = \mathbf{F}^{-1}\nabla_X \qquad [5.33]$$

Angular momentum balance demonstrates that the Cauchy stress is symmetric:

$$\boldsymbol{\sigma}^{\mathrm{T}} = \boldsymbol{\sigma} \qquad [5.34]$$

Reference configuration: total Lagrangian formulation

Transformation of the linear momentum balance of differential volume of Eqn [5.30] configuration yields

$$\int_V \boldsymbol{b}\,\mathrm{d}V + \int_V \boldsymbol{t}\,\mathrm{d}A = \int_{V_X} \boldsymbol{b}J\,\mathrm{d}V_X + \int_{A_X} \boldsymbol{\sigma}\cdot J\mathbf{F}^{-T}\boldsymbol{n}_X\,\mathrm{d}A_X = \mathbf{0} \qquad [5.35]$$

Application of divergence theorem yields

$$\int_{V_X} (\boldsymbol{b}J + \nabla_X \cdot \boldsymbol{P})\mathrm{d}V_X = \mathbf{0} \qquad [5.36]$$

Enforcing that Eqn [5.36] hold for all V_0 yields

$$\boldsymbol{b}_X + \nabla_X \cdot \boldsymbol{P} = \mathbf{0} \qquad [5.37]$$

where $\boldsymbol{b}_X = \boldsymbol{b}J$.

Angular momentum balance demonstrates that the first Piola–Kirchhoff stress is not symmetric:

$$\mathbf{F}\boldsymbol{P}^{\mathrm{T}} = \boldsymbol{P}\mathbf{F}^{\mathrm{T}} \qquad [5.38]$$

5.2.3 Small-deformation formulations

When the deformation is small, the deformed configuration V_X is not significantly different from that of the undeformed V_x. Then, derivatives with respect to $\boldsymbol{X}$ and $\boldsymbol{x}$ are equal. Furthermore, the second-order term of the strain in Eqns [5.10] and [5.16] is negligible, which implies that strain may be defined as

$$\boldsymbol{\varepsilon} = \frac{1}{2}(\mathbf{d} + \mathbf{d}^{\mathrm{T}}) \qquad [5.39]$$

Table 5.2 Comparison of total, updated Lagrangian and small-deformation formulations

	Total Lagrangian	Updated Lagrangian	Small deformation
Governing equation	$\nabla_X \cdot \boldsymbol{P} + \boldsymbol{b}_x = \boldsymbol{0}$	$\nabla_x \cdot \boldsymbol{\sigma} + \boldsymbol{b} = \boldsymbol{0}$	$\nabla \cdot \boldsymbol{\sigma} + \boldsymbol{b} = \boldsymbol{0}$
Primary field	$\boldsymbol{u}(\boldsymbol{X}, t)$	$\boldsymbol{u}(\boldsymbol{x}, t)$	$\boldsymbol{u}(\boldsymbol{x}, t)$
Secondary field	$\boldsymbol{E} = \frac{1}{2}(\mathbf{D} + \mathbf{D}^{\mathrm{T}}) + \frac{1}{2}\mathbf{D}\mathbf{D}^{\mathrm{T}}$	$\mathbf{A} = \frac{1}{2}(\mathbf{d} + \mathbf{d}^{\mathrm{T}}) - \frac{1}{2}\mathbf{d}\mathbf{d}^{\mathrm{T}}$	$\varepsilon = \frac{1}{2}(\mathbf{d} + \mathbf{d}^{\mathrm{T}})$
Essential boundary condition	$\boldsymbol{t}_x = \boldsymbol{P} \cdot \boldsymbol{n}_x$ $\boldsymbol{u} = \boldsymbol{u}^{\mathrm{p}}$ on A_{X_u}	$\boldsymbol{t} = \boldsymbol{\sigma} \cdot \boldsymbol{n}$ $\boldsymbol{u} = \boldsymbol{u}^{\mathrm{p}}$ on A_u	$\boldsymbol{t} = \boldsymbol{\sigma} \cdot \boldsymbol{n}$ $\boldsymbol{u} = \boldsymbol{u}^{\mathrm{p}}$ on A_u
Auxiliary boundary condition	$\boldsymbol{t}_X = \boldsymbol{t}_X^{\mathrm{p}}$ on A_{X_t}	$\boldsymbol{t} = \boldsymbol{t}^{\mathrm{p}}$ on A_t	$\boldsymbol{t} = \boldsymbol{t}^{\mathrm{p}}$ on A_t

Table 5.2 lists a summary of the measures for strain, stress, and corresponding equilibrium equation for the updated Lagrangian, total Lagrangian and small-deformation formulations.

5.2.4 Constitutive response

In weld modeling, an elastic–plastic constitutive response is typically assumed. During non-active yielding, the stress is a function of strain only and follows the same path during loading and unloading:

$$\boldsymbol{S} = \mathbf{C}\mathbf{E} \qquad [5.40]$$

where $\mathbf{C}$ is the fourth-order material stiffness tensor. Equation [5.40] can be written in matrix vector form as follows:

$$\bar{\boldsymbol{S}} = \mathbf{C}\boldsymbol{\varepsilon} \qquad [5.41]$$

where $\bar{\boldsymbol{S}}$ is the second Piola–Kirchhoff stress written in a vector form and $\mathbf{C}$ is the material stiffness matrix.

During active yielding, the yield condition, flow rule and hardening rule are used to calculate the rate of plastic strain. Plasticity formulations have been derived using additive decomposition of total strain to a plastic and an elastic component or multiplicative decomposition of the total deformation gradient into elastic and plastic components.

5.2.5 Isotropic hardening plasticity

In this section, a summary of the isotropic hardening plasticity formulations are presented assuming additive strain decomposition. The derivations for a multiplicative decomposition have been presented previously[23–25].

Formulations in a deviatoric space

Assuming thermoelastoplasticity, and using engineering notation, the additive decomposition of the total strain vector is applied as

$$\boldsymbol{\varepsilon} = \boldsymbol{\varepsilon}_{\mathrm{e}} + \boldsymbol{\varepsilon}_{\mathrm{p}} + \boldsymbol{\varepsilon}_{\mathrm{th}} \qquad [5.42]$$

where $\boldsymbol{\varepsilon}_{\mathrm{e}}$, $\boldsymbol{\varepsilon}_{\mathrm{p}}$ and $\boldsymbol{\varepsilon}_{\mathrm{th}}$ are the elastic, plastic and thermal strain vectors respectively in engineering notation.

Using Eqn [5.42], the stress–strain relationship is

$$\boldsymbol{\sigma} = \mathbf{C}\boldsymbol{\varepsilon}_{\mathrm{e}} = \mathbf{C}(\boldsymbol{\varepsilon} - \boldsymbol{\varepsilon}_{\mathrm{p}} - \boldsymbol{\varepsilon}_{\mathrm{th}}) \qquad [5.43]$$

where $\mathbf{C}$ is the material stiffness matrix.

For the isotropic material, the thermal strain vector is

$$\boldsymbol{\varepsilon}_{\mathrm{th}} = \varepsilon_{\mathrm{th}}\boldsymbol{j} \qquad [5.44]$$

$$\varepsilon_{\mathrm{th}} = \alpha(T - T^{\mathrm{ref}}) \qquad [5.45]$$

$$\boldsymbol{j} = [1 \quad 1 \quad 1 \quad 0 \quad 0 \quad 0]^{\mathrm{T}} \qquad [5.46]$$

where α is the thermal expansion coefficient and T^{ref} is the reference temperature.

From the von Mises yield criterion, the yield function is

$$f = \sigma_{\mathrm{M}} - \sigma_{\mathrm{Y}}(\varepsilon_{\mathrm{q}}, T) \qquad [5.47]$$

$$= \frac{1}{2^{1/2}}\left[(\sigma_x - \sigma_y)^2 + (\sigma_y - \sigma_z)^2 + (\sigma_z - \sigma_x)^2 + 6(\tau_{xy}^2 + \tau_{yz}^2 + \tau_{zy}^2)\right]^{1/2} - \sigma_{\mathrm{Y}}(\varepsilon_{\mathrm{q}}, T) \qquad [5.48]$$

$$= \left(\frac{3}{2}\right)^{1/2}(\boldsymbol{s}^{\mathrm{T}} \cdot \mathbf{L}\boldsymbol{s})^{1/2} - \sigma_{\mathrm{Y}}(\varepsilon_{\mathrm{q}}, T) \qquad [5.49]$$

where σ_{M} is the von Mises stress, σ_{Y} is the yield stress and $\boldsymbol{s}$ denotes the deviatoric stress vector which is

$$\boldsymbol{s} = \boldsymbol{\sigma} - \boldsymbol{\sigma}_{\mathrm{h}} \qquad [5.50]$$

$$\boldsymbol{\sigma}_{\mathrm{h}} = \sigma_{\mathrm{h}}\boldsymbol{j} \qquad [5.51]$$

$$\sigma_{\mathrm{h}} = \frac{1}{3}(\sigma_x + \sigma_y + \sigma_z) \qquad [5.52]$$

where $\boldsymbol{\sigma}_{\mathrm{h}}$ is the hydrostatic or volumetric stress vector.

The strain vector also can be divided by volumetric and deviatoric vectors, and the relationships between stress and strain are the following:

$$\boldsymbol{\varepsilon} = \boldsymbol{\varepsilon}_{\mathrm{h}} + \boldsymbol{e} \quad [5.53]$$

$$\boldsymbol{\varepsilon}_{\mathrm{h}} = \varepsilon_{\mathrm{h}} \boldsymbol{j} = \frac{1}{3} \boldsymbol{j} \cdot \boldsymbol{j}^{\mathrm{T}} \cdot \boldsymbol{\varepsilon} \quad [5.54]$$

$$\varepsilon_{\mathrm{h}} = \frac{1}{3}(\varepsilon_x + \varepsilon_y + \varepsilon_z) \quad [5.55]$$

$$\boldsymbol{\sigma}_{\mathrm{h}} = 3k\boldsymbol{\varepsilon}_{\mathrm{h}} \quad [5.56]$$

$$\boldsymbol{s} = 2G\mathbf{L}^{-1}\boldsymbol{e} \quad [5.57]$$

$$G = \frac{E}{2(1+\nu)} \quad [5.58]$$

$$k = \frac{E}{3(1-2\nu)} \quad [5.59]$$

where $\boldsymbol{\varepsilon}_{\mathrm{h}}$ and $\boldsymbol{e}$ are the volumetric and deviatoric strain vectors respectively, and G and k are the shear modulus and bulk modulus respectively.

In conjunction with Eqns [5.47]–[5.49], the Prandtl–Reuss flow rules are

$$\dot{\boldsymbol{\varepsilon}}_{\mathrm{p}} = \dot{\boldsymbol{\varepsilon}}_{\mathrm{q}} \boldsymbol{a} \quad [5.60]$$

$$\boldsymbol{a} = \left(\frac{\partial f}{\partial \boldsymbol{\sigma}}\right)^{\mathrm{T}} = \frac{\partial}{\partial \boldsymbol{s}}\left(\frac{3}{2}\boldsymbol{s}^{\mathrm{T}} \cdot \mathbf{L}\boldsymbol{s}\right)^{1/2} = \frac{3}{2\sigma_{\mathrm{M}}}\mathbf{L}\boldsymbol{s} \quad [5.61]$$

$$\dot{\boldsymbol{\varepsilon}}_{\mathrm{q}} = \left(\frac{2}{3}\right)^{1/2} (\dot{\boldsymbol{\varepsilon}}_{\mathrm{p}}^{\mathrm{T}} \cdot \mathbf{L}^{-1} \dot{\boldsymbol{\varepsilon}}_{\mathrm{p}})^{1/2} = \left(\frac{2}{3}\right)^{1/2} (\dot{\boldsymbol{e}}_{\mathrm{p}}^{\mathrm{T}} \cdot \mathbf{L}^{-1} \dot{\boldsymbol{e}}_{\mathrm{p}})^{1/2} \quad [5.62]$$

where $\boldsymbol{a}$ is the flow vector and Eqn [5.62] is valid because there is no volumetric plastic evolution in the von Mises plasticity. The following equation can be derived from Eqns [5.50]–[5.59]:

$$\mathbf{C}\boldsymbol{a} = \frac{3G}{\sigma_{\mathrm{M}}}\boldsymbol{s} \quad [5.63]$$

The radial return algorithm

For simplicity, superscripts for the current increment ${}^{n}(\cdot)$ will be dropped but those for the previous increment will be kept, i.e. all quantities without left superscript are for the current increment in this section.

The stress increment for the current iteration is

$$\Delta\boldsymbol{\sigma} = \Delta\mathbf{C}({}^{n-1}\boldsymbol{\varepsilon} - {}^{n-1}\boldsymbol{\varepsilon}_{\mathrm{p}} - {}^{n-1}\boldsymbol{\varepsilon}_{\mathrm{th}}) + \mathbf{C}(\Delta\boldsymbol{\varepsilon} - \Delta\boldsymbol{\varepsilon}_{\mathrm{p}} - \Delta\boldsymbol{\varepsilon}_{\mathrm{th}}) \qquad [5.64]$$

$\Delta\boldsymbol{\varepsilon}$ is the total strain increment corresponding to the current displacement increment:

$$\Delta\boldsymbol{\varepsilon} = \boldsymbol{B}(\boldsymbol{U} - {}^{n-1}\boldsymbol{U}) = \boldsymbol{B}\sum\delta\boldsymbol{U} \qquad [5.65]$$

For isotropic materials, the thermal strain vector is

$$\Delta\boldsymbol{\varepsilon}_{\mathrm{th}} = \Delta\varepsilon_{\mathrm{th}}\boldsymbol{j} \qquad [5.66]$$

$$\Delta\varepsilon_{\mathrm{th}} = \alpha(T - T^{\mathrm{ref}}) - {}^{n-1}\alpha({}^{n-1}T - T^{\mathrm{ref}}) \qquad [5.67]$$

Now, the only unknown on the right-hand side of Eqn [5.64] is $\Delta\boldsymbol{\varepsilon}_{\mathrm{p}}$. So let us define the elastic predictor $\boldsymbol{\sigma}_B$ and corresponding yield function f_B:

$$\boldsymbol{\sigma}_B = \boldsymbol{\sigma}\big|_{\Delta\varepsilon_{\mathrm{p}}=0} = {}^{n-1}\boldsymbol{\sigma} + \Delta\boldsymbol{\sigma}\big|_{\Delta\varepsilon_{\mathrm{p}}=0} \qquad [5.68]$$

$$f_B = \sigma_{\mathrm{M}}(\boldsymbol{\sigma}_B) - \sigma_{\mathrm{Y}}({}^{n-1}\varepsilon_{\mathrm{q}}, T) \qquad [5.69]$$

If f_B is smaller than zero, $\boldsymbol{\sigma}$ is equal to $\boldsymbol{\sigma}_B$ because there is no plastic evolution. If f_B is larger than 0 there is plastic evolution. From Eqns [5.50], [5.61], [5.64], [5.63] and [5.68] the relationship between $\boldsymbol{\sigma}$ and $\boldsymbol{\sigma}_B$ can be rewritten as

$$\begin{aligned}\boldsymbol{\sigma} &= \boldsymbol{\sigma}_B - \boldsymbol{C}\Delta\boldsymbol{\varepsilon}_{\mathrm{p}} \\ &= \boldsymbol{\sigma}_B - \Delta\varepsilon_{\mathrm{q}}\boldsymbol{C}\boldsymbol{a}_B \\ &= \boldsymbol{\sigma}_B - \frac{3G\Delta\varepsilon_{\mathrm{q}}}{\sigma_{B\mathrm{M}}}\boldsymbol{s}_B \\ &= \boldsymbol{\sigma}_{B\mathrm{h}} + \boldsymbol{s}_B - \frac{3G\Delta\varepsilon_{\mathrm{q}}}{\sigma_{B\mathrm{M}}}\boldsymbol{s}_B \\ &= \boldsymbol{\sigma}_{B\mathrm{h}} + \beta\boldsymbol{s}_B\end{aligned} \qquad [5.70]$$

$$\boldsymbol{\sigma} = \boldsymbol{\sigma}_{\mathrm{h}} + \boldsymbol{s} \qquad [5.71]$$

$$\beta = 1 - \frac{3G\Delta\varepsilon_{\mathrm{q}}}{\sigma_{B\mathrm{M}}} \qquad [5.72]$$

where $\boldsymbol{\sigma}_{B\mathrm{h}}$ and $\boldsymbol{s}_B$ are the hydrostatic and deviatoric stress vectors respectively of $\boldsymbol{\sigma}_B$. Thus we obtain the following relationships:

$$\boldsymbol{\sigma}_{\mathrm{h}} = \boldsymbol{\sigma}_{B\mathrm{h}} \qquad [5.73]$$

$$\boldsymbol{s} = \beta\boldsymbol{s}_B \qquad [5.74]$$

$$\boldsymbol{a} = \boldsymbol{a}_B \qquad [5.75]$$

From Eqns [5.47]–[5.49], [5.72] and [5.74] the current yield function f is computed as

$$\begin{aligned} f &= \sigma_M(\boldsymbol{s}) - \sigma_Y(\varepsilon_q, T) \\ &= \sigma_M(\beta \boldsymbol{s}_B) - \sigma_Y({}^{n-1}\varepsilon_q + \Delta\varepsilon_q, T) \\ &= \beta\sigma_M(\boldsymbol{s}_B) - \sigma_Y({}^{n-1}\varepsilon_q + \Delta\varepsilon_q, T) \\ &= \sigma_M(\boldsymbol{s}_B) - 3G\Delta\varepsilon_q - \sigma_Y({}^{n-1}\varepsilon_q + \Delta\varepsilon_q, T) = 0 \end{aligned} \qquad [5.76]$$

Now the current yield function is a function of only $\Delta\varepsilon_q$. In order to obtain the $\Delta\varepsilon_q$ iteratively, let us expand Eqn [5.76] in a Taylor series:

$$\begin{aligned} f^{J+1} &= f^J + \frac{\partial f}{\partial(\Delta\varepsilon_q)}\delta(\Delta\varepsilon_q) \\ &= f^J - \left(3G + \frac{\partial\sigma_Y}{\partial\varepsilon_q}\frac{\partial\varepsilon_q}{\partial(\Delta\varepsilon_q)}\right)\delta(\Delta\varepsilon_q) \\ &= f^J - (3G + H^J)\delta(\Delta\varepsilon_q) = 0 \end{aligned} \qquad [5.77]$$

$$\delta(\Delta\varepsilon_q) = \frac{f^J}{3G + H^J} \qquad [5.78]$$

$$\Delta\varepsilon_q^{J+1} = \Delta\varepsilon_q^J + \delta(\Delta\varepsilon_q) \qquad [5.79]$$

where the superscript J denotes the plastic evolution iteration number and H is the isotropic plastic hardening which is a function of ε_q and T:

$$H^J = H(\varepsilon_q^J, T) \qquad [5.80]$$

Once $\Delta\varepsilon_q$ is evaluated, all other unknowns can be evaluated as follows:

$$\boldsymbol{s} = \frac{2}{3}\sigma_M \mathbf{L}^{-1}\boldsymbol{a} = \sigma_Y \boldsymbol{m} \qquad [5.81]$$

$$\boldsymbol{m} = \frac{2}{3}\mathbf{L}^{-1}\boldsymbol{a} = \frac{1}{\sigma_{BM}}\boldsymbol{s}_B \qquad [5.82]$$

$$\boldsymbol{\sigma} = \boldsymbol{\sigma}_h + \boldsymbol{s} = \boldsymbol{\sigma}_{Bh} + \sigma_Y \boldsymbol{m} \qquad [5.83]$$

$$\boldsymbol{\varepsilon}_p = {}^{n-1}\boldsymbol{\varepsilon}_p + \Delta\varepsilon_q\,\boldsymbol{a} \qquad [5.84]$$

$$\boldsymbol{\varepsilon}_e = {}^{n-1}\boldsymbol{\varepsilon}_e + \Delta\boldsymbol{\varepsilon} - \Delta\boldsymbol{\varepsilon}_{th} - \Delta\varepsilon_q\,\boldsymbol{a} \qquad [5.85]$$

$$\varepsilon_q = {}^{n-1}\varepsilon_q + \Delta\varepsilon_q \qquad [5.86]$$

Tangent modular matrix

In order to calculate the next iterative displacement increment $\delta\mathbf{u}$, the tangent modular matrix needs $d\sigma/d\varepsilon$ to be calculated. When an incremental algorithm such as the radial return method is used to compute the plastic strain increment $\Delta\varepsilon_p$, a consistent method with the computation of $\Delta\varepsilon_p$ needs to be used instead of the continuous tangent[26] $\mathbf{C}_t$. In this case, the tangent modular matrix is termed the consistent tangent matrix and is denoted as

$$\mathbf{C}_{tc} = \frac{d\boldsymbol{\sigma}}{d\boldsymbol{\varepsilon}} \quad [5.87]$$

Equation [5.87] is rewritten as

$$\dot{\boldsymbol{\sigma}} = \mathbf{C}_{tc}\dot{\boldsymbol{\varepsilon}} \quad [5.88]$$

All the strain vectors can be divided by volumetric and deviatoric vectors and there is no volumetric plastic strain evolution in the von Mises plasticity:

$$\begin{aligned}\boldsymbol{\varepsilon} &= \boldsymbol{\varepsilon}_e + \boldsymbol{\varepsilon}_p = (\boldsymbol{\varepsilon}_{eh} + \boldsymbol{e}_e) + (\boldsymbol{\varepsilon}_{ph} + \boldsymbol{e}_p)\\ &= (\boldsymbol{\varepsilon}_{eh} + \boldsymbol{\varepsilon}_{ph}) + (\boldsymbol{e}_e + \boldsymbol{e}_p)\\ &= \boldsymbol{\varepsilon}_h + \boldsymbol{e}\end{aligned} \quad [5.89]$$

$$\dot{\boldsymbol{\sigma}}_h = 3k\dot{\boldsymbol{\varepsilon}}_{eh} = 3k(\dot{\boldsymbol{\varepsilon}}_h - \dot{\boldsymbol{\varepsilon}}_{ph}) = 3k\dot{\boldsymbol{\varepsilon}}_h \quad [5.90]$$

In order to formulate the relationship between $\dot{\boldsymbol{s}}$ and $\dot{\boldsymbol{e}}$, let us differentiate Eqn [5.74]. Recall that $\boldsymbol{\varepsilon}_B$ does not have a plastic strain evolution but total strain increments ($\Delta\boldsymbol{\varepsilon}$) are the same for both $\dot{\boldsymbol{s}}$ and $\dot{\boldsymbol{s}}_B$:

$$\dot{\boldsymbol{s}} = \beta\dot{\boldsymbol{s}}_B + \dot{\beta}\boldsymbol{s}_B \quad [5.91]$$

$$\dot{\boldsymbol{s}}_B = 2G\mathbf{L}^{-1}\dot{\boldsymbol{e}}_{Be} = 2G\mathbf{L}^{-1}(\dot{\boldsymbol{e}}_B - \dot{\boldsymbol{e}}_{Bp}) = 2G\mathbf{L}^{-1}\dot{\boldsymbol{e}}_B = 2G\mathbf{L}^{-1}\dot{\boldsymbol{e}} \quad [5.92]$$

From Eqn [5.72],

$$\dot{\beta} = -\frac{3G}{\sigma_{BM}}\Delta\dot{\varepsilon}_q + \frac{3G}{\sigma_{BM}^2}\Delta\varepsilon_q\,\dot{\sigma}_{BM} \quad [5.93]$$

From Eqn [5.76],

$$\Delta\varepsilon_q = \frac{\sigma_{BM} - \sigma_Y}{3G} \quad [5.94]$$

$$\dot{f} = \dot{\sigma}_{BM} - 3G\,\Delta\dot{\varepsilon}_q - H\,\Delta\dot{\varepsilon}_q = 0 \quad [5.95]$$

$$\Delta\dot{\varepsilon}_q = \frac{\dot{\sigma}_{BM}}{3G + H} \quad [5.96]$$

From Eqns [5.49] and [5.92]

$$\dot{\sigma}_{BM} = \dot{\sigma}_M(\boldsymbol{s}_B) = \frac{3}{2}\frac{\boldsymbol{s}_B^T \cdot \mathbf{L}\dot{\boldsymbol{s}}_B}{\sigma_{BM}} = \frac{3G}{\sigma_{BM}}\boldsymbol{s}_B^T \cdot \dot{\boldsymbol{e}} \quad [5.97]$$

Combining Eqns [5.93]–[5.97], we obtain the following relationship:

$$\dot{\beta} = \frac{3G}{\sigma_{BM}^2}\left[-\frac{3G}{3G+H} + \left(1 - \frac{\sigma_Y}{\sigma_{BM}}\right)\right]\boldsymbol{s}_B^{\mathrm{T}} \cdot \dot{\boldsymbol{e}} \quad [5.98]$$

From Eqns [5.72] and [5.94], we can simplify β as follows:

$$\beta = 1 - \frac{3G}{\sigma_{BM}}\frac{\sigma_{BM} - \sigma_Y}{3G} = \frac{\sigma_Y}{\sigma_{BM}} \quad [5.99]$$

From Eqns [5.91], [5.92], [5.98] and [5.99],

$$\begin{aligned}\dot{\boldsymbol{s}} &= \frac{\sigma_Y}{\sigma_{BM}} 2G\mathbf{L}^{-1}\dot{\boldsymbol{e}} + \frac{3G}{\sigma_{BM}^2}\left[-\frac{3G}{3G+H} + \left(1 - \frac{\sigma_Y}{\sigma_{BM}}\right)\right]\boldsymbol{s}_B \cdot \boldsymbol{s}_B^{\mathrm{T}} \cdot \dot{\boldsymbol{e}} \\ &= \left[2G\frac{\sigma_Y}{\sigma_{BM}}\mathbf{L}^{-1} + \left(\frac{3GH}{3G+H} - 3G\frac{\sigma_Y}{\sigma_{BM}}\right)\frac{\boldsymbol{s}_B}{\sigma_{BM}}\left(\frac{\boldsymbol{s}_B}{\sigma_{BM}}\right)^{\mathrm{T}}\right]\dot{\boldsymbol{e}} \\ &= [2G_{\mathrm{eff}}\mathbf{L}^{-1} + (H_{\mathrm{eff}} - 3G_{\mathrm{eff}})\boldsymbol{m}\cdot\boldsymbol{m}^{\mathrm{T}}](\dot{\mathbf{e}} - \dot{\mathbf{e}}_{\mathrm{h}})\end{aligned} \quad [5.100]$$

where

$$G_{\mathrm{eff}} = G\frac{\sigma_Y}{\sigma_{BM}} \quad [5.101]$$

$$H_{\mathrm{eff}} = \frac{3GH}{3G+H} \quad [5.102]$$

From Eqns [5.54], [5.82], [5.90] and [5.100],

$$\begin{aligned}\dot{\boldsymbol{\sigma}} &= \dot{\boldsymbol{\sigma}}_{\mathrm{h}} + \dot{\boldsymbol{s}} \\ &= 3k\frac{1}{3}\boldsymbol{j}\cdot\boldsymbol{j}^{\mathrm{T}}\cdot\dot{\boldsymbol{\varepsilon}} \\ &\quad + [2G_{\mathrm{eff}}\mathbf{L}^{-1} + (H_{\mathrm{eff}} - 3G_{\mathrm{eff}})\boldsymbol{m}\cdot\boldsymbol{m}^{\mathrm{T}}]\left(\dot{\boldsymbol{\varepsilon}} - \frac{1}{3}\boldsymbol{j}\cdot\boldsymbol{j}^{\mathrm{T}}\cdot\dot{\boldsymbol{\varepsilon}}\right) \\ &= \left(\frac{3k - 2G_{\mathrm{eff}}}{3}\boldsymbol{j}\cdot\boldsymbol{j}^{\mathrm{T}} + 2G_{\mathrm{eff}}\mathbf{L}^{-1} + (H_{\mathrm{eff}} - 3G_{\mathrm{eff}})\boldsymbol{m}\cdot\boldsymbol{m}^{\mathrm{T}}\right)\dot{\boldsymbol{\varepsilon}}\end{aligned} \quad [5.103]$$

Equation [5.103] can be obtained using the following equalities:

$$\mathbf{L}^{-1}\boldsymbol{j}\cdot\boldsymbol{j}^{\mathrm{T}} = \boldsymbol{j}\cdot\boldsymbol{j}^{\mathrm{T}} \quad [5.104]$$

$$\begin{aligned}s_x + s_y + s_z &= (\sigma_x - \sigma_{\mathrm{h}}) + (\sigma_y - \sigma_{\mathrm{h}}) + (\sigma_z - \sigma_{\mathrm{h}}) \\ &= (\sigma_x + \sigma_y + \sigma_z) - 3\sigma_{\mathrm{h}} \\ &= 0\end{aligned} \quad [5.105]$$

$$(\boldsymbol{m}\cdot\boldsymbol{m}^{\mathrm{T}}\cdot\boldsymbol{j}\cdot\boldsymbol{j}^{\mathrm{T}}) = \frac{1}{\sigma_m^2}(\boldsymbol{s}\cdot\boldsymbol{s}^{\mathrm{T}}\cdot\boldsymbol{j}\cdot\boldsymbol{j}^{\mathrm{T}}) \quad [5.106]$$

$$= \frac{1}{\sigma_M^2}(s_x + s_y + s_z)\begin{bmatrix} s_x & s_y & s_z & 0 & 0 & 0 \\ s_x & s_y & s_z & 0 & 0 & 0 \\ s_x & s_y & s_z & 0 & 0 & 0 \\ 0 & 0 & 0 & 0 & 0 & 0 \\ 0 & 0 & 0 & 0 & 0 & 0 \\ 0 & 0 & 0 & 0 & 0 & 0 \end{bmatrix}$$

$$= \mathbf{0} \tag{5.107}$$

Finally from Eqns [5.88] and [5.103],

$$\mathbf{C}_{\mathrm{tc}} = \lambda_{\mathrm{eff}}\boldsymbol{j}\cdot\boldsymbol{j}^{\mathrm{T}} + 2G_{\mathrm{eff}}\mathbf{L}^{-1} + (H_{\mathrm{eff}} - 3G_{\mathrm{eff}})\boldsymbol{m}\cdot\boldsymbol{m}^{\mathrm{T}} \tag{5.108}$$

$$\lambda_{\mathrm{eff}} = \frac{3k - 2G_{\mathrm{eff}}}{3} \tag{5.109}$$

5.3 Modeling methods

5.3.1 Lagrangian formulations

Total Lagrangian

A weak formulation of the momentum balance is derived by multiplying Eqns [5.37] and [5.21] by a kinematic admissible function $\hat{\boldsymbol{u}}$, integrating over the reference volume V_0 and surface $A_{X\mathrm{bt}}$ respectively and adding them together. Integration by parts and application of the divergence theorem leads to the following weak formulation:

$$\int_{V_X} (\nabla_X \hat{\boldsymbol{u}} : \boldsymbol{P} - \hat{\boldsymbol{u}} \cdot \mathbf{b}_X)\mathrm{d}V_X - \int_{A_{X\mathrm{t}}} \hat{\boldsymbol{u}} \cdot \boldsymbol{t}_X^{\mathrm{p}}\, \mathrm{d}A_X = 0 \tag{5.110}$$

Considering that the first Piola–Kirchhoff stress is not symmetric, Eqn [5.110] is rewritten by using the second Piola–Kirchhoff stress, using $\boldsymbol{P} = \mathbf{F}\boldsymbol{S}$:

$$\begin{aligned} &\int_{V_X} (\nabla_X \hat{\boldsymbol{u}} : \mathbf{F}\boldsymbol{S} - \hat{\boldsymbol{u}} \cdot \boldsymbol{b}_X)\mathrm{d}V_X - \int_{A_{X\mathrm{t}}} \hat{\boldsymbol{u}} \cdot \boldsymbol{t}_X^{\mathrm{p}}\, \mathrm{d}A_X = 0 \\ &\int_{V_X} [\mathrm{sym}(\mathbf{F}^{\mathrm{T}}\nabla_X \hat{\boldsymbol{u}}) : \boldsymbol{S} - \hat{\boldsymbol{u}} \cdot \boldsymbol{b}_X]\mathrm{d}V_X - \int_{A_{X\mathrm{t}}} \hat{\boldsymbol{u}} \cdot \boldsymbol{t}_X^{\mathrm{p}}\, \mathrm{d}A_X = 0 \end{aligned} \tag{5.111}$$

Finite-element discretization is applied to the variational equations to obtain the expressions for the element residual in Eqn [5.110] as follows:

$$\mathcal{R}(\mathcal{U}) = \sum_{\text{elements}} \boldsymbol{R}(\boldsymbol{U}) = \mathbf{0} \quad \text{on } V_0 \text{ and } A_{0\mathrm{t}} \tag{5.112}$$

where $\mathcal{R}$ and $\mathcal{U}$ are the global residual and nodal displacement vectors respectively. The element residual $\boldsymbol{R}(\boldsymbol{U})$ is evaluated as follows:

$$\boldsymbol{R}(\boldsymbol{U}) = \sum_{V_{0\text{GaussPoint}}} (\mathbf{B}_{\text{nl}}^{\text{T}}\overline{\boldsymbol{S}} - \mathbf{N}^{\text{T}}\boldsymbol{b}_X) wJ_X - \sum_{A_{0\mathbf{t}\text{GaussPoint}}} \mathbf{N}^{\text{T}}\boldsymbol{t}_X^{\text{p}} wj_X \quad [5.113]$$

where $\overline{\boldsymbol{S}}$ is the second Piola–Kirchhoff stress written in vector form and $\boldsymbol{U}$ is the nodal displacement vector for each element. Equation [5.113] is presented in matrix-vector form. The matrices $\mathbf{N}$ and $\mathbf{B}_{\text{nl}}$ are the shape function and strain operators respectively:

$$\boldsymbol{u} = \mathbf{N}\boldsymbol{U} \quad [5.114]$$

$$\mathbf{L}\overline{\text{sym}(\mathbf{F}^{\text{T}}\nabla_X\hat{\boldsymbol{u}})} = \mathbf{B}_{\text{nl}}\hat{\boldsymbol{U}} \quad [5.115]$$

Substitution of $\mathbf{F} = \mathbf{I} + \mathbf{D}$ and $\nabla_X \cdot \boldsymbol{u} = \mathbf{D}$ into Eqn [5.115] yields

$$\begin{aligned}\mathbf{B}_{\text{nl}}\hat{\boldsymbol{U}} &= \mathbf{L}\overline{\text{sym}\left[(\mathbf{D}+\mathbf{I})^{\text{T}}\hat{\mathbf{D}}\right]} \\ &= \mathbf{L}\overline{\text{sym}(\mathbf{D}^{\text{T}}\hat{\mathbf{D}}+\hat{\mathbf{D}})} \\ &= \mathbf{L}\overline{\text{sym}(\mathbf{D}^{\text{T}}\hat{\mathbf{D}})} + \mathbf{L}\overline{\text{sym}(\hat{\mathbf{D}})} \quad [5.116] \\ &= \mathbf{B}_l\hat{\boldsymbol{U}} + \mathbf{B}_n\hat{\boldsymbol{U}} \quad [5.117]\end{aligned}$$

where $\mathbf{B}_{\text{l}}$ is the operator that computes the linear component of the Green strain in engineering format and $\mathbf{B}_{\text{n}}$ is twice the operator that computes the nonlinear component of the Green strain in engineering format (see Eqn [5.10]).

From Eqns [5.10] and [5.117] the Green strain in engineering format is computed as

$$\begin{aligned}\bar{\boldsymbol{\varepsilon}} &= \mathbf{L}\frac{1}{2}\overline{\text{sym}(\mathbf{D}^{\text{T}}\mathbf{D})} + \mathbf{L}\overline{\text{sym}(\mathbf{D})} \\ &= \boldsymbol{B}_{\text{l}}\boldsymbol{U} + \frac{1}{2}\boldsymbol{B}_{\text{n}}\boldsymbol{U} = \boldsymbol{B}_{\text{nl2}}\boldsymbol{U} \quad [5.118]\end{aligned}$$

The residual of Eqn [5.112] is minimized by solving for the global displacement vector $\mathcal{U}$ iteratively. An initial starting solution $\mathcal{U}^1$ is assumed and the Newton–Raphson method is applied to determine a corrected displacement field. Typically, the initial solution is assumed as zero $\mathcal{U}^1 = \mathbf{0}$. Application of the Newton–Raphson linearization yields the following iterative scheme for the computation of displacement field:

$$\delta\mathcal{U} = -\left[\left.\frac{\mathrm{d}\mathcal{R}}{\mathrm{d}\mathcal{U}}\right|_{\mathcal{U}^I}\right]^{-1} \mathcal{R}(\mathcal{U})|_{\mathcal{U}^I} \quad [5.119]$$

$$\mathcal{U}^{I+1} = \mathcal{U}^I + \delta\mathcal{U} \quad [5.120]$$

The displacement field is updated until the solution converges and the residual becomes very small. The tangent stiffness $\mathrm{d}\mathcal{R}/\mathrm{d}\mathcal{U}$ for Eqn [5.120] is calculated by differentiating the residual with respect to the nodal displacement vector $\boldsymbol{U}$:

$$\frac{\mathrm{d}\mathcal{R}}{\mathrm{d}\mathcal{U}} = \sum_{\text{elements}} \frac{\mathrm{d}\boldsymbol{R}}{\mathrm{d}\boldsymbol{U}} \qquad [5.121]$$

where $\mathrm{d}\boldsymbol{R}/\mathrm{d}\boldsymbol{U}$ is the element stiffness which is derived from

$$\mathrm{d}\boldsymbol{P} = \mathrm{d}\mathbf{F}\boldsymbol{S} + \mathbf{F}\,\mathrm{d}\boldsymbol{S} \qquad [5.122]$$

Substitution into Eqn [5.110] along with finite-element discretization yields

$$\begin{aligned}\frac{\mathrm{d}\boldsymbol{R}}{\mathrm{d}\boldsymbol{U}} &= \sum_{V_{X\text{GaussPoint}}} \left(\mathbf{B}_{\mathrm{nl}}^{\mathrm{T}} \frac{\mathrm{d}\overline{\boldsymbol{S}}}{\mathrm{d}\boldsymbol{\varepsilon}} \mathbf{B}_{\mathrm{nl}} + \boldsymbol{G}^{\mathrm{T}}\tilde{\boldsymbol{S}}\boldsymbol{G}\right) wJ_X \\ &\quad - \sum_{V_{X\text{GaussPoint}}} \left(\mathbf{N}^{\mathrm{T}} \frac{\mathrm{d}\boldsymbol{b}_X}{\mathrm{d}\boldsymbol{U}}\right) wJ_X - \sum_{A_{X\mathbf{t}\text{GaussPoint}}} \mathbf{N}^{\mathrm{T}} \frac{\mathrm{d}\boldsymbol{t}_X^{\mathrm{p}}}{\mathrm{d}\boldsymbol{U}} wj_X \end{aligned} \qquad [5.123]$$

where the following have been used:

$$\frac{\mathrm{d}\overline{\boldsymbol{S}}}{\mathrm{d}\boldsymbol{U}} = \frac{\mathrm{d}\overline{\boldsymbol{S}}}{\mathrm{d}\overline{\boldsymbol{\varepsilon}}} \frac{\mathrm{d}\overline{\boldsymbol{\varepsilon}}}{\mathrm{d}\boldsymbol{U}} = \frac{\mathrm{d}\overline{\boldsymbol{S}}}{\mathrm{d}\overline{\boldsymbol{\varepsilon}}} B_{\mathrm{nl}} \qquad [5.124]$$

and

$$\begin{aligned}\frac{\mathrm{d}\overline{\boldsymbol{\varepsilon}}}{\mathrm{d}\boldsymbol{U}} &= \frac{\mathrm{d}}{\mathrm{d}\boldsymbol{U}}\left[\overline{\mathbf{L}\mathrm{sym}(\mathbf{D})} + \frac{1}{2}\overline{\mathbf{L}\mathrm{sym}(\mathbf{D}^{\mathrm{T}}\mathbf{D})}\right] \\ &= \overline{\mathbf{L}\mathrm{sym}\left(\frac{\mathrm{d}\mathbf{D}}{\mathrm{d}\boldsymbol{U}}\right)} + \frac{1}{2}\overline{2\mathbf{L}\mathrm{sym}\left(\mathbf{D}^{\mathrm{T}}\frac{\mathrm{d}\mathbf{D}}{\mathrm{d}\boldsymbol{U}}\right)} \end{aligned} \qquad [5.125]$$

$$\begin{aligned} &= \mathbf{B}_{\mathrm{l}} + \mathbf{B}_{\mathrm{n}} \\ &= \mathbf{B}_{\mathrm{nl}} \end{aligned} \qquad [5.126]$$

and

$$\mathrm{d}\mathbf{F} = \mathrm{d}(\mathbf{D} + \mathbf{I}) = \mathrm{d}\mathbf{D} \qquad [5.127]$$

and

$$\hat{\mathbf{D}}:(\mathbf{D}\boldsymbol{S}) = \hat{\boldsymbol{U}}^{\mathrm{T}}\boldsymbol{G}^{\mathrm{T}}\tilde{\boldsymbol{S}}\boldsymbol{G}\boldsymbol{U} \qquad [5.128]$$

where the operator $\boldsymbol{G}$ computes the displacement gradient in a vector form:

$$\overline{\mathbf{D}} = \boldsymbol{G}\cdot\boldsymbol{U} \qquad [5.129]$$

and

$$\tilde{\boldsymbol{S}} = \begin{bmatrix} \boldsymbol{S} & \mathbf{0} & \mathbf{0} \\ \mathbf{0} & \boldsymbol{S} & \mathbf{0} \\ \mathbf{0} & \mathbf{0} & \boldsymbol{S} \end{bmatrix} \qquad [5.130]$$

where $\mathbf{0}$ is a 3×3 zero matrix:

$$\mathbf{0} = \begin{bmatrix} 0 & 0 & 0 \\ 0 & 0 & 0 \\ 0 & 0 & 0 \end{bmatrix} \tag{5.131}$$

The first component of the stiffness matrix of Eqn [5.123] is called the material stiffness:

$$\left.\frac{\mathrm{d}\boldsymbol{R}}{\mathrm{d}\boldsymbol{U}}\right|_{\mathrm{mat}} = \sum_{V_{0\mathrm{GaussPoint}}} \left(\mathbf{B}_{\mathrm{nl}}^{\mathrm{T}} \frac{\mathrm{d}\bar{\boldsymbol{S}}}{\mathrm{d}\bar{\boldsymbol{\varepsilon}}} \mathbf{B}_{\mathrm{nl}} \right) w J_X \tag{5.132}$$

The second component of the stiffness matrix of Eqn [5.123] is called the geometric of stress stiffness:

$$\left.\frac{\mathrm{d}\boldsymbol{R}}{\mathrm{d}\boldsymbol{U}}\right|_{\mathrm{geom}} = \sum_{V_{0\mathrm{GaussPoint}}} (\boldsymbol{G}^{\mathrm{T}} \boldsymbol{S} \boldsymbol{G}) w J_X \tag{5.133}$$

The remaining components of the stiffness matrix of Eqn [5.123] account for the dependences of the body force and traction on the deformation:

$$\left.\frac{\mathrm{d}\boldsymbol{R}}{\mathrm{d}\boldsymbol{U}}\right|_{\mathrm{load}} = - \sum_{V_{0\mathrm{GaussPoint}}} \left(\mathbf{N}^{\mathrm{T}} \frac{\mathrm{d}\boldsymbol{b}_X}{\mathrm{d}\boldsymbol{U}} \right) w J_X - \sum_{A_{0\mathbf{t}\mathrm{GaussPoint}}} \mathbf{N}^{\mathrm{T}} \frac{\mathrm{d}\boldsymbol{t}_X^{\mathrm{p}}}{\mathrm{d}\boldsymbol{U}} w j_x \tag{5.134}$$

Updated Lagrangian

The weak formulation for the updated Lagrangian approach can be derived similarly to the approach for the total Lagrangian formulation by multiplying Eqns [5.32] and [5.20] by a kinematic admissible function $\hat{\boldsymbol{u}}$, integrating over the current volume V and surface A_{bt} respectively, adding them together and finally applying integration by parts and the divergence theorem.

An alternative approach is to consider the current configuration as the reference configuration which will lead to the following substitutions:

$$\mathbf{F} \rightarrow \mathbf{I} \tag{5.135}$$

$$\boldsymbol{S} \rightarrow \boldsymbol{\sigma} \tag{5.136}$$

$$\boldsymbol{E} \rightarrow \boldsymbol{A} \tag{5.137}$$

$$\boldsymbol{\varepsilon} \rightarrow \boldsymbol{\varepsilon}^{\mathrm{A}} \tag{5.138}$$

$$\nabla_X \rightarrow \nabla_X \tag{5.139}$$

$$\mathbf{B}_{\mathrm{nl}} \rightarrow \mathbf{B}_{\mathrm{l}} \tag{5.140}$$

$$V_0 \rightarrow V \tag{5.141}$$

$$A_0 \rightarrow A \tag{5.142}$$

$$\boldsymbol{t}_X \rightarrow \boldsymbol{t} \tag{5.143}$$

$$\boldsymbol{b}_X \rightarrow \boldsymbol{b} \tag{5.144}$$

Then, the element residual for the updated Lagrangian formulation is computed as follows:

$$\boldsymbol{R}(\boldsymbol{U}) = \sum_{V_{\text{GaussPoint}}} (\mathbf{B}_{\text{l}}^{\text{T}}\bar{\boldsymbol{\sigma}} - \mathbf{N}^{\text{T}}\boldsymbol{b})wJ_x - \sum_{A_{t\text{GaussPoint}}} \mathbf{N}^{\text{T}}\boldsymbol{t}_X^{\text{P}}wj_x \tag{5.145}$$

where $\bar{\boldsymbol{\sigma}}$ is the Cauchy stress written in a vector form.

The element stiffness is computed as follows:

$$\frac{\mathrm{d}\boldsymbol{R}}{\mathrm{d}\boldsymbol{U}} = \sum_{V_{\text{GaussPoint}}} \left(\mathbf{B}_{\text{l}}^{\text{T}} \frac{\mathrm{d}\bar{\boldsymbol{\sigma}}}{\mathrm{d}\bar{\boldsymbol{\varepsilon}}^{\text{A}}} \mathbf{B}_{\text{l}} + \boldsymbol{g}^{\text{T}}\bar{\boldsymbol{\sigma}}\boldsymbol{g} \right) wJ_x - \sum_{V_{\text{GaussPoint}}} \left(\mathbf{N}^{\text{T}} \frac{\mathrm{d}\boldsymbol{b}}{\mathrm{d}\boldsymbol{U}} \right) wJ_x - \sum_{A_{t\text{GaussPoint}}} \mathbf{N}^{\text{T}} \frac{\mathrm{d}\boldsymbol{t}^{\text{P}}}{\mathrm{d}\boldsymbol{U}} wj_x \tag{5.146}$$

where the operator $\boldsymbol{g}$ computes the displacement gradient in a vector form:

$$\bar{\mathbf{d}} = \boldsymbol{g} \cdot \boldsymbol{U} \tag{5.147}$$

5.3.2 Eulerian formulation

This section presents a finite-element formulation for quasistatic thermo-elastoplastic systems. The formulation is suitable for modeling material processes such as welding and laser surfacing. For example, in welding, a torch is passed over a body (Fig. 5.2). A reference frame fixed to the material configuration, $\boldsymbol{r}$, denotes a Lagrangian frame. In the Lagrangian frame, the torch moves with a steady velocity and the state of the body changes in time. If the velocity and heat input are constant, the body is infinitely long in the direction of the velocity and its shape is uniform along that direction; then, in a reference frame $\boldsymbol{x}$ fixed to the torch, the state of the body does not change in time. Therefore, in the moving reference frame, the problem becomes static. In practice, the body has finite dimensions but, if the torch is sufficiently far from the edges, in view of St Venant's principle, the edge effects can be ignored. Thus it can be assumed that the solution in the new reference frame for an infinite continuum is the same as that for the finite body. Such a reference frame, which moves relative to the body, is similar to a control volume with an influx and outflux of material and denotes an Eulerian frame.

Welding is typically assumed to be a weakly coupled thermomechanical

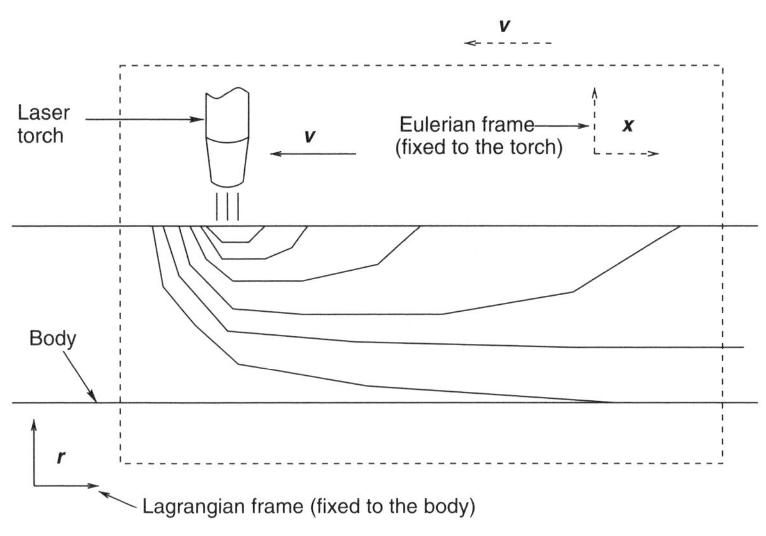

5.2 Eulerian and Lagrangian frames.

process. The temperature profile is assumed to be independent of stresses and strains. Thus a heat transfer analysis is performed initially, and the results are imported for the stress analysis. Modeling a history-dependent heat transfer problem in a Lagrangian frame requires a transient analysis. The numerical implementation involves an incremental scheme with several small time increments. The solution at a given time increment is obtained by using the solution at the previous time increment as an initial condition. This problem has been addressed in detail previously[4,27–29]. The subsequent history-dependent stress analysis is performed by modeling the stress problem as a quasistatic process[1,2,9,11,30–35]. Similar to the heat transfer analysis, the numerical implementation of the quasistatic analysis involves an incremental scheme with several small static increments. The solution at a given time interval is obtained by using the solution at the previous time increment as an initial condition. Each static analysis involves finding the plastified regions and updating the stresses and strains by integrating the incremental plasticity evolution equations. The consistent stiffness for these plastified regions has been developed by Simo and Taylor[26].

The solution field of a quasistatic process in an Eulerian frame for the heat transfer problem becomes steady state and for the stress problem static. This concept was originally proposed by Nguyen and Rahimian[36]. Thus the solution can be obtained by a single steady-state or static analysis. For an incremental analysis in the Lagrangian frame, a mesh with a uniformly high density in the direction along the velocity of the torch is required whereas, in an Eulerian frame, the mesh may be graded to lower densities away from the active regions. This advantage is more significant for three-dimensional problems as both longitudinal grading and transverse grading are feasible. Thus, for quasistatic processes such as welding and laser surfacing, the computational requirements can be lower for an Eulerian analysis than for a Lagrangian analysis by two or three orders of magnitude, depending on the size of the problem[37]. These savings become even more important in an optimization routine which performs several iterations with various parameters to obtain the parameters to be used for desired stress and distortion results.

The steady-state heat transfer problem in an Eulerian frame has been addressed previously[37–40]. For the elastoplastic problems, stress and strain histories need to be integrated to obtain the respective values at a given point. In flow problems, the history is often embedded in streamlines. Dawson[41] and Caswell and Viriyauthakorn[42] used a viscoplastic model and have introduced the concept of integrating along streamlines. These are determined by a distinct analysis before the integration is carried out. These formulations are velocity based and require further analysis to incorporate the effect of elastic strains[41,43].

Small-deformation Lagrangian formulation

Finite-element formulations for quasistatic thermoelastoplastic processes in Lagrangian reference frames have been widely used[1,2,11,26,30,32–35]. For a stationary reference frame $\boldsymbol{r}$, the stress equilibrium equation is given by

$$\nabla_r \boldsymbol{\sigma}(\boldsymbol{r},t) + \boldsymbol{b}(\boldsymbol{r},t) = 0 \qquad \text{in } V_r \tag{5.148}$$

where $\boldsymbol{\sigma}$ is the stress and $\boldsymbol{b}$ the body force. The boundary conditions are

$$\boldsymbol{u}(\boldsymbol{r},t) = \bar{\boldsymbol{u}}(\boldsymbol{r},t) \quad \text{on surface } A_r^u \tag{5.149}$$

$$\boldsymbol{\sigma}(\boldsymbol{r},t)\cdot\boldsymbol{n}(\boldsymbol{r},t) = \bar{\boldsymbol{t}}(\boldsymbol{r},t) \quad \text{on surface } A_r^t \tag{5.150}$$

where $\bar{\boldsymbol{u}}(\boldsymbol{r}, t)$ are the prescribed displacements on surface A_r^u, $\bar{\boldsymbol{t}}$ are the prescribed tractions on surface A_r^t, and $\boldsymbol{n}$ is the unit outward normal to the surface A_r^t.

The total strain is the Green strain:

$$\boldsymbol{\varepsilon}(\boldsymbol{r},t) = \frac{1}{2}\left\{\nabla_r \boldsymbol{u}(\boldsymbol{r},t) + [\nabla_r \boldsymbol{u}(\boldsymbol{r},t)]^{\mathrm{T}}\right\} \tag{5.151}$$

Assume that $\boldsymbol{\varepsilon}$ is applied as

$$\boldsymbol{\varepsilon} = \boldsymbol{\varepsilon}_{\mathrm{e}} + \boldsymbol{\varepsilon}_{\mathrm{p}} + \boldsymbol{\varepsilon}_{\mathrm{th}} \tag{5.152}$$

where $\boldsymbol{\varepsilon}_{\mathrm{e}}$, $\boldsymbol{\varepsilon}_{\mathrm{p}}$ and $\boldsymbol{\varepsilon}_{\mathrm{th}}$ are the elastic, plastic and thermal strains respectively. The thermal strain $\boldsymbol{\varepsilon}_{\mathrm{th}}$ is evaluated using

$$\boldsymbol{\varepsilon}_{\text{th}} = [\varepsilon_{\text{th}} \quad \varepsilon_{\text{th}} \quad \varepsilon_{\text{th}} \quad 0 \quad 0 \quad 0]^{\text{T}} \qquad [5.153]$$

$$\varepsilon_{\text{th}} = \alpha^{\text{T}}(T - T_{\text{ref}}) - \alpha^{T_{\text{amb}}}(T_{\text{amb}} - T_{\text{ref}}) \qquad [5.154]$$

where α^{T} and $\alpha^{T_{\text{amb}}}$ are the thermal expansion coefficients at temperatures T and T_{amb} respectively. T_{ref} is the reference temperature.

The initial conditions are:

$$\boldsymbol{u}(\boldsymbol{r}, t_0) = \boldsymbol{u}^0(\boldsymbol{r}) \qquad [5.155]$$

$$\boldsymbol{\varepsilon}_{\text{p}}(\boldsymbol{r}, t_0) = \boldsymbol{\varepsilon}_{\text{p}}^0(\boldsymbol{r}) \qquad [5.156]$$

$$\varepsilon_{\text{q}}(\boldsymbol{r}, t_0) = \varepsilon_{\text{q}}^0(\boldsymbol{r}) \qquad [5.157]$$

where ε_{q} is the equivalent plastic strain and t_0 is the time for the previous time increment.

Using Eqn [5.151], the stress–strain relationship is

$$\boldsymbol{\sigma} = \boldsymbol{C}(\boldsymbol{\varepsilon} - \boldsymbol{\varepsilon}_{\text{p}} - \boldsymbol{\varepsilon}_{\text{th}}) \qquad [5.158]$$

where $\boldsymbol{C}$ is a tensor. Using the associative J_2 plasticity[44], the yield function f and equivalent plastic strain ε_{q} are

$$f = \sigma_{\text{e}} - (\sigma_{y0} + H\varepsilon_{\text{q}}) \qquad [5.159]$$

$$\dot{\varepsilon}_{\text{q}} = \left(\frac{2}{3}\right)^{1/2} \|\dot{\boldsymbol{\varepsilon}}_{\text{p}}\| \qquad [5.160]$$

where σ_{e} and H are the von Mises stress and the hardening coefficient respectively. The evolution equation for active yielding is

$$\dot{f} = 0 \qquad [5.161]$$

Differentiating Eqn [5.159], using Eqn [5.160] for the definition of equivalent plastic strain and differentiating Eqn [5.158] according to

$$\dot{\boldsymbol{\sigma}} = \boldsymbol{C}(\dot{\boldsymbol{\varepsilon}} - \dot{\boldsymbol{\varepsilon}}_{\text{p}} - \dot{\boldsymbol{\varepsilon}}_{\text{th}}) + \dot{\boldsymbol{C}}(\boldsymbol{\varepsilon} - \boldsymbol{\varepsilon}_{\text{p}} - \boldsymbol{\varepsilon}_{\text{th}}) \qquad [5.162]$$

the plasticity evolution equations become

$$\dot{\varepsilon}_{\text{q}} = \frac{\gamma}{H}[2\mu\boldsymbol{a}\cdot(\dot{\boldsymbol{\varepsilon}} - \dot{\boldsymbol{\varepsilon}}_{\text{p}}) + 2\dot{\mu}\boldsymbol{a}\cdot(\boldsymbol{\varepsilon} - \boldsymbol{\varepsilon}_{\text{p}}) - \dot{H}\varepsilon_{\text{q}} - \sigma_{\dot{y}0}] \qquad [5.163]$$

$$\dot{\boldsymbol{\varepsilon}}_{\text{p}} = \boldsymbol{a}\dot{\varepsilon}_{q} \qquad [5.164]$$

where $\boldsymbol{a}$ is the flow vector and μ is the shear modulus of the material. γ is a multiplier which depends on the yield function f as follows:

$$\gamma = \begin{cases} 1 & \text{if } f \geqslant 0 \\ 0 & \text{otherwise} \end{cases} \qquad [5.165]$$

Since the plastic strain and equivalent plastic strains evolve in time, the rate equations can be written as time rate equations, i.e.

$$\frac{\partial \varepsilon_q}{\partial t} = \frac{\gamma}{H}\left[2\mu \boldsymbol{a}\cdot\left(\frac{\partial \boldsymbol{\varepsilon}}{\partial t} - \frac{\partial \boldsymbol{\varepsilon}_p}{\partial t}\right) + 2\frac{\partial \mu}{\partial t}\boldsymbol{a}\cdot(\boldsymbol{\varepsilon} - \boldsymbol{\varepsilon}_p) - \frac{\partial H}{\partial t}\varepsilon_q - \frac{\partial \sigma_{y0}}{\partial t}\right] \quad [5.166]$$

$$\frac{\partial \boldsymbol{\varepsilon}_p}{\partial t} = \boldsymbol{a}\frac{\partial \varepsilon_q}{\partial t} \quad [5.167]$$

For a quasistatic analysis, the time rate equations become incremental equations:

$$\Delta\varepsilon_q = \frac{\gamma}{H}[2\mu \boldsymbol{a}\cdot(\Delta\boldsymbol{\varepsilon} - \Delta\boldsymbol{\varepsilon}_p) + 2\Delta\mu \boldsymbol{a}\cdot(\boldsymbol{\varepsilon} - \boldsymbol{\varepsilon}_p) - \Delta H\varepsilon_q - \Delta\sigma_{y0}] \quad [5.168]$$

$$\Delta\boldsymbol{\varepsilon}_p = \boldsymbol{a}\,\Delta\varepsilon_q \quad [5.169]$$

where Δ represents the change for each increment. These incremental equations are solved at each Gauss point for each increment[19].

The finite-element formulation for the Lagrangian reference frame is derived using the Galerkin method, where the weight function is the kinematicaly admissible function $\hat{\boldsymbol{u}}$. Integration by parts and application of the divergence theorem on Eqn [5.148] yields the following variational form of the initial-boundary value problem:

$$\int_{V_r} [\hat{\boldsymbol{\varepsilon}}\boldsymbol{C}(\boldsymbol{\varepsilon} - \boldsymbol{\varepsilon}_p - \boldsymbol{\varepsilon}_{th}) - \hat{\boldsymbol{u}}\cdot\boldsymbol{b}]\mathrm{d}V - \int_{A_r^t} \hat{\boldsymbol{u}}\cdot\bar{\boldsymbol{t}}\,\mathrm{d}A = 0 \quad [5.170]$$

along with the boundary conditions in Eqns [5.149] and [5.150], and the initial conditions in Eqns [5.155]–[5.157]. In the finite-element formulation of Eqn [5.170], the primary field is the displacement field $\boldsymbol{u}$, while the total strain $\boldsymbol{\varepsilon}$, plastic strain $\boldsymbol{\varepsilon}_p$ and equivalent plastic strain ε_q are secondary fields computed through a strong enforcement of Eqns [5.151], [5.158], [5.168] and [5.169] at each Gauss point. The displacement boundary condition in Eqn [5.149] is strongly enforced. The equilibrium equation, Eqn [5.148] and the traction boundary condition in Eqn [5.150] are weakly enforced.

Transformation to an Eulerian frame

A reference frame $\boldsymbol{x}$ moving with respect to the material configuration is defined. The Eulerian frame $\boldsymbol{x}$ is a control volume through which the material configuration flows. Specifically,

$$\boldsymbol{x} = \boldsymbol{r} - \boldsymbol{v}t \quad [5.171]$$

where $\boldsymbol{v}$ is a steady velocity. A field $\boldsymbol{\Psi}$ in $\boldsymbol{r}$ transforms to $\bar{\boldsymbol{\Psi}}$ in $\boldsymbol{x}$ as follows:

$$\boldsymbol{\Psi}(\boldsymbol{r}, t) = \bar{\boldsymbol{\Psi}}(\boldsymbol{x}, t) \quad [5.172]$$

$$\nabla_r \boldsymbol{\Psi}(\boldsymbol{r},t) = \nabla_x \overline{\boldsymbol{\Psi}}(\boldsymbol{x},t)\frac{\partial \boldsymbol{x}}{\partial \boldsymbol{r}} \tag{5.173}$$

$$\frac{\partial \boldsymbol{\Psi}(\boldsymbol{r},t)}{\partial t} = \frac{\partial \overline{\boldsymbol{\Psi}}(\boldsymbol{x},t)}{\partial \boldsymbol{x}}\frac{\partial \boldsymbol{x}}{\partial t} + \frac{\partial \overline{\boldsymbol{\Psi}}(\boldsymbol{x},t)}{\partial t} \tag{5.174}$$

For the transformation assumed in Eqn [5.171], the Jacobian $\partial \boldsymbol{x}/\partial \boldsymbol{r}$ is a unit matrix and the quantity $\partial \boldsymbol{x}/\partial \boldsymbol{t}$ is the velocity $-\boldsymbol{v}$. If the loading, shape and material properties are independent of $\boldsymbol{x}$ in the Eulerian frame, as described before, the solution is static and the quantity $\partial \overline{\boldsymbol{\Psi}}(\boldsymbol{x},t)/\partial \boldsymbol{r}$ can be dropped. The field $\boldsymbol{\Psi}$ thus will be a function of $\boldsymbol{x}$ only. The transformation is thus

$$\boldsymbol{\Psi}(\boldsymbol{r},t) = \overline{\boldsymbol{\Psi}}(\boldsymbol{x}) \tag{5.175}$$

$$\nabla_r \boldsymbol{\Psi}(\boldsymbol{r},t) = \nabla_x \overline{\boldsymbol{\Psi}}(\boldsymbol{x}) \tag{5.176}$$

$$\frac{\partial \boldsymbol{\Psi}(\boldsymbol{r},t)}{\partial t} = -\frac{\partial \overline{\boldsymbol{\Psi}}(\boldsymbol{x})}{\partial \boldsymbol{x}}\cdot \boldsymbol{v} \tag{5.177}$$

The quantity $\partial \overline{\boldsymbol{\Psi}}(\boldsymbol{x})/\partial \boldsymbol{x}.\ \boldsymbol{v}$ is the spatial derivative of $\overline{\boldsymbol{\Psi}}$ along the streamline and will henceforth be denoted by $\partial \overline{\boldsymbol{\Psi}}/\partial \boldsymbol{x}_i$. In the Eulerian frame $\boldsymbol{x}$, the surfaces $\boldsymbol{v}\cdot\boldsymbol{n} > 0$ and $\boldsymbol{v}\cdot\boldsymbol{n} < 0$ are called the inlet and outlet faces, where the material enters and leaves respectively the control volume.

Application of the transformations of Eqns [5.175]–[5.177] in Eqns [5.148], [5.151], [5.152], [5.158], [5.166] and [5.167] results in the following boundary value problem:

$$\nabla_x \boldsymbol{\sigma} + \boldsymbol{b} = 0 \quad \text{in } V_x \tag{5.178}$$

$$\boldsymbol{\varepsilon} = \frac{1}{2}\left[\nabla_x \boldsymbol{u} + (\nabla_x \boldsymbol{u})^{\mathrm{T}}\right] \tag{5.179}$$

$$\boldsymbol{\varepsilon} = \boldsymbol{\varepsilon}_{\mathrm{e}} + \boldsymbol{\varepsilon}_{\mathrm{p}} + \boldsymbol{\varepsilon}_{\mathrm{th}} \tag{5.180}$$

$$\boldsymbol{\sigma} = \boldsymbol{C}(\boldsymbol{\varepsilon} - \boldsymbol{\varepsilon}_{\mathrm{p}} - \boldsymbol{\varepsilon}_{\mathrm{th}}) \tag{5.181}$$

$$\frac{\partial \varepsilon_{\mathrm{q}}}{\partial \boldsymbol{x}_i} = \frac{\gamma}{H}\left[2\mu \boldsymbol{a}\left(\frac{\partial \boldsymbol{\varepsilon}}{\partial \boldsymbol{x}_i} - \frac{\partial \boldsymbol{\varepsilon}_{\mathrm{p}}}{\partial \boldsymbol{x}_i}\right) + 2\frac{\partial \mu}{\partial \boldsymbol{x}_i}\boldsymbol{a}\cdot(\boldsymbol{\varepsilon} - \boldsymbol{\varepsilon}_{\mathrm{p}}) - \frac{\partial H}{\partial \boldsymbol{x}_i}\varepsilon_{\mathrm{q}} - \frac{\partial \sigma_{y0}}{\partial \boldsymbol{x}_i}\right] \tag{5.182}$$

$$\frac{\partial \boldsymbol{\varepsilon}_{\mathrm{p}}}{\partial \boldsymbol{x}_i} = \boldsymbol{a}\frac{\partial \varepsilon_{\mathrm{q}}}{\partial \boldsymbol{x}_i} \tag{5.183}$$

where all the fields are now functions of $\boldsymbol{x}$ only.

The boundary conditions on the inlet face of the Eulerian control volume are the initial conditions in the Lagrangian problem. Either the initial displacement or the initial traction and plastic strain (Eqn [5.184]) can be prescribed.

$$\left.\begin{aligned} \boldsymbol{\varepsilon}_{\rm p} &= 0 \\ \varepsilon_{\rm q} &= 0 \\ \boldsymbol{\sigma}\cdot\boldsymbol{n} &= 0 \end{aligned}\right\} \quad \text{on the inlet face} \tag{5.184}$$

Since the fields are steady in the Eulerian frame, the outlet face has a prescribed stress or displacement. This prescribed value is not directly known but depends on the solution field itself. The Eulerian control volume is taken to be large enough for the solution fields to be uniform at the outlet. Hence the following boundary conditions are specified at the outlet:

$$\left.\begin{aligned} \frac{\partial \boldsymbol{\varepsilon}_{\rm p}}{\partial \boldsymbol{x}_i} &= 0 \\ \frac{\partial \varepsilon_{\rm q}}{\partial \boldsymbol{x}_i} &= \boldsymbol{0} \\ \frac{\partial \boldsymbol{\sigma}}{\partial \boldsymbol{x}_i} &= 0 \end{aligned}\right\} \quad \text{on the outlet face} \tag{5.185}$$

The outlet boundary condition is implemented by prescribing nodal forces at the outlet to be same as those immediately upstream[40].

The remaining prescribed displacements and tractions in the Lagrangian frame transform to the Eulerian frame as follows:

$$\begin{aligned} \boldsymbol{u}(\boldsymbol{x}) &= \bar{\boldsymbol{u}}(\boldsymbol{r},t) \quad \text{on surface } A_x^u \\ \boldsymbol{t}(\boldsymbol{x}) &= \boldsymbol{t}(\boldsymbol{r},t) \quad \text{on surface } A_x^t \end{aligned} \tag{5.186}$$

Finite-element formulation

History dependence in stress problems in an Eulerian frame can be handled by using either Galerkin methods or streamline integration[41,42,45–47]. In the streamline approach, the finite-element formulation is similar to Eqn [5.170] where $\boldsymbol{u}$ is the primary solution field. The residual and stiffness contributions of each Gauss point are evaluated, by integrating the plastic strains and equivalent plastic strains accumulated from Eqns [5.168] and [5.169], starting at the inlet conditions in Eqn [5.184] and up to that Gauss point along the streamline on which it lies. Thus, the element residuals and stiffnesses are evaluated on an 'upstream first' basis so that the upstream stresses and strains can be used for the downstream evaluations. This has two major consequences.

1. Elaborate bookkeeping is required, and the sequencing needs to be changed for every new problem.
2. Because of this spatial dependence, the solution $\boldsymbol{u}$ for each element depends on the solution $\boldsymbol{u}$ for every other element upstream.

The consistent global stiffness matrix for such a formulation is thus a full matrix representing global field dependences. This arrangement is not practical, since for large problems it translates to extremely high computation times and storage requirements. Another issue with streamline integration is that interpolation across elements leads to an inconsistent history communication if the mesh is graded downstream in a direction transverse to the flow[47]. This forces the mesh to have higher densities downstream than necessary.

Balagangadhar *et al.*[40] and Paul *et al.*[15] developed a displacement-based mixed formulation. In a mixed formulation approach, along with the displacement degrees of freedom $\boldsymbol{u}$, the plastic strain $\boldsymbol{\varepsilon}_{\mathrm{p}}$ and equivalent plastic strain ε_{q} are included as primary solution variables for the steady-state Eulerian analysis. The evolution equations are weakly enforced using the Galerkin method. Thus, at the expense of additional solution variables, interelement spatial dependences are avoided, resulting in a banded global stiffness matrix that reduces storage requirements and computation time. Moreover, there is no need for any sequencing for evaluating residuals and stiffnesses, and the mesh can be made coarse downstream as required. Spurious oscillations resulting for the hyperbolic nature of evolution equations are suppressed by use of the streamline upwind Petrov–Galerkin (SUPG) method developed by Hughes[48]. The SUPG method is a modification of the Galerkin method in which the weighting function introduces a discontinuity in the test field.

Integration by parts and application of the divergence theorem. $\hat{\boldsymbol{u}}$, $\hat{\boldsymbol{\varepsilon}}_{\mathrm{p}}$ and $\hat{\varepsilon}_{\mathrm{q}}$ are any chosen kinematically admissible functions, and the resulting variational form is

$$\mathbf{0} = \int_{V_{\boldsymbol{x}}} [\hat{\boldsymbol{\varepsilon}}\boldsymbol{C}(\boldsymbol{\varepsilon} - \boldsymbol{\varepsilon}_{\mathrm{p}} - \boldsymbol{\varepsilon}_{\mathrm{th}}) - \hat{\boldsymbol{u}}\cdot\boldsymbol{b}]\mathrm{d}V - \int_{A_{\boldsymbol{x}}^{t}} \hat{\boldsymbol{u}}\cdot\bar{\boldsymbol{t}}\,\mathrm{d}A \qquad [5.187]$$

$$\mathbf{0} = \int_{V_x}\left(\hat{\varepsilon}_{\mathrm{q}} + \tau\frac{\partial\hat{\varepsilon}_{\mathrm{q}}}{\partial\boldsymbol{x}_i}\right)\left\{\frac{\partial\varepsilon_{\mathrm{q}}}{\partial\boldsymbol{x}_i} - \gamma\frac{1}{H}\left[2\mu\boldsymbol{a}\left(\frac{\partial\boldsymbol{\varepsilon}}{\partial\boldsymbol{x}_i} - \frac{\partial\boldsymbol{\varepsilon}_{\mathrm{p}}}{\partial\boldsymbol{x}_i}\right)\right.\right.$$
$$\left.\left. + 2\frac{\partial\mu}{\partial\boldsymbol{x}_i}\boldsymbol{a}\cdot(\boldsymbol{\varepsilon} - \boldsymbol{\varepsilon}_{\mathrm{p}}) - \frac{\partial H}{\partial\boldsymbol{x}_i}\varepsilon_{\mathrm{q}} - \frac{\partial\sigma_{y0}}{\partial\boldsymbol{x}_i}\right]\right\}\mathrm{d}V \qquad [5.188]$$

$$\mathbf{0} = \int_{V_x}\left(\hat{\boldsymbol{\varepsilon}}_p + \gamma\frac{\partial\hat{\boldsymbol{\varepsilon}}_p}{\partial\boldsymbol{x}_i}\right)\left(\frac{\partial\hat{\boldsymbol{\varepsilon}}_p}{\partial\boldsymbol{x}_i} - \boldsymbol{a}\frac{\partial\varepsilon_{\mathrm{q}}}{\partial\boldsymbol{x}_i}\right)\mathrm{d}V \qquad [5.189]$$

Owing to the Eulerian formulation, active yielding occurs when $f \geq 0$ and $\dot{f} \geq 0$. Thus γ in Eqn [5.188] is evaluated as follows:

$$\gamma = \begin{cases} 1 & \text{if } f \geqslant 0 \text{ and } \dfrac{\partial f}{\partial \boldsymbol{x}_i} \geqslant 0 \\ 0 & \text{otherwise} \end{cases} \qquad [5.190]$$

Finite-element discretization of Eqns [5.187]–[5.189] results in the formulation of the global residual which is solved iteratively by using the Newton–Raphson method. Details of the formulations of the element residual and tangent matrix have been presented previously[15].

5.4 Modeling material properties

The modeling of the material behavior from room temperature up to melting temperature poses a challenge. This becomes even more difficult in the presence of phase transformations. Thus the aspects of material modeling that must be handled in an appropriate way in order to obtain an accurate solution are as follows:

1. Modeling elastic and plastic deformation in combination with a cut-off temperature.
2. Accounting for the effect of solid–solid, solid–liquid transformations or other microstructural changes in the material behavior.

All models used are based on assuming thermoelastic material behavior. Young's modulus E and the thermal dilatation (expansion) ε_{th} are the most important parameters. Poisson's ratio ν has a smaller influence[49] on the residual stresses and deformations. A general discussion of material modeling has been given by Lindgren[49] and Oddy and Lindgren[50].

The rate-independent deviatoric plasticity model with the von Mises yield condition and the associated flow rule has been used with success in many welding simulations[49]. Some work has also used viscoplastic models[2,51,52] or combined rate-independent plasticity at lower temperatures with viscoplastic models at higher temperatures[53]. The hardening behavior at lower temperatures is important for the residual stresses. The material near the weld experiences cyclic loading and choosing isotropic or kinematic hardening will affect the stresses in this region[54].

For temperatures that exceed the melting point the modulus of elasticity and yield limit have zero values. Viscous flow models are more appropriate at these temperatures. However, elastoplastic or elastoviscoplastic models may still be used in conjunction with the use of a cut-off temperature. This technique is used for two reasons, the properties at high temperatures are usually unknown and a too soft weld metal in the model may give numerical problems. In this technique, during the mechanical analysis, temperatures that exceed a cut-off value are reset to the cut-off temperature. An appropriate choice of cut-off temperature will not affect the residual stresses[3,11,49].

The very low value of yield strength along with the use of elastoplastic models may result into artificial hardening and lead to higher than the actual residual stresses. To account for the annealing effect that naturally occurs in metals, either a creep model may be used or all accumulated plastic strains may set to zero above a critical temperature[8]. Solid-phase transformations are important to include in the model in order to obtain accurate deformations but not for residual stresses according to Leblond[55].

Solid-phase transformations that occur during the thermal cycle produced by welding lead to irreversible plastic deformation known as transformation plasticity. This phenomenon is driven by the volume change during solid-state phase transformations[5,56–59]. Oddy et al.[10] decomposed the total strain into its elastic, plastic, thermal, volumetric change and transformation plasticity (a term proportional to the deviatoric stress tensor) components. They reported that transformation plasticity significantly affects longitudinal and transverse residual stress distributions. Early work on weld modeling modifying the thermal expansion coefficient to take into account the volumetric change during phase transformations has been described by Argyris *et al.*[2] and Papazoglou and Masubuchi[3]. Argyris *et al.*[2] also modified the yield strength during cooling to reflect the austenite, ferrite and pearlite formation. In their numerical simulations, Argyris *et al.* concluded that the individual stress components are affected by the phase transformations during the heat cycle; however, the effects of phase transformations are dampened upon cooling of the weld. They also concluded that the equivalent von Mises stress is not affected by the phase transformations.

5.5 References

1. Hibbitt, H., and Marcal, P.V., 'A numerical, thermo-mechanical model for the welding and subsequent loading of a fabricated structure', *Comput. Structs*, **3**, 1145–1174, 1973.
2. Argyris, J.H., Szimmat, J., and Willam K.J., 'Computational aspects of welding stress analysis', *Comput. Meth. Appl. Mech. Engng*, **33**, 635–666, 1982.
3. Papazoglou, V.J., and Masubuchi, K., 'Numerical analysis of thermal stresses during welding including phase transformation effects', *J. Pressure Vessel Technol.*, **104**, 198–203, 1982.
4. Goldak, J., Chakravarti, A., and Bibby, M., 'A new finite element model for welding heat sources', *Metall. Trans. B*, **15B**, 299–305, 1984.
5. Leblond, J.B., and Devaux, J., 'A new kinetic model for anisothermal metallurgical transformations in steels including effect of austenite grain size', *Acta Metall.*, **32**(1), 137–146, 1984.
6. Watt, D.F., Coon, L., Bibby, M.J., Goldak, J., and Henwwod, C., 'An algorithm for modeling microstructural development in weld heat-affected zones', *Acta Metall.*, **36**(11), 3029–3035, 1988.
7. Henwood, C., Bibby, M., Goldak, J., and Watt, D., 'Coupled transient heat transfer–microstructure weld computations, Part B', *Acta Metall.*, **36**(11), 3037–3046, 1988.
8. Michaleris, P., Feng, Z., and Campbell, G., 'Evaluation of 2D and 3D FEA models for predicting residual stress and distortion', *Proc. Pressure Vessel and Piping Conf.,* American Society of Mechanical Engineers, New York, 1997, pp. 91–102.

9. Michaleris, P., and DeBiccari, A., 'Prediction of welding distortion', *Weld. J.*, **76**(4), 172–180, 1997.
10. Oddy, A.S., Goldak, J.A., and McDill, J.M.J., 'Numerical analysis of transformation plasticity in 3D finite element analysis of welds', *Eur. J. Mech., A/Solids*, **9**(3), 253–263, 1990.
11. Tekriwal, P., and Mazumder, J., 'Transient and residual thermal strain–stress analysis of GMAW', *J. of Engng. Mater. Technol.*, **113**, 336–343, 1991.
12. Brown, S.B., and Song, H, 'Implications of three-dimensional numerical simulations of welding of large structures', *Weld. J.*, **71**(2), 55-s–62-s, 1992.
13. Michaleris, P, Tortorelli, D.A., and Vidal, C.A., 'Analysis and optimization of weakly coupled thermo-elasto-plastic systems with applications to weldment design', *Int. J. Numer. Meth. Engng*, **38**(8), 1259–1285, 1995.
14. Michaleris, P., Dantzig, J.A., and Tortorelli, D.A., 'Minimization of welding residual stress and distortion in large structures', *Weld. J.*, **78**(11), 361-s–366-s, 1999.
15. Paul, S., Michaleris, P., and Shanghvi, J., 'Optimization of thermo-elasto-plastic finite element analysis using an Eulerian formulation', *Int. J. Numer. Meth. Engng*, **56**, 1125–1150, 2003.
16. Shanghvi, J., and Dydo, J., 'A transient thermal tensioning process for mitigating distortion in stiffened structures', *Proc. 39th Annual Technical Meet. of Society of Engineering Science*, State College, Pennsylvania, USA, 2002, Society of Engineering Science, State College, Pennsylvania, 2002.
17. ESI Group, 'SYSWELD: engineering simulation solution for heat treatment, welding, and welding assembly', ESI Group, Paris, http://www.esi-group.com/Products/Welding, 2003.
18. Bathe, K.J., *Finite Element Procedures in Engineering Analysis*, Prentice-Hall, Englewood Cliffs, New Jersey, 1982.
19. Crisfield., M.A., *Non-linear Finite Element Analysis of Solids and Structures*, John Wiley, New York, 1991.
20. Belytscho, T., Liu, W.K., and Moran, B., *Nonlinear Finite Elements for Continua and Structures*, John Wiley, New York, 2000.
21. Owen, D.R.J., and Hinton, E., *Finite Elements in Plasticity Theory and Practice*, Pineridge Press, Swansea, 1980.
22. Zienkiewicz, O.C., and Taylor, R.L., *The Finite Element Method, Vol. 2, Solid and Fluid Mechanics, Dynamics and Non-linearity*, 4th edition, McGraw-Hill, New York, 1981.
23. Lubliner, J., 'Normality rules in large deformation plasticity', *Mech. Mater.*, **5**, 29–34, 1986.
24. Simo, J.C., 'A framework for finite strain elastoplasticity based on maximum plastic dissipation and the multiplicative decomposition', *Comput. Meth. Appl. Mech. Engng*, **66**, 199–219, 1988.
25. Simo., J.C., 'A framework for finite strain elastoplasticity based on maximum plastic dissipation and the multiplicative decomposition, Part ii: computational aspects', *Comput. Meth. Appl. Mech. Engng*, **68**, 1–31, 1988.
26. Simo, J.C., and Taylor, R.L., 'Consistent tangent operators for rate-independent elasto-plasticity', *Comput. Meth. Appl. Mech. Engng*, **48**, 101–118, 1985.
27. Goldak, J., and Bibby, M., 'Computational thermal analysis of welds: current status and future directions', *Modeling of Casting and Welding Processes IV*, (eds A.F. Giamei and G.J. Abbaschian) Palm Coast, Florida, USA, 1988, The Minerals & Materials Society, Warrendale, Pennsylvania, pp. 153–166.

28. Watt, D.F., Coon, L., Bibby, M., and Henwood, C., 'Coupled transient heat transfer-microstructure weld computations (Part b)', *Acta Metall.*, **36**(11), 3037–3046, 1988.
29. Tekriwal, P., and Mazumder, J., 'Finite element analysis of three-dimensional transient heat transfer in GMA welding', *Weld. J. Res. Suppl.*, **67**, 150-s–156-s, 1988.
30. Braudel, H.J., Abouaf, M., and Chenot, J.L., 'An implicit and incremental formulation for the solution of elastoplastic problems by the finite element method', *Comput. Structs*, **22**(5), 801–814, 1986.
31. McDill, J.M.J., Oddy, A.S., and Goldak, J.A., 'Consistent strain fields in 3D finite element analysis of welds', *Trans. ASME, J. Pressure Vessel Technol.*, **112**(3), 309–311, 1990.
32. Oddy, A.S., Goldak, J.A., and McDill, J.M.J., 'Numerical analysis of transformation plasticity in 3D finite element analysis of welds', *Eur. J. Mech., A/solids*, **9**, 1–11, 1990.
33. Rybicki, E.F., and Stonesifer, R.B., 'Computation of residual stresses due to multipass welds in piping systems', *J. Pressure Vessel Technol.*, **101**, 149–154, 1979.
34. Shim, Y., Feng, Z., Lee, S., Kim, D.S., Jaeger, J., Paparitan, J.C., and Tsai, C.L., 'Determination of residual stress in thick-section weldments', *Weld. J.*, **71**, 305-s–312-s, 1992.
35. Bertram, L.A., and Ortega, A.R, 'Automated thermomechanical modeling of welds using interface elements for 3D metal deposition', *Proc. ABAQUS User's Conf.*, Oxford, 1991, Hibbit, Karlsson and Sorensen, Pawtuchet, Rhode Island, 1991.
36. Nguyen, Q.S., and Rahimian, M., 'Mouvement permanent d'une fissure en milieu elastoplastique', *J. Mec. Appl.*, **5**, 95–120, 1981.
37. Rajadhyaksha, S., and Michaleris, P., 'Optimization of thermal processes using an Eulerian formulation and application in laser hardening', *Int. J. Numer. Meth. Engng*, **47**, 1807–1823, 2000.
38. Goldak J., Gu M., and Hughes, E., 'Steady state thermal analysis of welds with filler metal addition', *Can. Metall. Q.*, **32**(1), 49–55, 1993.
39. Ohmura, E., Takamachi, Y., and Inoue, K., 'Theoretical analysis of laser transformation hardening process of hypoeutectoid steel based on kinetics', *Trans. Japan Soc. of Mech. Engng (Ser. A)*, **56**, 1496–1503, 1990.
40. Balagangadhar, D., Dorai, G.A., and Tortorelli, D.A., 'A displacement-based Eulerian steady-state formulation suitable for thermo-elasto-plastic material models', *Int. J. Solids Structs*, **36**(16), 2397–2416, 1999.
41. Dawson, P.R., 'Viscoplastic finite element analysis of steady-state processes including strain history and stress flux dependence', *Applics Num. Meth. Forming Processes*, **28**, 55–67, 1978.
42. Caswell, B., and Viriyauthakorn, M., 'Finite element simulation of viscoelastic flow.', *J. Non-Newtonian Fluid Mech.*, **6**, 245–267, 1980.
43. Onate, E., Zienkiewicz, O.C., and Jain, P.C., 'Flow of solids during forming and extrusion: some aspects of numerical solution', *Int. J. Solids Structs*, **14**, 15–38, 1978.
44. Lubliner, J., *Plasticity Theory*, 1st edition, Macmillan, London, 1990.
45. Zienkiewicz, O.C., Thompson, E.G., and Pittman, J.F.T., 'Some integration techniques for the analysis of viscoelastic flows', *Int. J. Numer. Meth. Fluids*, **3**, 165–177, 1983.

46. Pont, D., Bergheau J.M., and Leblond, J.B., 'Three-dimensional simulation of a laser surface treatment through steady state computation in the heat source's comoving frame', *Proc. IUTAM Symp. on Mechanical Effects of Welding*, Springer, Berlin, 1992, pp. 85–92.
47. Gu, M., and Goldak, J.A., 'Steady-state formulation for stress and distortion of welds', *Trans. ASME, J. Engng Industry*, **116**, 467–474, 1994.
48. Hughes, T.J.R., 'A theoretical framework for Petrov–Galerkin methods with discontinuous weighting functions: application to the streamline upwind procedure', *Finite Element in Fluids*, Vol. 4 (eds R.H. Gallagher, D.H. Norrie, J.T. Oden and O.L. Zienkiewicz), John Wiley, Chichester, West Sussex, 1982.
49. Lindgren, L.-E., 'Finite element modelling and simulation of welding, Part 2, improved material modelling', *J. Thermal Stresses*, **24**, 195–231, 2001.
50. Oddy, A. S., and Lindgren, L.-E., 'Mechanical modeling and residual stresses', *Modeling in Welding, Hot Powder Forming, and Casting*, (ed. L. Karlsson) ASM International, Materials Park, Ohio, 1997, pp. 31–59.
51. Myhr, O., 'Modeling of microstructure evolution and residual stresses in processing and welding of 6082 and 7108 aluminium alloys', *Proc. 5th Int Conf. on Trends in Welding Research* (eds H.D. Brody and D. Apelian), Pine Mountain, Georgia, USA, ASM International, Materials Park, Ohio, 1998.
52. Wang, Z., and Inoue, T., 'Viscoplastic constitutive relation incorporating phase transformation – application to welding', *Mater. Sci. Technol.*, **1**, 899–903, 1985.
53. Goldak. J, 'Thermal stress analysis in solids near the liquid region in welds', *Proc. 3rd Int. Cont. on Mathematical Modelling of Weld Phenomena*, Graz, Austria, 1997.
54. Bammann, D., and Ortega, A., 'The influence of the Bauschinger effect and yield definition on the modeling of welding processes', *Proc. 6th Int. Conf. on Modeling of Casting, Welding and Advance Solidification Processes* (eds. T.S. Piwonka, V. Voller and L. Katgerman), Palm Coast, Florida, USA, 1993, The Minerals & Materials Society, Warrendale, Pennsylvania, 1993, pp. 543–551.
55. Leblond, J., 'Metallurgical and mechanical consequences of phase transformations in numerical simulations of welding processes', *Modeling in Welding, Hot Powder Forming, and Casting* (ed. L. Karlsson), 1997, pp. 61–89.
56. Leblond, J.B., 'Mathematical modelling of transformation plasticity in steels II: Coupling with strain hardening phenomena', *Int. J. Plasticity*, **5**, 573–591, 1989.
57. Leblond, J.B., Mottet, G., and Devaux, J.C., 'A theoretical and numerical approach to the plastic behavior of steels during phase transformations – I. Derivation of general relations', *J. Mech. Phys. Solids*, **34**(4), 395–409, 1985.
58. Leblond, J.B., Mottet, G., and Devaux, J.C., 'A theoretical and numerical approach to the plastic behaviour of steels during phase transformations – 2. Study of classical plasticity for ideal-plastic phases', *J. Mech. Phys. Solids*, **34**(4), 411–432, 1986.
59. Magee, G.L., and Davies, R.G., 'On the volume expansion accompanying the f.c.c. to b.c.c. transformation in ferrous alloys', *Acta Metall.*, **20**, 1031–1043, 1972.

Part II

Applications

6 Measuring welding-induced distortion

Y. J. CHAO, University of South Carolina, USA

6.1 Introduction

In welding, plastic thermal strain develops near the weld region because of heating, melting and cooling of the weldment and the nearby base metals. The plastic strain results in permanent deformation of the welded structure after the welding. This is called welding-induced distortion. The mode of welding-induced distortion could show up as in-plane deformation such as stretching and out-of-plane deformation such as bending, rotation or buckling. Typical welding-induced distortion[1] is shown in Fig. 6.1.

In contrast with the welding-induced residual stress that is mostly localized in the neighborhood of the weldment, welding-induced distortion is to the entire welded structure. Therefore, in assembling welded components into another structure, tremendous fit-up problems could be present because of the dimensional change of the substructures due to welding. A typical example is in shipbuilding where large welded panels are to be assembled step by step into the final shape of a ship using welding. Welding-induced buckling distortion is often the most challenging problem in shipbuilding because of the small thickness-to-in-plane dimensions of the panel. Various industrial and military codes and specifications that cover welding fabrication provide standards for allowable distortions, e.g. NAVYSHIPS 0900-0060-4010 and ASME Pressure Vessel and Piping Code Section III. However, even when the code requirements are met, welding-induced distortion could affect the performance of the welded structure in service as well. For example, welding-induced distortion in a pressure vessel head can serve as the geometrical imperfection for potential buckling failure of the vessel head.

Measurement of welding-induced distortion is normally performed to provide the dimensions of the welded structure during or after welding and in some cases to provide the actual data for examining the accuracy of the mathematical, numerical or empirical predictions. In this chapter, we briefly review the techniques used in distortion measurement (Section 6.2)

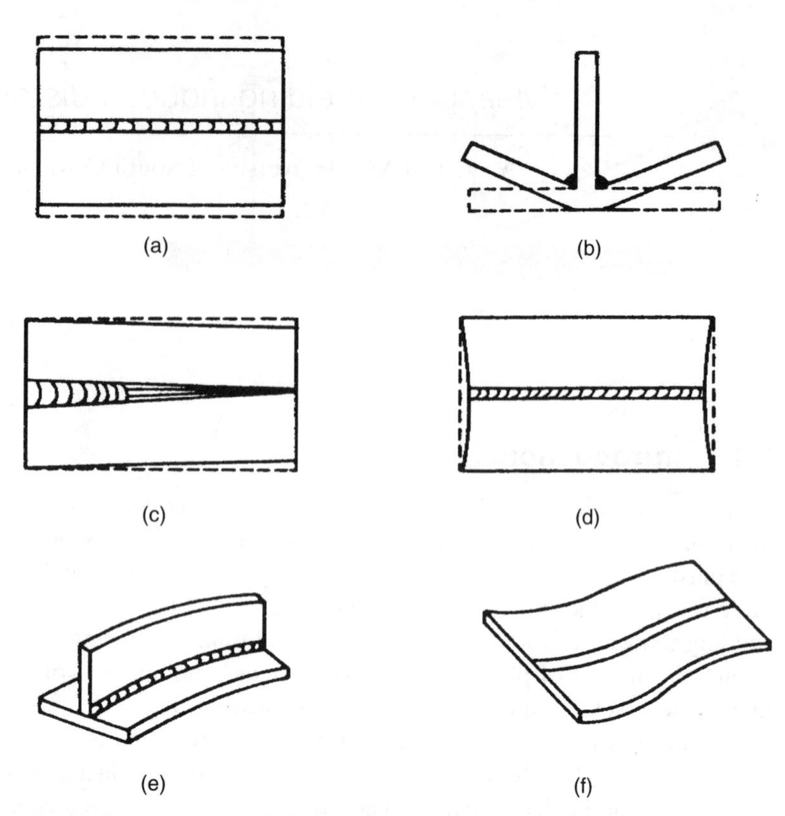

6.1 Various types of welding-induced distortion[1]: (a) transverse shrinkage; (b) angular change; (c) rotational distortion; (d) longitudinal shrinkage; (e) longitudinal bending distortion; (f) buckling distortion.

followed by examples in small-scale local distortion measurement (Section 6.3) and large-scale global distortion measurement (Section 6.4). Special consideration in interpretation of welding-induced distortion is discussed in Section 6.5.

6.2 Techniques in distortion measurement

In principle, any dimensional measuring instrument can be used for measuring the welding induced distortion in the post-weld cooled state. For example, as shown in Fig. 6.2(a), the angular distortion of a T-joint can be measured easily by a ruler as the welded plate is situated on a flat reference plane. In Fig. 6.2(b) a mechanical dial gage is used to measure the bending distortion continuously as it moves across the welded beam. A ruler can also be used to measure the continuous angular shrinkage of a panel welded with stiffeners as shown in Fig. 6.2(c).

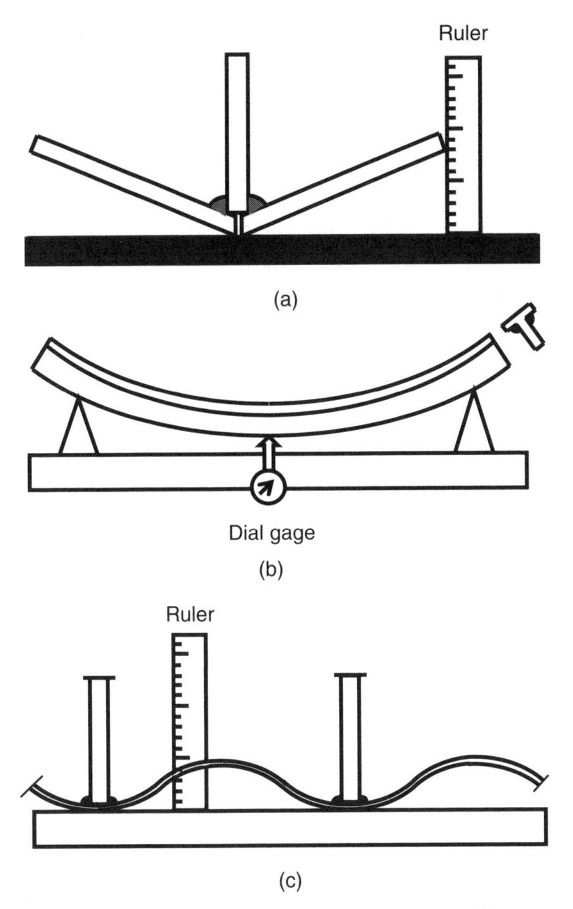

6.2 Distortion measurement in the post-weld cooled state: (a) angular distortion; (b) bending distortion; (c) angular shrinkage of panels welded with stiffeners.

Note that this type of measurement determines the shape of the structure after welding relative to a reference plane. The deformation of the structure due to welding is the difference between the initial shape (before welding) and the final shape (after welding). The initial shape of the structure may or may not be flat. Therefore, the choice of the reference plane needs careful consideration before the measurement is performed.

The instrumentation used in measuring the dimensions or shape of a structure in general can be classified as contact and noncontact, point and full field, or stationary and transient. The contact instrument is normally mechanical such as a dial gage, coordinate-measuring machine or linear variable-differential transformer (LVDT). The noncontact method is typically optical such as laser displacement sensors and digital image correlation (DIC). Each dial gage, LVDT or laser displacement sensor measures the distortion at a point. On the other hand, the DIC is a full-field method, which adopts the principle of stereovision or photogrammetry and could conceptually determine the deformation or shape of an entire structure with one system. The transient distortion during welding can also be determined by the LVDT or DIC as long as the data and images can be acquired, buffered and stored with sufficient speed.

Since the welding process involves high temperature, any contact-measuring instrument if placed near the weldment for transient distortion measurement must be able to tolerate the high temperature in the welding process. In addition, shielding of the light from the welding arc may be used if an optical system is used for the transient deformation measurement.

6.3 Small-scale local distortion measurement

For localized measurement, an LVDT or laser displacement sensor is rather convenient. An LVDT is an excellent device for converting the mechanical displacement into an electrical signal[2]. The sensitivity of a commercial LVDT varies from 0.003 to 0.25 V/mm of displacement per volt of excitation. The excitation could be either alternating current or direct current with 3 V excitation typical. The ranges of displacement that can be measured by an LVDT vary from 0.125 mm to 0.64 m, which covers most of the distortion encountered in welding. The error of the LVDT is typically about 0.5% of the maximum linear output. The frequency response from the LVDT is not high because it is a mechanical device. However, it far exceeds the frequency in transient welding distortion measurement since welding deformation is nearly quasistatic relative to other dynamic events such as impact.

An example of using the LVDT for transient welding distortion measurement is presented here. As shown in Fig. 6.3, the well-designed experiment for demonstrating the development of transient and permanent distortion, residual stress and transient temperature history during welding was performed by Masubuchi[3]. In the experiment, a rectangular plate made of aluminum alloy 5052-H32 was used. Gas–metal arc welding was applied with a moving torch applied at the top longitudinal edge of the plate of 1220 mm (48 in) long, 152 cm (6 in) wide and 12.5 mm (0.5 in) thick. The plate is simply supported at both ends. Strain gages and thermocouples were mounted on the plate as shown in Fig. 6.3(a) to record the transient strain and temperature. An LVDT was used to measure the transient deflection of the beam at the lower midpoint of the plate. The traveling speed of the welding arc is 7.34 mm/s (0.289 in/s) which yields a total welding time of 166 s. The welding current and arc voltage are 260 A and 23 V respectively. Such a geometry and welding arrangement was chosen to produce a

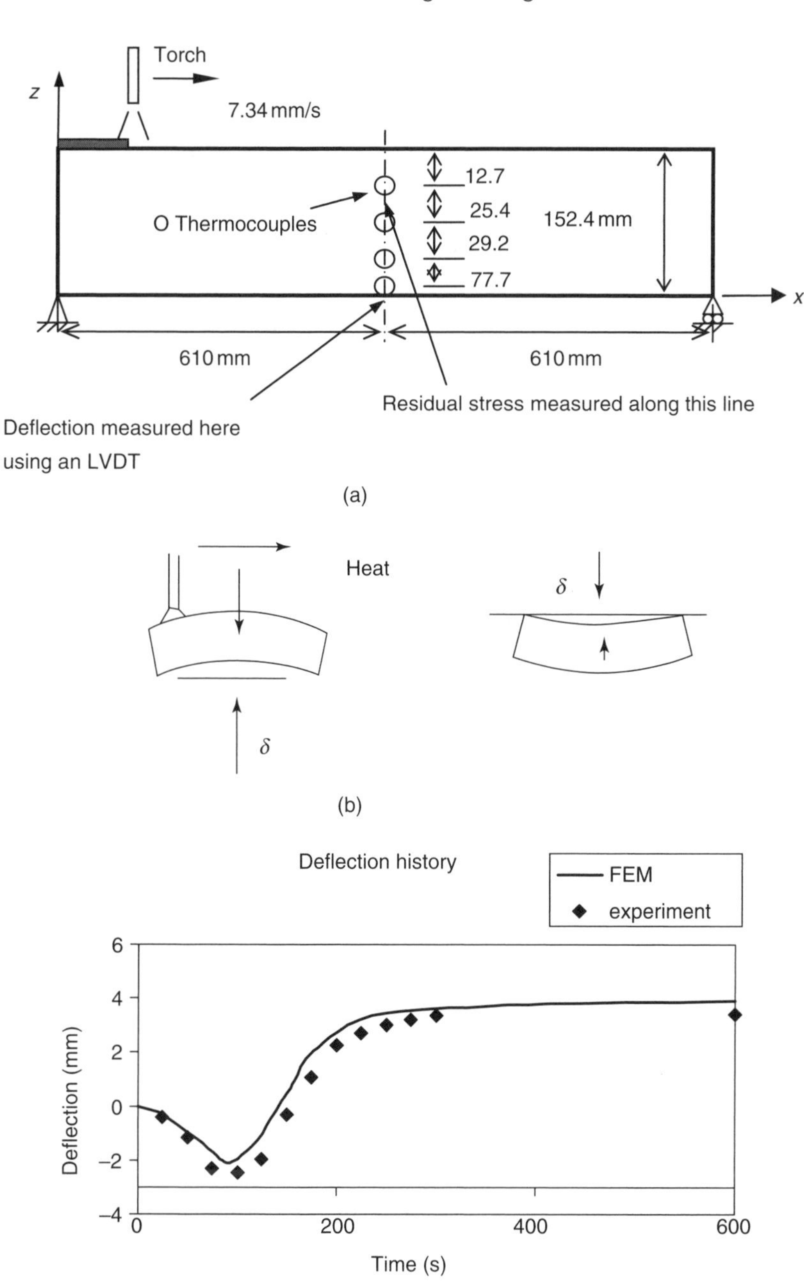

6.3 Transient welding-induced deformation[3,4]: (a) workpiece dimensions and the LVDT at the low center point; (b) anticipated displacement during heating and cooling-down phases; (c) measured results from the LVDT as well as the finite element analysis.

complete set of experimental data that include transient temperature, transient strain, distortion and residual stress at various locations in the plate. As such, it provides excellent information to validate results from any computer simulation. More detailed technical information for the welding can be found in the book by Masubuchi[3].

An interesting feature of the welding design shown in Fig. 6.3 is the trend of the transient distortion of the plate. As heat is applied to the top edge during welding, the upper part of the plate expands due to thermal gradients relative to the lower part of the plate. Consequently, the plate is bent upward as shown in Fig. 6.3(b). During the cool-down phase after welding is completed, the weldment shrinks. The longitudinal shrinkage of the top edge makes the plate bent downward. The transient deflection at the lower center point as measured by the LVDT, therefore, first experiences upward and then downward displacement and finally leaves a downward displacement as the permanent (plastic) distortion at the post-weld cooled state, shown in Fig. 6.3(c). In Fig. 6.3(c), results from numerical simulation by Chao *et al.*[4] are also included for comparison. The numerical simulation predicts the transient and post-welding distortion well.

6.4 Large-scale global distortion measurement

In this section, a noncontact technique appropriate for large-scale global distortion measurement in either transient or post-welding condition is discussed. Specifically, the principle of the DIC method developed at the University of South Carolina is reviewed. Examples are then given for a bead-on-plate gas–metal arc welding (GMAW) test and a T-joint welding. Distortion results are presented and compared with data obtained from an LVDT.

The concept of DIC was first introduced by Peters and Ranson[5] and Chu *et al.*[6] in the early 1980s for two-dimensional (2D) in-plane deformation measurement. After a decade-long development, the 2D technique was eventually extended to the three-dimensional (3D) technique by Luo *et al.*[7,8] in the early 1990s. A comprehensive review of the principles, milestone development and applications of the DIC in both 2D and 3D conditions can be found in the work by Sutton *et al.*[9].

The basis of two-dimensional digital image correlation (DIC-2D) for the measurement of surface deformation is the matching of one point from an image of an object's surface before loading (the undeformed image) to a point in an image of the object's surface taken at a later time or loading (the deformed image). Assuming a one-to-one correspondence between the deformations in the image recorded by the camera and the deformations of the surface of the object, an accurate point-to-point mapping from the undeformed image to the deformed image will allow the displacement of the object's surface to be measured. Three main requirements must be met for the successful use of the DIC-2D. First, in order to provide features for the numerical matching process, the surface of the object must have a pattern that produces varying intensities of diffusely reflected light from its surface. This pattern may be applied to the object or it may occur naturally. Secondly, the imaging camera must be positioned so that its sensor plane is parallel to the surface of the planar object. Thirdly, the displacement, including both the deformation and the rigid-body motion, between the two images must be in plane or perpendicular to the optical axis of the imaging system. The third requirement apparently presents a severe limitation when applied to measuring welding distortion of panels.

Three-dimensional digital image correlation (DIC-3D) uses two video cameras to obtain 2D images of a specimen surface from two different perspectives, as shown in Fig. 6.4. In principle, the stereovision model is similar to human depth perception. By comparing the images mathematically using a correlation algorithm the *profile* of the specimen surface can be determined. The surface profiles obtained before and after the surface has been deformed or loaded are then compared with the results in 3D deformation of the 3D object. Note that this operation requires accurate information about the placement and the operating characteristics of the cameras being used. To obtain this information, a camera calibration process must be performed prior to the test. In addition, in the case of dynamic deformation measurements, the two cameras have to be synchronized. The synchronization can be achieved by controlling the cameras from a single computer with two image acquisition boards. Data and images can be obtained at the

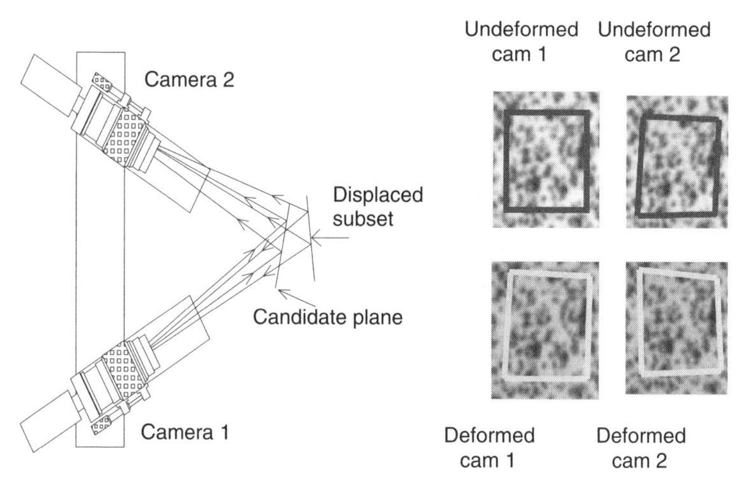

6.4 3D deformation measurement using a stereoscopic video imaging system.

framing rate of the video cameras; thus dynamic information can be acquired during a thermomechanical event such as the GMAW process. In the following, two examples are provided to illustrate the application of DIC-3D to the measurement of transient welding-induced distortion[10–12].

6.4.1 Experimental setup

The imaging system consisted of two digital Pulnix cameras (Fig. 6.5). The Pulnix cameras were connected to a Bitflow Raptor board that permitted simultaneous acquisition of 768 pixel × 480 pixel × 8 bit images at rates of up to 16 frames/s.

In order to generate speckle patterns on the specimen surface for the image correlation calculation, a ceramic paint manufactured by XYZ, Inc. was used. Ceramic paints allow accurate measurements of deformations using DIC at temperatures of up to 650 °C. The paint used in this investigation consisted of titanium dioxide particles suspended in water. When dry, the particles would rigidly move with the specimen surface, causing the pattern to distort with deformations equivalent to the specimen surface. To

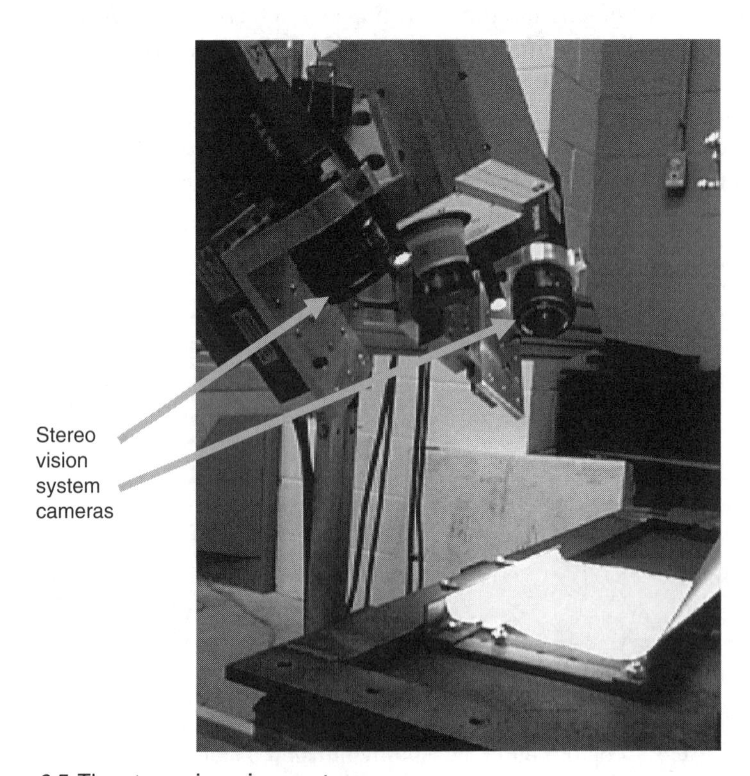

6.5 The stereo imaging system.

maintain cohesion between the paint and the specimen surface, the surface had to be roughened with 600 grit sandpaper and then degreased.

The first application shown was a bead-on-plate test. A 150mm square ASTM A36 steel plate with thicknesses of 3.18mm was welded in a bead-on-plate configuration (Fig. 6.6(a) and Fig. 6.6(b)). A Miller GMAW machine was used with E70s-6 welding wire and Ar+2% O_2 shielding gas to weld the specimens at torch travel rates of 11 mm/s and wire feed rates of 80.4 mm/s. Specimens were placed on top of four ceramic balls located at each corner of the specimen to minimize heat conduction from the plate. Also, two 6.35 mm × 20 mm screws contacted each side of the specimens in order to conduct current and to keep the specimen in place.

For this experiment, images were obtained on the same side as the welding process. To prevent the light generated by the welding process from oversaturating the stereoscopic video imaging system and to protect the system from expelled weld material, a special shield was constructed (Fig. 6.7). To obtain the maximum field of view, the shielding was located 15 mm from the center of the weld line. A piece of heat-resistant cloth was also placed between the shield and the specimen to seal the gap, and the shield was loosely attached to a pair of slots in order to permit out-of-plane motion.

The second application was welding a 500 mm × 250 mm × 12.7 mm T-section to a 500 mm × 500 mm × 12.7 mm baseplate. Figure 6.8 shows the experimental setup. The T-section was initially tag welded to the baseplate at both edges and the center of the plate on either side and then mounted to a frame by four screws. The frame also supported the travel rail of the welding machine. The cameras are mounted on computer-controlled translation stages and the stages are mounted to a rigid aluminum bar. In the configuration shown, measurements were made on the bottom of the steel plate, while the T-section was welded to the top of the plate. For this application, no special shielding was necessary, as the arc could not be seen by the cameras. A side view of a specimen sitting on the frame and showing the angular distortion generated by the welding process is shown in Fig. 6.9.

6.4.2 Results

Weld bead on plate

Measurements of the u, v and w displacement fields (corresponding to displacement in x, y and z directions) obtained from the surface of the plate 40 s after the start of the GMAW process are presented in Fig. 6.10(a), Fig. 6.10(b) and Fig. 6.10(c) respectively. The displacements were measured with an accuracy of approximately 40 μm over a 100 mm × 50 mm area. The x axis corresponds approximately to the location of the weldment, with welding

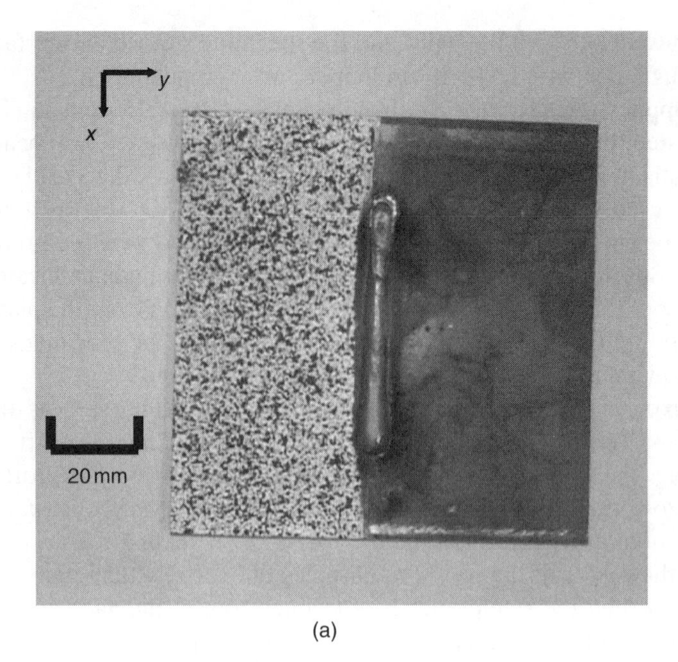

(a)

Welding direction

150 mm

10 mm

75 mm

y

x

150 mm

(b)

6.6 Bead-on-plate welded ASTM A36 steel plate.

6.7 Shielding for the imaging system used during the GMAW process.

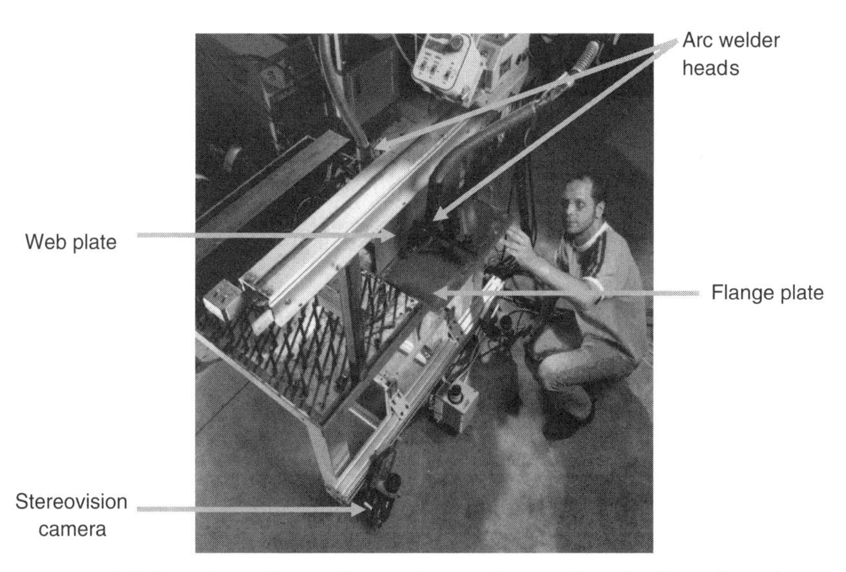

6.8 Overview of experimental setup for welding the T-section plate.

6.9 Side view of the T-section welded plate.

along the positive x direction. Over most of the plate, the in-plane displacements u and v are fairly negligible.

The w displacement field corresponds to the bending-induced out-of-plane distortions that are more pronounced and need to be mitigated in the GMAW process. These distortions exhibit significant curvature in both the x and the y directions. Therefore, it may be reasonable to assume that the distortions are due to bending moments occurring about both the x and the y axes. Figure 6.10(c) shows the shape of the plate at $t = 40$ s, a moment after cooling down. The data provided by the DIC-3D system was smoothed and resampled to a regular grid in the x–y plane. One can clearly see the saddle shape of the plate surface. Also shown in the plot is an anticlastic surface fit with a curvature of R^{-1} about the y axis and νR^{-1} about the x axis, where R is the principal radius of curvature and ν is Poisson's ratio. For this fit, ν was set to 0.3 and only the radius R was used as a variable for the regression. The maximum deviation of the fit from the data was $\pm 8\,\mu$m for the smoothed data shown and $\pm 40\,\mu$m for the raw data. This result indicates that the bending of the plate is mainly due to a one-dimensional bending moment around the y axis, which is generated by thermal residual stresses or thermal strain and shrinkage accumulated during the cooling of the weldment.

Welding of the T-section

For this experiment, the computer vision system was set up to view an area of approximately 250 mm × 200 mm on one side of the T-section, and LVDT

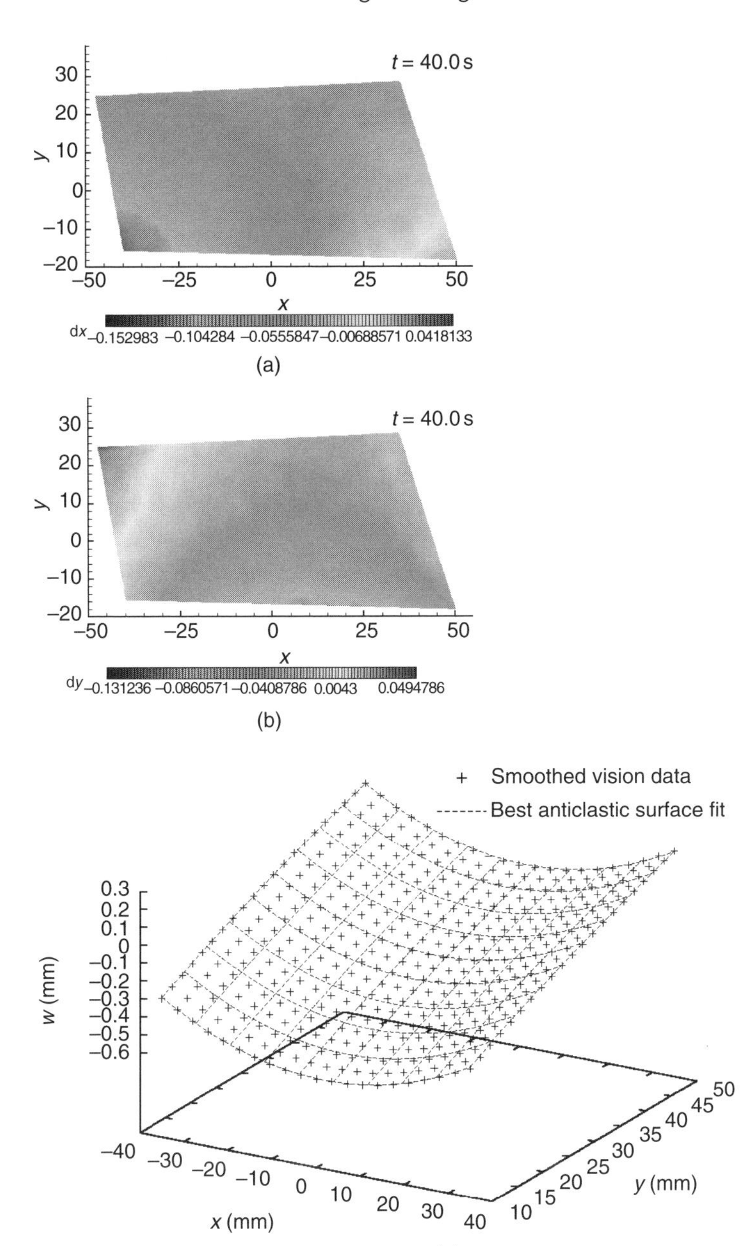

6.10 (a) The *u* displacement field, (b) the *v* displacement field and (c) the *w* displacement field measured 40 s after the start of the GMAW process. (c) includes a fit of an anticlastic surface to the *w* displacement field for the welded specimen using Poisson's ratio equal to 0.3.

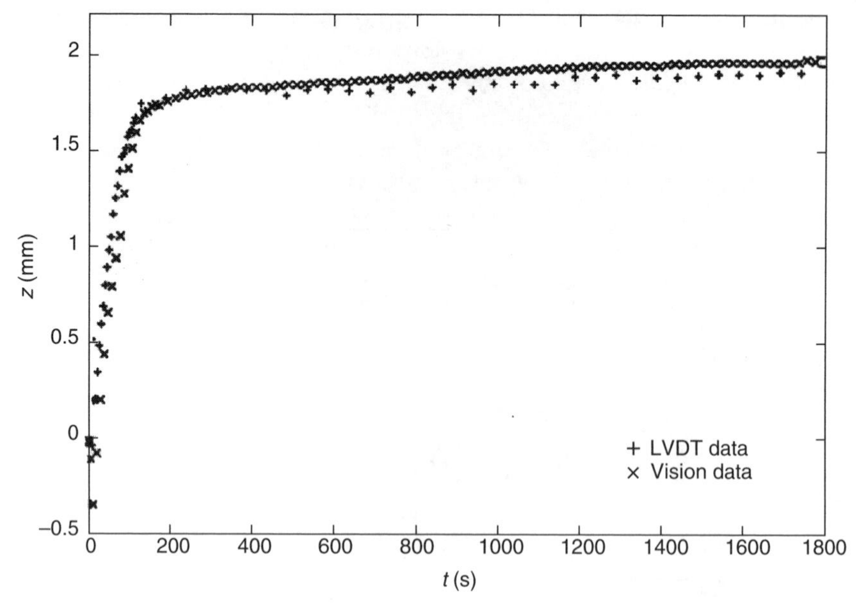

6.11 Comparison between LVDT and DIC *w* displacement data at a point.

sensors were placed on the opposite side. Figure 6.11 shows a comparison between LVDT data and vision data for two opposing points at the same distance from the centerline of the plate. The data from this particular experiment match very well, although a deviation on the order of ±0.2 mm was noted in other experiments. As profile measurements of the baseplate after welding show, the plate does not deform symmetrically, which explains the deviations between data taken on opposite sides of the welds. It is also noted that measurements between experiments with the same nominal process parameters vary on the same order in the *w* displacement component. This could be attributed to the slight nonsymmetry in the T-section as well as the imperfections of the panels before welding.

6.5 Special considerations

It should be emphasized that the experimental results from any measurement of welding distortion are based on a reference coordinate system or a reference plane. The measured data from a welded structure may include both rigid-body motion and the actual deformation of the structure due to welding. On the other hand, the displacement fields from any numerical simulation are often based on specific boundary conditions assigned in the model to eliminate the rigid-body motion. Therefore, special care must be taken when interpreting the welding-induced distortion or comparing the measured data with results from numerical modeling. This can be done by applying appropriate boundary conditions (e.g. clamping) in the tests and/or providing a reference plane for easy interpretation of the distortion. In addition, in many applications, the detailed shape of the initial configuration before welding is also important and warrants measurement effort. For example, maximum variation in flatness on the order of 1.2 mm for a panel of 1220 mm × 101 mm × 4.76 mm (length × width × thickness) before welding could result in buckling distortion when the panel is welded to form a T-section[13].

6.6 Acknowledgments

The author wishes to thank Dr Michael A. Sutton, Dr Steve R. McNeil, Dr Wally H. Peters and Dr William F. Ranson at the University of South Carolina for their insightful discussion on the DIC method. In addition, the support from the National Science Foundation to the author through Grant CMS0116238 and Program Director Dr Kenneth Chong during the course of this writing is deeply appreciated.

6.7 References

1. *Welding Handbook*, Vol. I, 8th edition, American Welding Society, Miami, Florida, 1987.
2. Dally, J.W., and Riley, W.F., *Experimental Stress Analysis*, 3rd edition, College House Enterprises, LLC, Knoxville, Tennessee, 2002.
3. Masubuchi, K., *Analysis of Welded Structures*, Pergamon, Oxford, 1980.
4. Chao, Y.J., Zhu, X. K., and Qi, X., 'WELDSIM – A WELDing SIMulation code for the determination of transient and residual temperature, stress and distortion', *Advances in Computational Engineering and Sciences*, Vol. II (eds S.N. Atluri and F.W. Brust) Tech Science Press, Irvine, California, 2000, pp. 1206–1211.
5. Peters, W.H., and Ranson, W.F., 'Digital image techniques in experimental stress analysis', *Opt. Engng*, **21**(3), 427–432, 1982.
6. Chu, T.C., Ranson, W.F., Sutton, M.A., and Peters, W.H., 'Applications of digital image correlation techniques to experimental mechanics', *Expl. Mech.*, **25**(3), 232–244, 1985.
7. Luo, P.F., Chao, Y.J., Sutton, M.A., and Peters, W.H., 'Accurate measurement of three dimensional deformations in deformable and rigid bodies using computer vision', *Expl Mech.*, **33**(2), 123–132, 1993.
8. Luo, P.F., Chao, Y.J., and Sutton, M.A., 'Application of stereo vision to 3D deformation analysis in fracture mechanics', *Opt. Engng*, **33**(3), 981–990, 1994.
9. Sutton, M.A., McNeill, S.R., Helm, J.D., and Chao, Y.J., 'Advances in two-dimensional and three-dimensional computer vision', *Photomechanics*, Topics in Applied Physics, Vol. 77, Springer, Berlin, 2000, pp. 323–372.
10. Bruck, H., Schreier, M.A., Sutton, Y.J., Chao, X.D., and Davoud, M., 'Distortions in GMAW of thin plates: temperature and 3-D deformation measurements

using high-speed thermal imaging and stereoscopic video imaging', *Trends in Welding Research, Proc. 5th Intl. Conf. on Trends in Welding Research* (eds J.M. Vitek, S.A. David, J.A. Johnson, H.B. Smartt and T. DebRoy), Callaway Garden Resort, Pine Mountain, Georgia, USA, 1998, American Welding Society, Miami, Florida, ASM International, Materials Park, Ohio, 1998, pp. 967–971.

11. Schreier, H.W., Sutton, M.A., Chao, Y.J., Bruck, H.A., and Dydo, J., 'Full-field temperature and three-dimensional displacement measurements in hostile environments', *Technical Papers of the North American Manufacturing Research Institution of Society of Manufacturing Engineers*, Society of Manufacturing Engineers, Dearborn, Michigan, 1999, pp. 75–79.
12. Bruck, H.A., Schreier, H., Sutton, M.A., and Chao, Y.J., 'Development of a measurement system for combined temperature and strain measurement during welding', *Proc. Taiwan Int. Welding Conf. on Technology Advancements and New Industrial Applications in Welding* (eds C.L. Tsai and H.L. Tsai), Taipei, 1998, Taiwan, 1998, pp. 523–526.
13. Qi, X., 'Numerical simulation of buckling in welding thin panels', PhD Dissertation, Department of Mechanical Engineering, University of South Carolina, Columbia, South Carolina, 2003.

7
Modeling distortion and residual stress during welding: practical applications

X. L. CHEN, Z. YANG, Caterpillar Inc., USA, and
F. W. BRUST, Battelle Memorial Institute, USA

7.1 Introduction

Welding distortion and residual stress have remained major challenges in manufacturing. Extensive experimental efforts have been made in the history of manufacturing to understand and control the welding distortions and residual stresses. Tremendous work has also been reported in the past two decades in using analytical and computational methods to predict and quantify the effect of the distortion and residual stress during the welding process. This chapter introduces some examples of industrial welding applications using computational methods to highlight the challenges and the benefits of using analysis and simulation tools to improve the quality and to reduce the time and cost of welding process developments and problem resolutions[1–3].

7.2 Challenges for industrial applications

Welding, along with thermal cutting, bending and machining, is usually referred to as one of the key manufacturing 'building blocks'. Most fabricated structures start with steel plates received from the steel supplier(s). The plates are cleaned, typically by shot blasting. Thermal cutting processes, such as oxyacetylene-fuel flame cutting, plasma cutting and recently laser cutting, are used to cut the parts from the plate into desired profiles and dimensions. Bending or forming processes could be used to change the parts into the desired shapes. The piece parts are then tacked together in a fixture and welded into an assembly. Machining may be used at either part or assembly level to achieve the tolerance of the critical dimensions. Each operation introduces manufacturing variability. The final quality of the assembly depends on the quality control of each process. In addition, the quality of one process could be affected by the upstream processes and will influence the downstream processes.

Process simulations have been reported extensively to understand the physics and science of the manufacturing processes, as illustrated in Fig. 7.1. The following is a brief description of the through-process simulations associated with each manufacturing building block[3].

1. *Steel-rolling simulation.* The model simulates the typical steel-plate-making processes focusing on the locked-in residual stresses. Such residual stresses could alter the plate response during the cutting process. Therefore, the residual stress produced during the steelmaking, such as rolling, is one of the key inputs to the downstream thermal cutting simulation.
2. *Shot-blasting simulation.* The model predicts the results of shot impact on the plate surface. It simulates the plastic deformation, residual stresses and oxide removal.
3. *Thermal cutting simulation.* The model simulates the distortions and residual stresses that result from thermal cutting processes. It predicts the thermal–mechanical response of the cutting process as well as the interaction of thermally induced distortion with the residual stresses from the steelmaking process. The simulation can improve the cutting accuracy by providing the critical information that is used to optimize the part nesting and cutting sequence.
4. *Bending simulation.* The model simulates the in-process mechanical response of the material. The simulation predicts the bending accuracy and spring-back based on the material variables and process parameters.

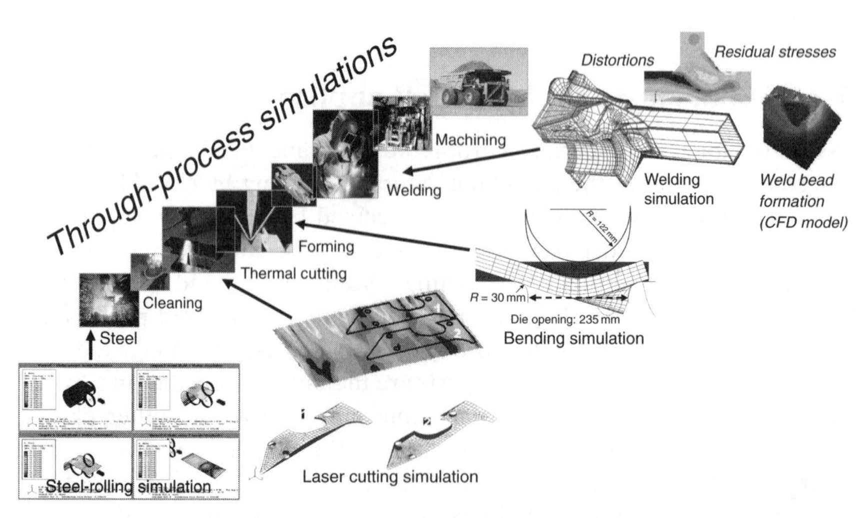

7.1 Through-process simulations, including steelmaking, thermal cutting, bending–forming and welding (CFD, computational fluid dynamics).

5. *Welding simulation.* The model simulates a wide range of welding phenomena including the thermodynamics of the weld pool, the metallurgical transformations in the weld and its heat-affected zone, and the thermal–mechanical response of the structure during the welding process. Simulation provides a unique tool for the welding engineer to design and optimize welding processes. The issues to be considered in a welding simulation are further illustrated in Fig. 7.2.

Based on the objectives of simulations, welding simulations could roughly be divided into two categories, namely a global model (three-dimensional (3D)) or a local model (two-dimensional (2D)). A global model is normally used for predicting the distortion of a large-scale and/or complicated component geometry. In a global model, the details of the welded joint are simplified to reduce the size of the model. Typical applications for global model are to explore the optimized welding sequencing to reduce welding distortions, to study the effects of fixture and clamping or to design the process parameters. On the other hand, a local model focuses on the welded-joint region that is a small portion of the component. The joints are usually modeled in great detail. Local models are commonly used to predict the thermal histories and the residual stresses of the welded joints. A detailed local model would even incorporate thermal fluid flow phenomena and metallurgical transformations. Figure 7.3 shows some sample industrial components in heavy manufacturing. The challenge of industrial applications lies in many aspects including the size of the components, geometric complexity, material variability, process variability and incompleteness of model theory as discussed in previous chapters.

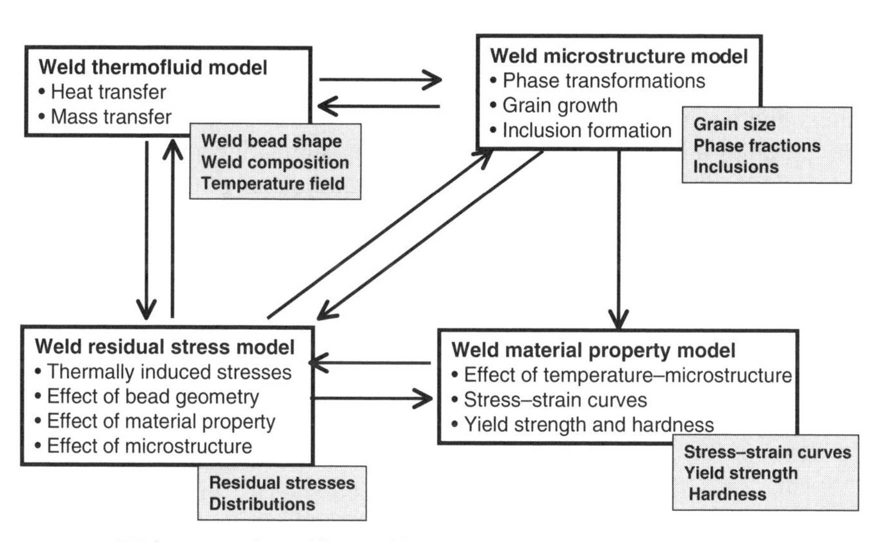

7.2 Issues to be addressed in welding simulation.

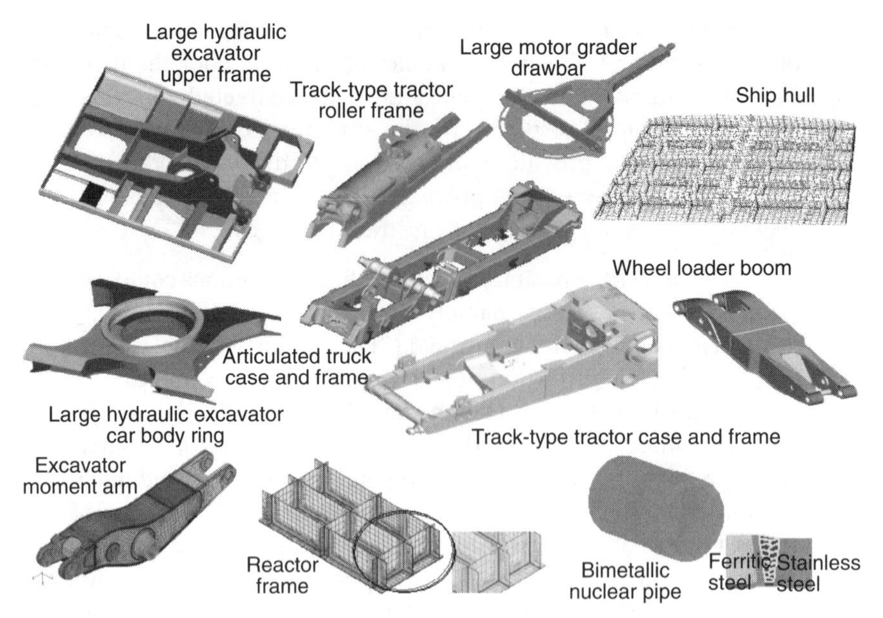

7.3 Examples of welding simulations for various industrial components.

Most welding simulations use numerical computational methods, such as the finite-element method, using either off-the-shelf code (with customization) or in-house development. Welding simulation of a large-size and/or complicated component imposes tremendous computational demands. Welding simulation usually requires transient analysis to capture the response of the component during the welding, such as the welding sequence, welding direction, cooling and fixture. Welding simulation is non-linear and usually difficult to converge because of the material behavior at the elevated temperature. Moreover, welding simulation requires quality finite-element meshing with sufficient mesh density along the welds and the heat-affected zone. This requires both time and expertise in finite-element meshing and usually results in a large number of degrees of freedom in the model. The computational cost quadruples with increase in the model size. Despite the fast-growing computer technology, the complexity of an industrial problem could easily make the simulation model infeasible to complete in an acceptable turnaround time. In many cases, decisions have to be made to balance between the size of the model that one could afford and the level of the detail that one would want to capture.

Figure 7.4 shows a multiple-pass welding mock-up sample of the lifting component for a mining machine. Thermocouples and laser displacement sensors were used to monitor the transient responses of the component

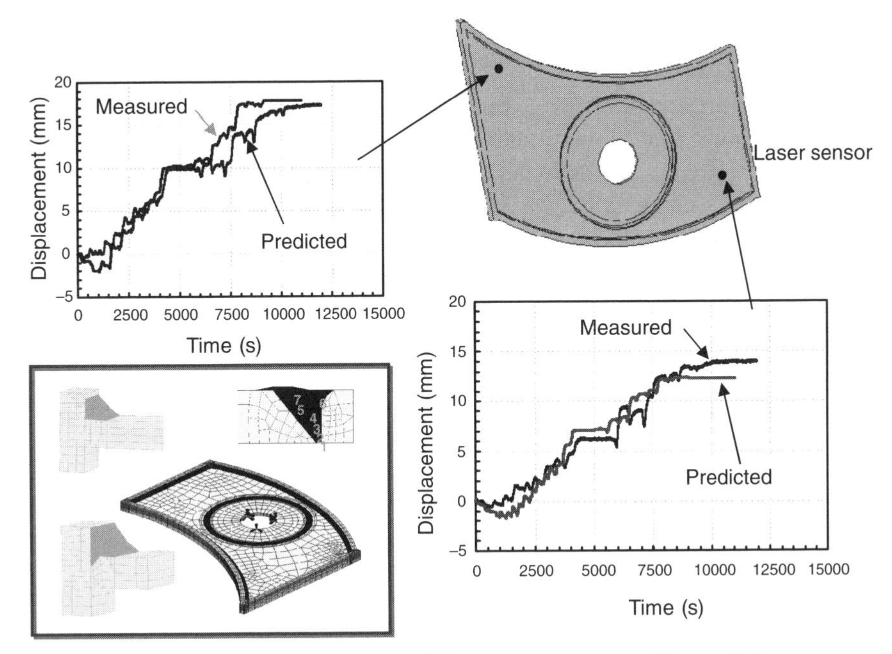

7.4 Mock-up sample of the lifting component used in a mining machine.

during the welding. The spikes in the temperature profiles and distortions represent roughly the responses of the structure to multiple passes of the welding. Although this is only a mock-up sample, the simulation faced significant challenges as follows.

1. The size of the sample is about 500 mm by 1000 mm. The total length of the weld is more than 20 m. Including the interpass cooling time, the total welding time is more than 5 h. Using an average 10 s time step and average 10 min computation time per step to reach the convergence, this model would need about 300 h to complete and a large disk space for storing the results.
2. Material melting at the elevated temperature has been one of the major challenges in welding simulation. Modeling melting and solidification, which cause the relaxation and recovery of elastic and plastic strains, are very complicated and not fully understood. Moreover, multiple-pass welding imposes the challenge of tracking material behavior during the remelting.
3. A hexagonal or brick element is the preferred element type for solid meshing. Brick elements would reduce the number of degrees of freedom and can avoid overstiffness (locking) of tetrahedral elements. However, building a brick element model for this mock-up sample is

not an easy task, although its geometry is considered fairly simple compared with other real components.

Lack of material properties at the elevated temperature is a well-known challenge in welding simulations. In addition, welding simulation for industrial application also needs to deal with the inevitable process variability in a shop environment. Different from a controlled laboratory experiment, welding on the shop floor could vary from time to time and from welder to welder. Usually the welding procedure specification represents the normal welding parameters while the true parameters in the shop could vary significantly, including the heat input, welding speed, weld size, welding sequence, interpass cooling and even part fit-up, enough to cause grief in a simulation.

7.3 Welding distortion simulation for large and complicated structures

Welding distortion simulation normally adopts sequentially coupled thermal structural analysis, i.e. the thermal analysis is performed first and then the structural analysis will be performed using the temperature predicted in the thermal analysis as the thermal load in conjunction with any additional mechanical loads or constraints. Material response in a welding process is very localized along the welds. For large fabricated structures, the simulations will be extremely computation intensive to consider millions of degrees of freedom and the highly nonlinear characteristics relating to the welding process. As such, coarse meshes have to be used for the global distortion predictions. However, numerical thermal predictions using such coarse meshes are inadequate, especially for capturing the thermal gradients and cooling rates during welding processes. For a reasonable turnaround time, developing an efficient and effective simulation procedure is crucial to the success of applying welding simulation in industrial problems.

The comprehensive thermal solution procedure (CTSP)[4,5] is an in-house thermal analysis code specifically developed by Caterpillar Inc. and Battelle Memorial Institute for global distortion prediction of production components. The code is an analytical solution based on the Rosenthal solution of a point heat source moving in an infinite domain in a constant direction and at a constant speed. Without additional treatment, the Rosenthal solution by itself could not be used for the temperature profiles of the industrial applications such as the example shown above. To simulate the surface of a component, the CTSP uses the imaginary heat sources reflected on the surface of the component to achieve the equivalent heat conduction. Meanwhile, the CTSP uses the 'negative' heat sources starting at the time of welding end to simulate the stop of the welding. Using these techniques, the CTSP is able to simulate typical weld joint types such as a groove joint, lap joint and T-fillet joint. The validation of the CTSP against the experimental measurement is shown in Fig. 7.5, and the central processing unit (CPU) time comparison between the CTSP and its corresponding thermal analysis using the commercial finite-element analysis (FEA) code ABAQUS® is shown in Fig. 7.6. The details of the CTSP have been well documented and reported in the literature[4,5]. Only its advantage and essential features are highlighted as follows (see also Chapter 8).

1. The computation time for the CTSP is much faster than that of a numerical solution. Depending on the complexity of the structure, a thermal calculation using the CTSP can be 10–100 times faster than an FEA. One reason for this significant speedup is that the CTSP avoids the calculation of the whole structure and only focuses on the local region around the heat-affected zone.

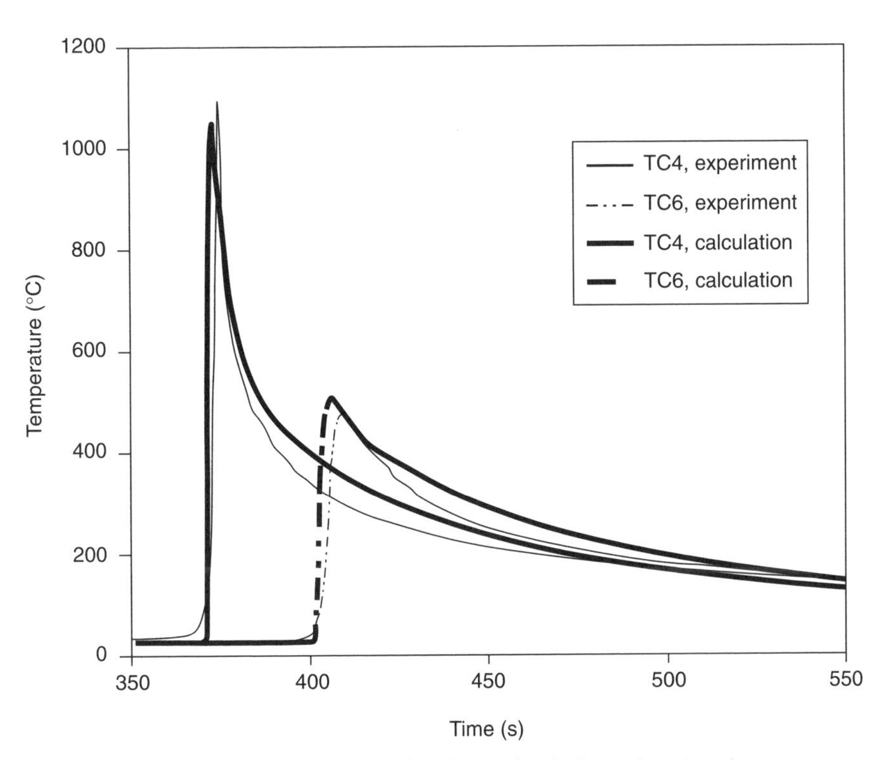

7.5 Validation of the CTSP for thermal solution, showing the comparison of the experimental and CTSP temperature–time histories in a T-joint: TC4, 0.5 mm from the weld toe; TC6, 4.5 mm from the weld toe. (TC = Thermocouple)

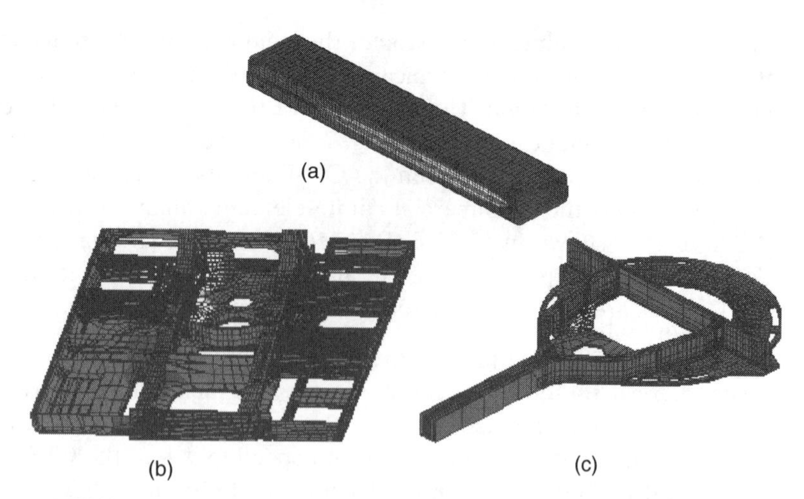

7.6 CPU time comparison between CTSP and FEA solution (the computations were performed on an HP 180 UNIX workstation): (a) 5000 shell elements, single-pass weld, FEA using ABAQUS (10 h) versus CTSP (3 min); (b) 37 000 elements, single-pass weld, FEA using ABAQUS (240 h) versus CTSP (2.5 h); (c) 15 000 elements, multiple-pass weld, FEA using ABAQUS (80 h) versus CTSP (3 h).

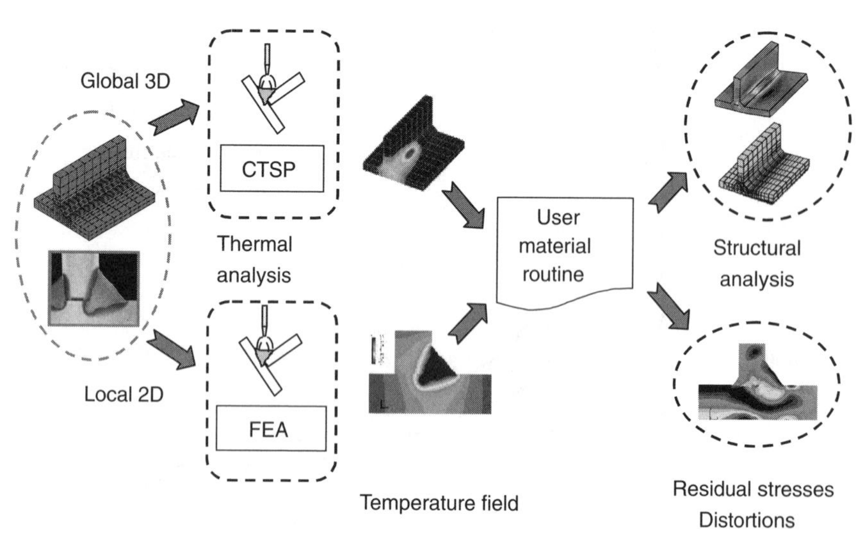

7.7 Typical simulation flow for both global and local modeling approaches.

2. The CTSP solutions for coarse meshes are very accurate for predicting the temperatures at the coarse node points. These temperatures provide accurate through-thickness temperatures, which are critical for providing accurate distortion prediction.

3. The input and analysis procedure is much simpler than those using numerical methods. This is especially true for large structures with multipass welds.

Another issue to be addressed in a welding simulation is the material behavior at an elevated temperature. Melting, solidification and phase transformation are important metallurgical phenomena associated with a welding process. Distortion is the primary interest of a global model and therefore the phase transformation model would need to be simplified to save computational time. Usually a user material routine would be sufficient to treat the material behavior in the weld zone including the heat-affected zone. Figure 7.7 shows the suggested simulation procedures for both global and local modeling approaches for industrial applications.

7.4 Distortion control and mitigation for individual components

Excessive welding distortion would result in expensive post-welding correction and poor quality of a welded structure. This section reports some examples of simulation-aided welding process development of fabricated components for mining and construction equipment.

7.4.1 Reducing welding distortion by alternative welding parameters

One of the major uses of welding simulation is to explore the alternative welding parameters for distortion control. The case study presented here is a real example of using the alternative weld size and sequence to minimize the welding distortion. Figure 7.8 shows a design of the rear frame of an articulated dump truck. The center portion of the rear frame is welded as the center subassembly. The inner bore (front and back) is machined to the dimensions after center subassembly. The center subassembly is then welded to the rails at an upper-level assembly. Distortions from the excessive heat cause the ovality of the bore to exceed the tolerance. The objective of the simulation is to develop a welding process maintaining the inner dimensions (front and back) of the bore within the tolerance during the upper-level welding assembly. In this example, the welds are 8 mm single pass and 10 mm three passes. Figure 7.9 shows the weld locations and welding sequence by the numbers in the angular brackets. Welding simulation was conducted using virtual fabrication technology (VFT^{TM})[6] developed by Caterpillar Inc. and Battelle Memorial Institute[7] (see also numerous papers by Brust *et al.* on weld modeling in this set of volumes). The simulation correlated well with the measurement (Fig. 7.10).

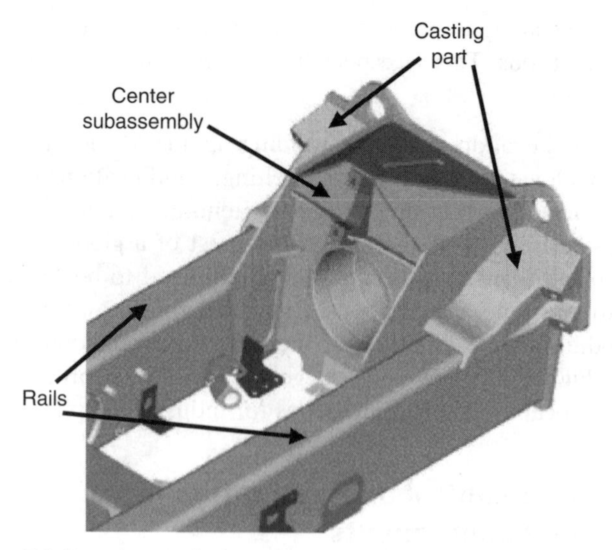

7.8 Rear frame design of an articulated dump truck.

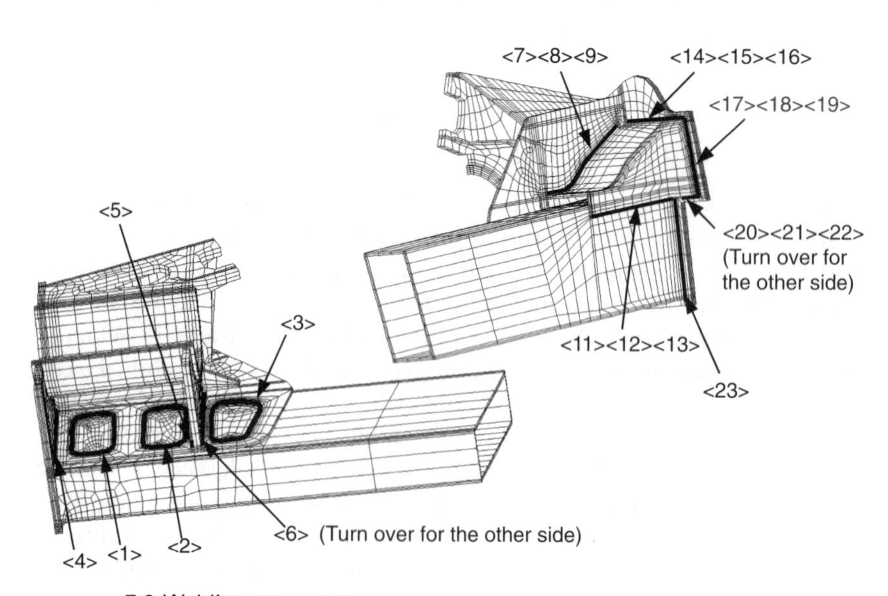

7.9 Welding sequence.

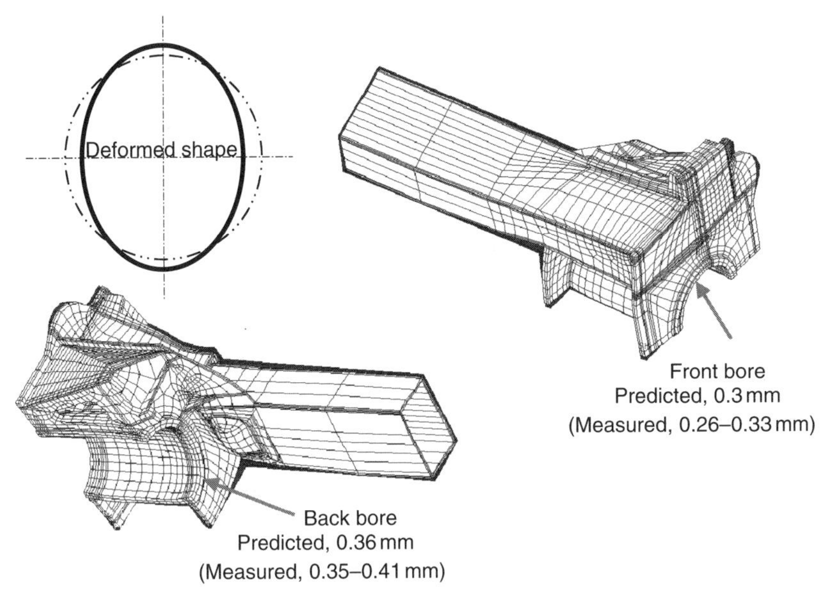

7.10 Welding distortion.

In this practice, the thermal response was calculated using the CTSP and the thermal mechanical analysis was performed using ABAQUS®, a general-purpose commercial FEA code with addition of a user-defined routine for welding material behavior. At first, the simulation used the full geometry of the rear frame. It was found that the turnaround time for each simulation would be more than 600 h on a HP240 UNIX workstation. The model was then simplified to a half-model using geometry symmetry. Despite the symmetry of the geometry, it was noted that the welding simulation model was not symmetric since the welding sequence from one side of the rear frame to the other side violated the symmetry requirement. The half-model was based on the assumption that the distortions and residual stresses of welding on one side of the rear frame were localized and would not affect those of the other side. However, additional time was added to the interpass cooling to simulate the delay of the welding when welding was performed on the other side of the rear frame, and therefore the total welding time remained the same. The turnaround time of the half-model was reduced to around 150 h. Note that the speedup was mainly due to the reduction in the degrees of freedom.

A turnaround time of 150 h would not be acceptable, especially for a fast-pace development project. One technique called lumped heating was used in this project to reduce further the computational time. In a transient welding simulation, time steps would typically be divided according to the size of the element along the welding path, i.e. the time step equals the length of the element divided by the welding arc travel speed along the welding path. For example, the time step will be 6s in a model of 30mm elements and a 5mm/sec welding arc travel speed. As a result, there are hundreds of tiny time steps in a simulation, and multiple substeps and iterations need to be performed for each time step for solution convergence. It is quite usual that a model would need a computational time of a few minutes to simulate physical welding for 1min, which leads to computation for many hours to complete the simulation. The lumped-heating method reduces the number of solution steps by grouping a few element-wise steps, typically either a segment of the weld or one weld pass, together into one lumped step. In this case study, each weld pass was lumped together and 22 physical weld passes resulted in a total of 44 steps (22 heating steps and 22 cooling steps) in the simulation. One challenge of lumped heating is to develop the equivalent heat input and heating duration for the lumped step. One approach is to match the heat-affected zone size by comparing the lumped heating with the ordinary moving-arc solution. Lumped-heating capability has also been implemented in the CTSP. The major advantage of the lumped heating in the CTSP is not the computational time but the feature of calculating heat input and duration of heating to match the heat-affected zone size automatically.

The lumped-heating model, after being validated against the moving-arc solution, was used to investigate the potential solutions for reducing the distortion. The turnaround time of the lumped-heating model was reduced within 24h. The first option explored was the alternative welding sequence. Instead of finishing the welds on one side and then moving to the other side, the alternative welding sequence spreads the passes from one side to the other to allow more interpass cooling time and less heat buildup. Some distortion reduction was observed; however, it was not enough to reduce the distortion within the tolerance. Further exploration of changing 8mm welds to 6mm welds reduced the distortion to 0.22mm for the back bore and 0.24mm for the front bore. Final reduction of the distortion to the tolerance required moving some welds from the upper-level assembly to the center subassembly, replacing single-pass 6mm welds with multiple-pass 3mm welds and alternating the welding from side to side.

7.4.2 Welding process development for a new product

In the example described here, the fabricated component was a new design for manufacturability and performance improvement. The component of interest to welding is the yoke assembly (draw bar) for a motor grader, a commonly used machine for ground leveling in the field. Figure 7.11 shows the design of the component that consists of a circular plate and some solid

7.11 Motor grader yoke assembly (drawbar).

bars welded to the top of the plate and some attachments welded at the bottom of the plate. One critical requirement is that the flatness of the yoke assembly be less than 1.5 mm at the six matting locations on the bottom surface of the plate. A traditional process would use the post-welding correction and then the machining of the bottom of the draw bar. The traditional process not only is expensive but also needs the plate thicker than the required thickness to allow the machining. The goal of the project was to develop a welding process that would require neither post-welding dimensional corrections nor machining.

Weld sequencing design was first considered as discussed in the previous example. It was hoped that the distortion would be balanced by altering the welding sequence appropriately. Figure 7.12 shows the distortion from one of the welding sequences using single-pass 10 mm fillet welds on the top of the plate and 6 mm fillets at the bottom of the plate. Unfortunately the distortion could not be controlled within the required tolerance (1.5 mm) after multiple sequences were explored. Manufacturing considerations also ruled out the other alternatives such as changing the weld size, breaking single pass into multiple passes or back-step welding.

Applying mechanical clamping or deformation is another commonly used distortion control method. A method called precambering was investigated. Precambering deforms the part in a fixture during the welding to compensate the weld distortion. The location and the amount of deformation at the precambered locations need to be determined so that the flatness of the welded part will be controlled within the design tolerance (1.5 mm) after the part is released from the fixture. Precambering designs could be performed by the traditional trial-and-error method. It was estimated in this application that it would require at least five iterations of welding trail in the shop and two iterations of fixture design and build,

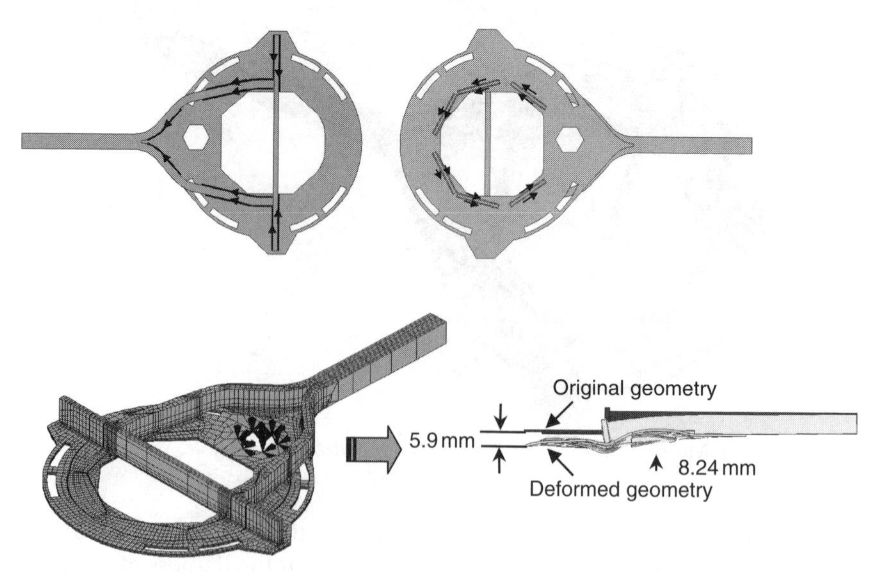

7.12 Welding distortion of a drawbar.

which would take a development time of more than 2 months. The team decided to use the simulation to develop the concept of the precambering process to save the time and cost of the process development. More than ten simulation iterations were performed using the VFT™ simulation package[6]. Since the process was designed for robot welding, welding fixture concept design was in parallel with welding simulation. Some precambering designs were recycled because of the accessibility problem identified by the robot simulation (Fig. 7.13). The final precambering scheme was reached, as shown in Fig. 7.14.

A welding fixture for precambering has to be designed to withstand the reaction forces from the mechanical deformation induced on the material and thermal distortions produced by the welding. In the past, the thermally induced reaction forces were usually ignored or not fully understood in the fixture design owing to the lack of knowledge in thermal stresses. As a result, a weld fixture might deform or even fracture during the welding, which significantly affected the quality and reliability of the welding process.

Typically in a welding simulation, the clamping locations are assumed fixed and infinitely rigid. In this case study, the reaction forces on the welding fixture at various precambering locations during and after welding are shown in Fig. 7.15(a)[8]. The mechanical forces due to precambering are shown at time zero. The values of the mechanical forces at various locations depend on the magnitudes and directions of precambering at the corresponding positions. Owing to the welding thermal stresses and the

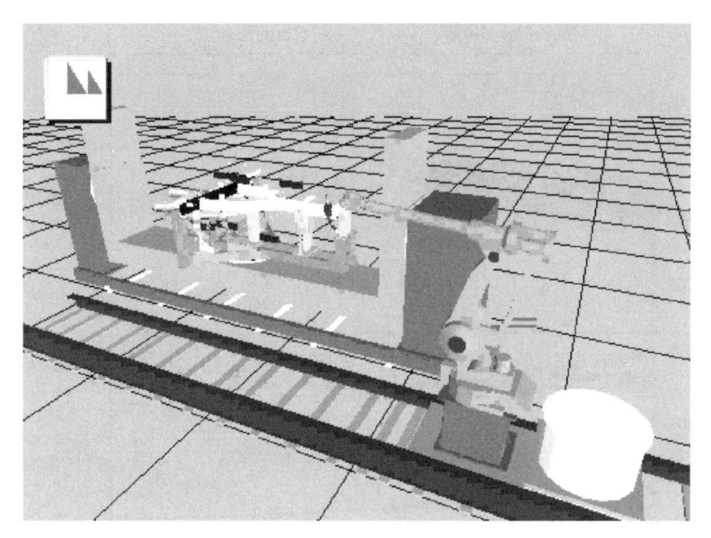

7.13 Robot simulation for the reach study.

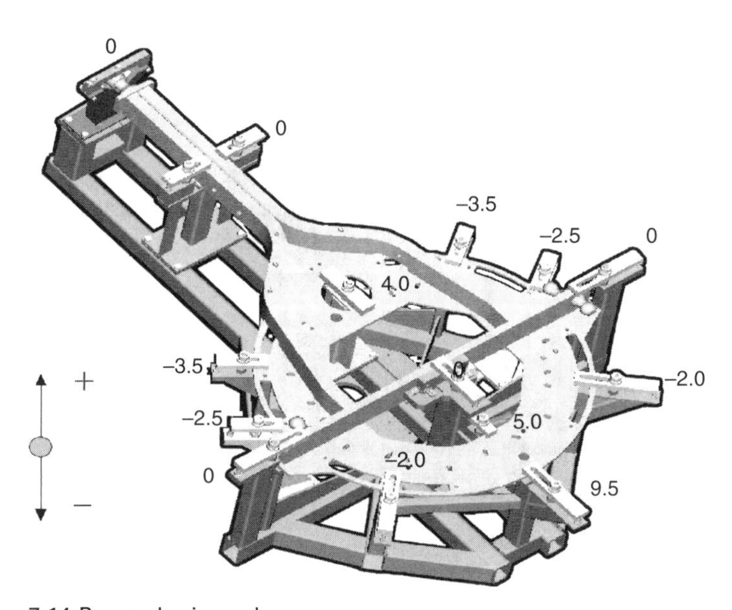

7.14 Precambering scheme.

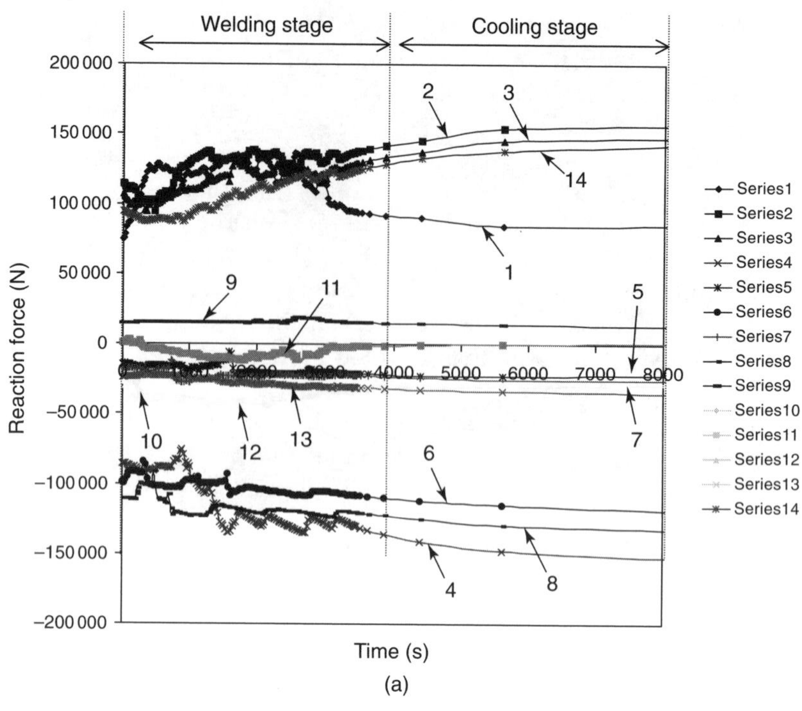

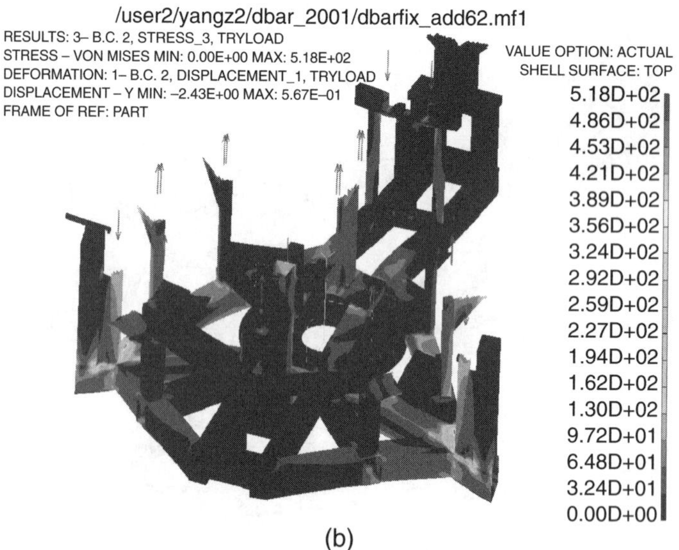

7.15 Reaction force of the fixture: (a) the reaction forces at clamping locations during welding process; (b) predicted distortion and stress in the weld fixture.

resulting residual stresses, the reaction forces vary significantly during the welding and in the subsequent cooling stage. It can be seen that the reaction forces at certain locations are almost doubled because of the welding-induced thermal stresses.

In this application, the stiffness of the fixture was examined through linear static analysis, in which the reaction forces from the welding simulation were applied to the fixture finite-element model. The results from the structural analysis suggest that the fixture would be significantly deformed, as shown in Fig. 7.15(b), and the maximum stress at the 'hot' spot in the model is 518 MPa, which is much higher than the yield strength of the material. This implies the weld fixture would have been underdesigned if weld thermal stresses were not considered.

The distortions at the critical points on the bottom plate of the welded parts were measured after the parts were cooled to room temperature and released from the fixture. The flatness check was made on seven parts, of which three were welded using the original weld fixture and four were welded using the modified weld fixture to sustain reaction force from the welding stresses, as shown in Fig. 7.16. It can be observed that the distortions in the three parts using the original fixture exceed the tolerance (1.5 mm). As mentioned previously, this is primarily due to the failure of the clamps of the fixture and the global deformation of the fixture during welding. In contrast, the distortions were significantly reduced after using the modified weld fixture that was stiff enough to support the precambering scheme. Almost all four parts meet the flatness requirement. In this development it was also found that the flatness of the welded drawbar was

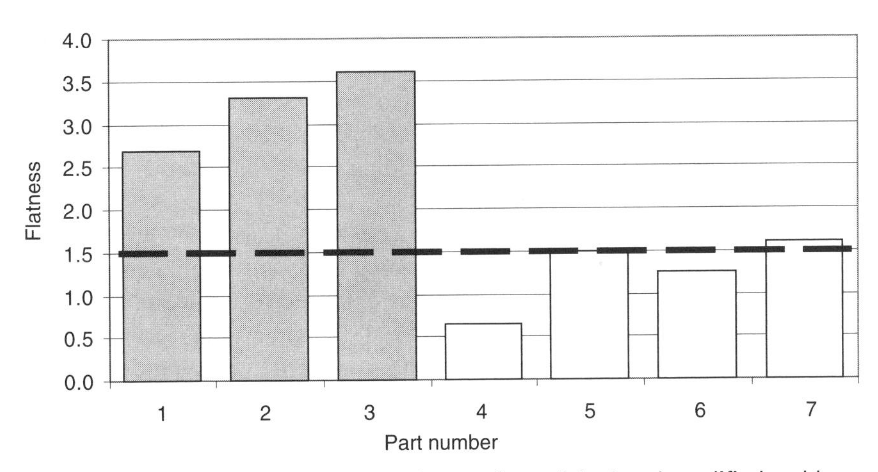

7.16 Distortions in weld products using original and modified weld fixtures (parts 1–3 used the original fixture and parts 4–7 used the modified fixture).

very sensitive to the flatness of the plate before welding, which was caused by the manufacture variability in the upstream operations. Steel-plate-rolling simulation and thermal cutting simulation were also performed in this project, which is not covered in here.

7.5 Residual stress prediction and mitigation for individual components

Residual stress is another important aspect of the welding simulation. The residual stress induced by the thermal cycle(s) and metallurgical transformation is localized around the weld and its heat-affected zone. In contrast with predicting the distortion of a large structure, investigating the residual stress at the welded joint normally requires the details of the weld joint. Residual stress analysis could use a sequential thermal and structural analysis approach assuming that the work energy generated in mechanical response is not high enough to cause significant temperature change. However, the thermal analysis would need to be coupled with the metallurgical model (if considered) since the latent heat of the phase transformation needs to be considered as a function of thermal history in the analysis. Figure 7.17 shows an example of the residual stress prediction of a bead-on-plate specimen[9].

7.5.1 Reducing welding residual stresses by welding sequencing

Cracking is one common premature failure for welded structures with thick plates and strong constraints. Residual stresses from welding could be high enough to cause cracking in welded joints without any applied loads. Figure 7.18 shows a crack at the weld toe of a car body structure fabricated with multiple-pass welding process. The constraint, defined as the trace of the stress tensor, is high with this type of welding. High constraint is known to reduce fracture resistance.

Welding residual stress can be reduced by selection of appropriate welding process, sequence, fixture–clamping and stress relieving. In this case, an alternative welding sequence was explored to reduce the residual stresses and to resolve the cracking problem. A simulation model was built with the axisymmetric elements for the ring structure considering the details of the welding passes and the sequences, as shown in Fig. 7.19. In the original sequence, all passes at the inner side were finished first and then the welds at the outer side were made, as shown in Fig. 7.19(a). In this sequence, high residual stress can be observed at the weld toe at the inner side of the ring, as shown in Fig. 7.19(c). The proposed new sequence alter-

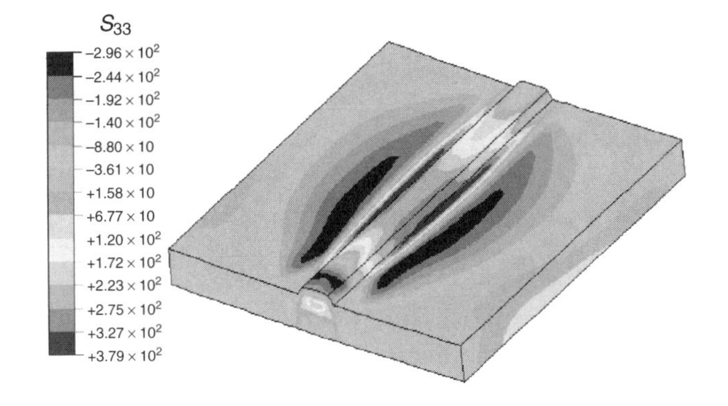

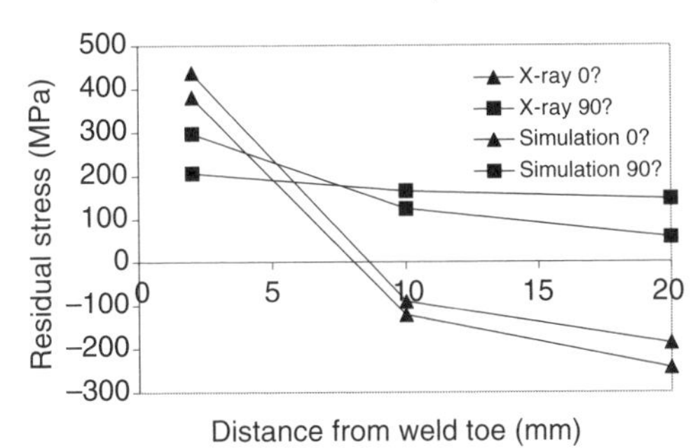

7.17 Welding residual stress prediction and comparison with the experimental results.

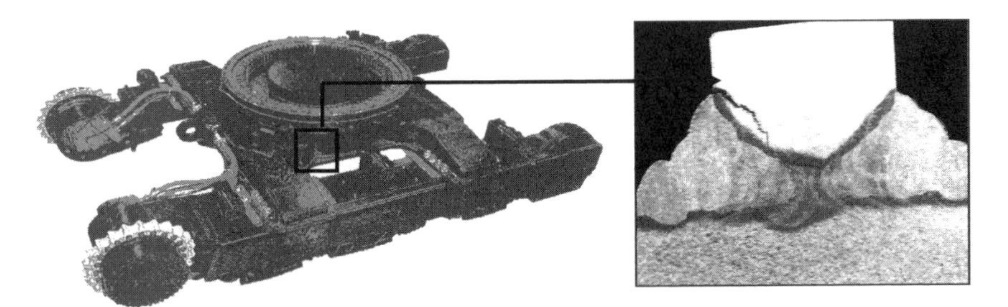

7.18 Crack at the weld toe in a car body structure due to residual stresses.

nates the passes between the inner and the outer sides, as shown in Fig. 7.19(b). The intent of alternating the weld passes was to avoid heat buildup and to allow more interpass cooling. The simulation results confirmed that the welding residual stresses at the weld toe would be significantly reduced, as shown in Fig. 7.20. The alternative welding sequence was then implemented in production to resolve the cracking problem.

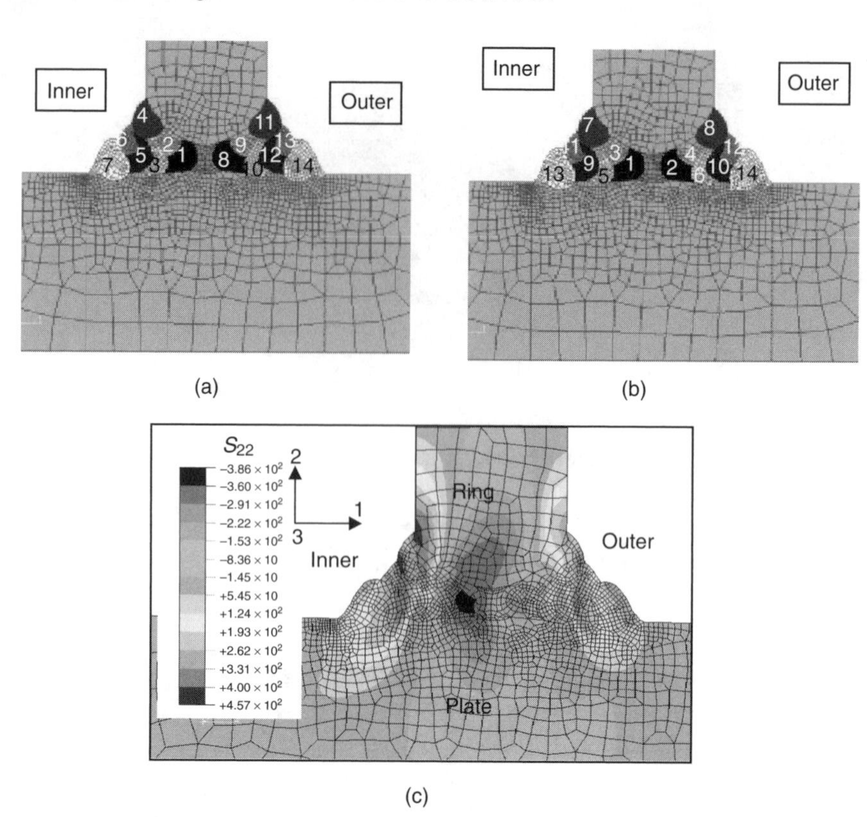

7.19 Axisymmetric model with welding passes and sequences: (a) original sequence; (b) alternative sequence; (c) residual stress distribution of the original welding sequence.

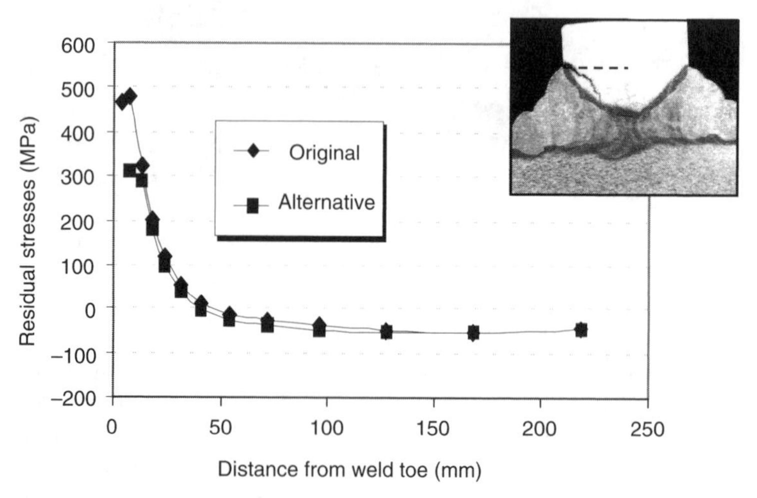

7.20 Comparison of welding residual stresses from different welding sequences.

7.5.2 Reducing welding residual stresses by using a special welding consumable

Welding residual stresses modifies the mean stress experienced by a welded joint under the fatigue loading. It is normally assumed that cyclic stresses near a weld are fully damaging because of the high pre-existing tensile residual stress. In T-fillet welds, weld fatigue failure in most cases occurs at the weld toe owing to a high stress concentration, high constraint and also high tensile residual stress in this region. To reduce the stress concentration, efforts have been made to achieve the favorable weld bead shape by welding process design, process control and/or post-weld treatment such as penning, grinding or even machining to eliminate the sharp-notch effect. Fabrication of Advanced Structures Using Intelligent and Synergistic Materials Processing (FASIP) is an Advanced Technology Program project Sponsored by the National Institute of Standards and Technology during 1995–1999 to develop the in-process control welding technology to achieve a welding strength improvement of ten times. One of FASIP's goals was to produce an as-welded bead shape to reduce the stress concentration at the weld toe.

In general, high tensile residual stresses are expected at the weld toe. This is particularly important for the welding of high-strength steels where higher tensile residual stresses at the weld toe are expected compared with those of the mild steels with lower yield strengths. In recent years, some reports from Japan have demonstrated that compressive residual stresses can be obtained at the weld toe by using a lower-temperature-transformation consumable. Compressive residual stresses are induced by the expansion of volume due to the lower-temperature martensite transformation. Experimental results also showed that weld fatigue strength could increase by a factor of about 2 owing to the introduction of compressive residual stress at the weld toe. A case study has been performed using a simulation model considering welding thermal, phase transformation and residual stress analysis. Figure 7.21(a) shows the residual stresses using a conventional welding consumable producing high tensile stresses close to yield strength level. Figure 7.21(b) is the prediction of the residual stresses of the same weld joint configuration using a low-temperature-transformation consumable. Low-temperature martensite transformation 'overrides' the tensile stress from the cooling process and results in compressive residual stresses at the weld toe. Laboratory experiments were conducted and confirmed the prediction of the simulation. In this study, the simulation model is able to predict the residual stress distribution quantitatively as a function of martensite transformation temperatures, weld geometry and plate thickness.

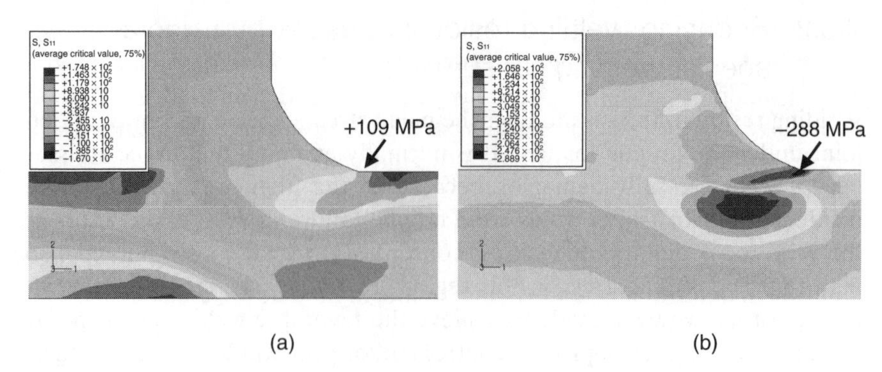

7.21 Residual stress distribution in T-fillet welds using (a) a conventional welding consumable and (b) a lower-temperature-transformation consumable.

7.6 Weld thermofluid modeling

The analytical thermal solution CTSP mentioned earlier is not able to incorporate nonlinear material properties unless an iterative procedure is used. It is also less suitable for integration with metallurgical transformation analysis. Heat transfer in the weld pool and the heat-affected zone is a very complicated phenomenon. A finite-element heat conduction model with sufficient weld details is able to overcome these limitations. However, an FEA approach still suffers from the fact that it is not able to simulate the fluid behavior of the molten metals. Therefore it is not possible to model physically the heat transfer within the weld pool and the interface between the solid and the fluid. More importantly, a finite-element heat conduction model is not able to predict the weld bead details such as the weld toe radius and penetration which are essential for predicting the stress concentration and the strength of the welded joint.

The weld bead profile is one of the critical factors in the specification for a welded-joint design since it has a direct influence on welded structure stresses, welded-joint strength and fatigue life. Weld bead geometry is determined by the heat transfer and fluid flow in the weld pool. In the past decade, significant progress has been made in understanding the effect of heat transfer and fluid flow on weld bead shape formation by mathematical modeling. It is crucial to establish an accurate thermofluid model to simulate the entire welding thermal and fluid processes to obtain the desired weld bead profile and thermal history for structurally sound and high-performance welded-joint designs.

Thermofluid dynamic modeling is well established and computational fluid dynamics (CFD) modeling has been widely used in many industrial applications including aerospace, appliances, automotive, biomedical, chemical process, electronics cooling, environmental, nuclear power, power generation, semiconductors and microelectromechanical systems. In the area of welding applications, most investigations of thermofluid flow in weld pools have concentrated on gas–tungsten arc welding processes. Little work has been done in the area of fluid flow and heat transfer in the gas–metal arc (GMA) welding pool because of the complexity of the process. During the GMA welding, weld metal is deposited into a weld pool from an electrode as a droplet stream. The metal transfer mechanisms play an important role in determining the resultant heat and fluid flow characteristics and the final weld profile. The heat content and impact force of droplets tend to induce a series of physical, chemical and metallurgical changes in the weld pool. Therefore, it is difficult to simulate the GMA welding process, mainly owing to the droplet impact and the large flow and temperature gradients associated with this process. A better understanding of droplet impact effects on weld heat transfer and fluid flow is critical to investigating welding pool phenomena and predicting weld bead profiles. The case studies presented here use a 3D transient weld pool dynamics model to simulate the GMA welding process. The governing equations, including continuity, momentum and energy equations, are discretized using the control volume method. The regular rectangle meshes are used in the whole domain, and the volume-of-fraction method is employed to treat free surfaces. The details have been described by Cao *et al.*[10]. CFD solutions can be used as a first step in identifying the bead shape to be used for a global distortion analysis and/or a local residual stress calculation. Knowledge of the bead shape and fusion zone is critical for such analyses. The bead shape can always be determined from experiments or experience (for a given set of weld parameters); however, the use of a CFD model to predict these shapes before a distortion analysis provides a virtual development environment for effective welding process design and more accurate thermal information for welding metallurgical and stress analysis. The examples presented here used a CFD model to predict a moving GMA weld pool by considering not only the heat transfer and fluid flow driven by surface tension gradients, electromagnetic forces, buoyancy and arc pressure, but also detailed information about droplet flow effects on weld pool, its solidification and weld bead shape. Because of its unique capacity in simulation of free surface, this model can be applied to various types of welded joint.

7.6.1 Bead on plate

Figure 7.22 shows the calculated temperature and fluid flow fields at a longitudinal section at different moments of time for a bead-on-plate weld case. The dark color at the middle of the weld pool indicates a temperature above the melting temperature, and the small white arrows inside the weld

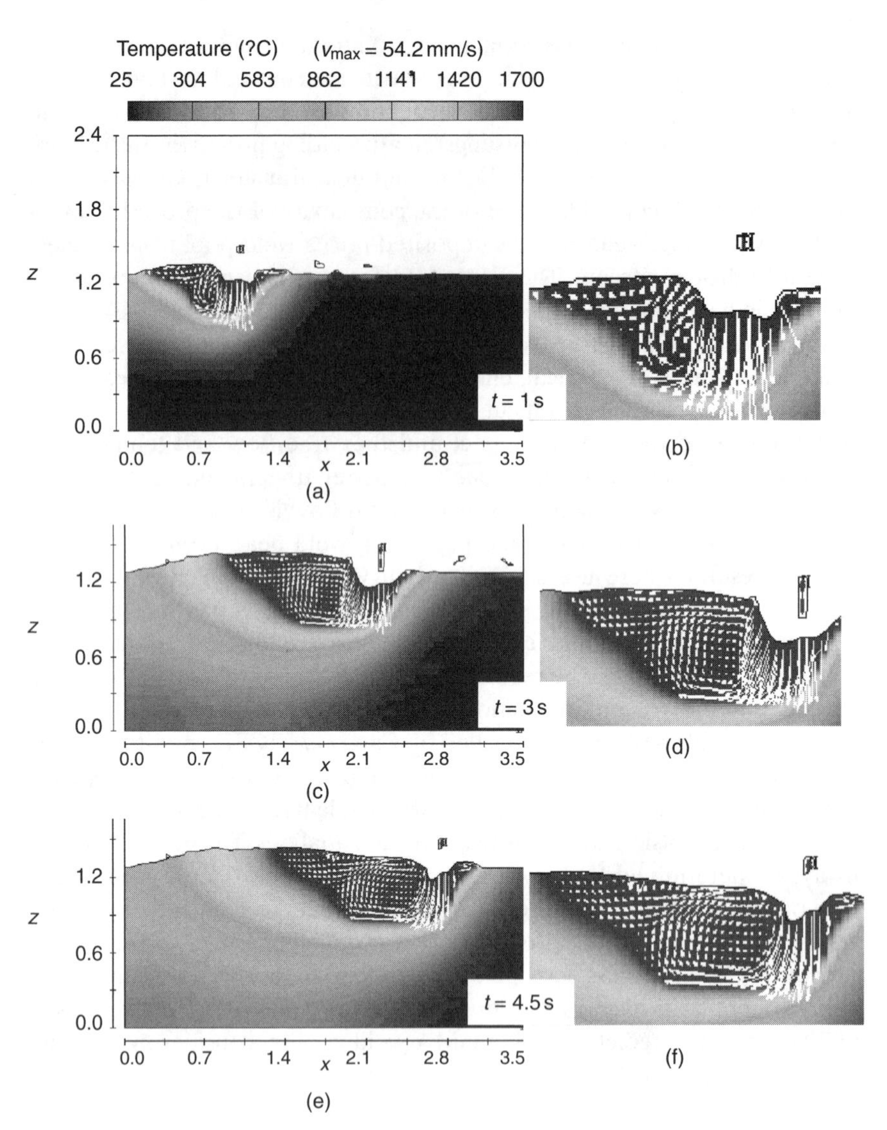

7.22 Temperature and fluid flow fields at the longitudinal section at different moments for the bead-on-plate case.[10]

pool represent the molten material flow direction and magnitude. It can be seen that the weld pool penetration is developed gradually as the welding time increases. As droplets enter the weld pool, both deep pool surface depression under the arc and high elevation away from the arc are clearly observed. The deposited material forms weld reinforcement behind the welding torch.

Figure 7.22(a), Fig. 7.22(c) and Fig. 7.22(e) show the temperature and fluid flow fields, and Fig. 7.22(b), Fig. 7.22(d) and Fig. 7.22(f) show the corresponding enlarged local views. It is seen that the weld pool depth does not change much after 3 s, which means that the majority of penetration happens in the first 3 s since the high-speed droplets penetrate the pool so quickly that this kind of penetration reaches equilibration after 3 s. As for fluid flow inside the weld pool, Fig. 7.22(a), Fig. 7.22(c) and Fig. 7.22(e) show a complex flow pattern in the weld pool. There exist two flow loops at the longitudinal cross-section at all three moments. The larger radial inward flow loop is dominated by the high-speed droplets and electromagnetic force. This is the driving force for GMA to form a deeply penetrated weld. The small radial outward flow loop behind the former flow loop is driven by the surface tension gradient, and it makes the weld pool longer.

Figure 7.23 shows the comparison of the predicted weld bead shape and the experiment coupon. The predicted width and depth are 1.24 mm and 0.39 mm respectively, and the measured width and depth are 1.29 mm and 0.41 mm respectively. Also, the finger-like penetration is captured from the prediction, and its penetration shape matches well the experimental shape.

7.6.2 T-fillet joint and plug weld

The T-fillet joint is one of the most common weld joint configurations. In this study, two welding positions, horizontal and flat, are simulated. Figure 7.24 shows the calculated temperature distribution and fluid flow field at transverse cross-sections and at different moments of time for the horizontal welding position. As shown in Fig. 7.24(a), finger-like penetration still occurs for this case because of the combination of the droplet impact force and electromagnetic force. The maximum velocity at this moment is 53.9 mm/s, which is slightly larger than the specified velocity of the metal droplet. When the arc moves away from this location, as shown in Fig. 7.24(b), the maximum velocity decreased to 7.29 mm/s. This is because the

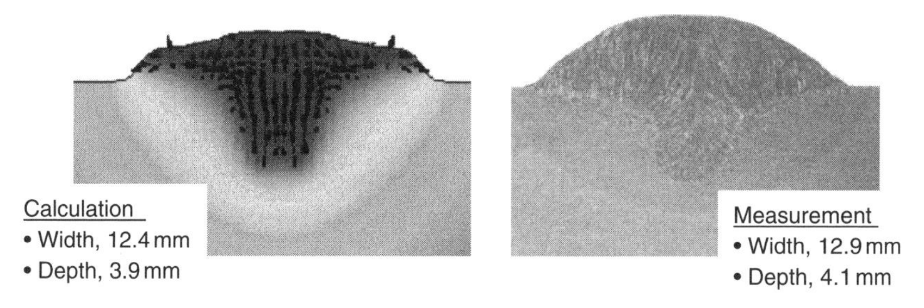

7.23 Comparison of predicted and actual bead-on-plate weld profiles.

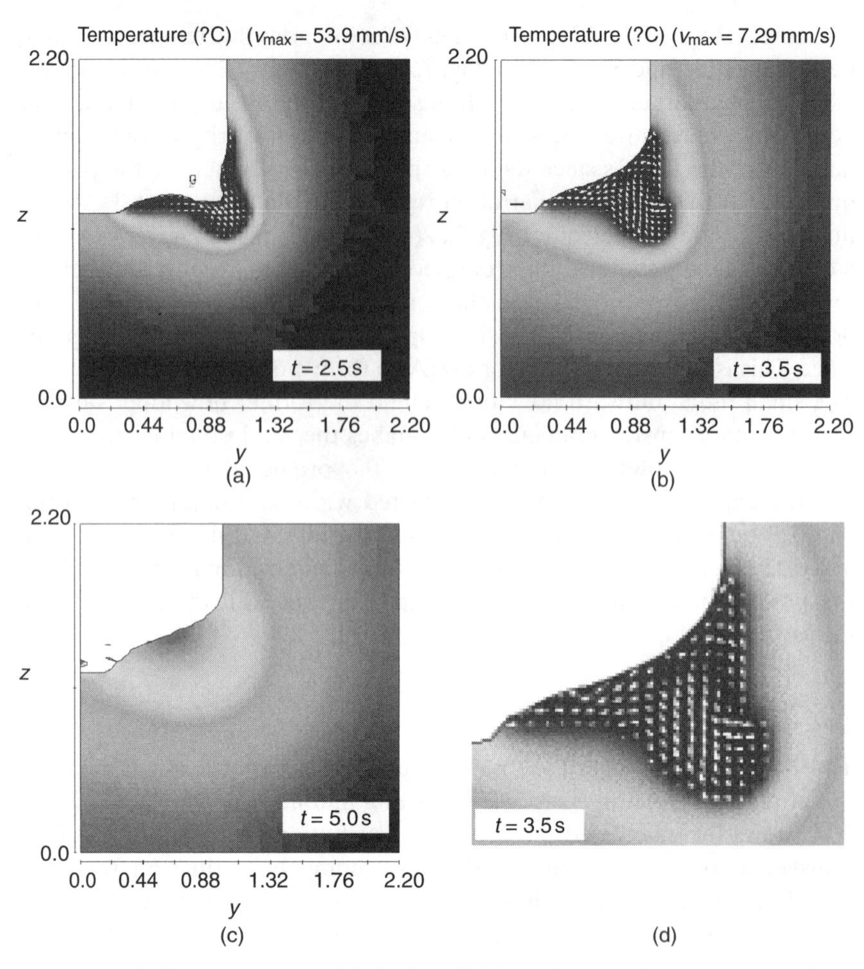

7.24 Temperature and fluid flow fields at the transverse cross-section at different moments for the horizontal T-fillet case.[10]

surface tension gradient force at this moment is dominant at this location. Thus, the fluid near the pool surface flows from the center outward. In addition, more fluid is formed near the horizontal plate owing to gravity. As a result, the leg length of the weld on the horizontal side is larger than that on the vertical side. The predicted final weld bead shape after solidification is shown in Fig. 7.24(c). The 3D temperature distribution for this case is shown in Fig. 7.25. It can be observed that there is severe spatter at the beginning of the weld, and some spatter is located on the side of the weld on the horizontal plate. However, the situation is improved when time increases. The weld bead shape can be observed clearly. Figure 7.26 compares the predicted and actual weld bead shapes from both horizontal-

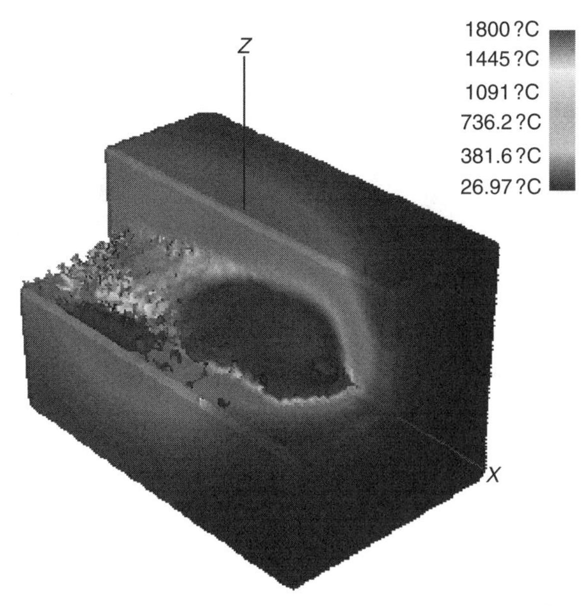

7.25 Three-dimensional temperature distribution for the horizontal T-fillet case.[10]

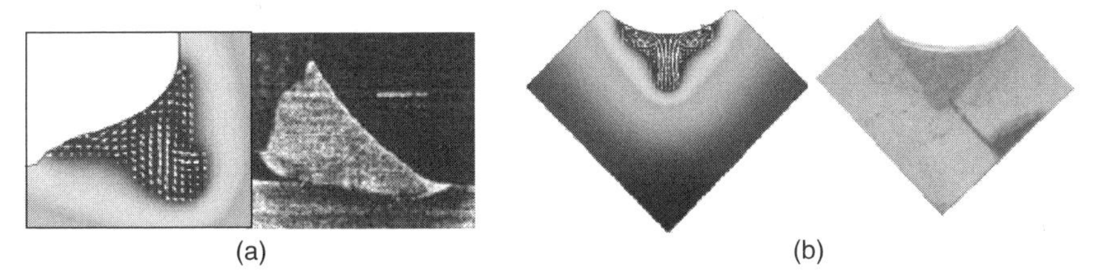

7.26 Comparison of predicted and actual weld bead shapes from (a) horizontal-position welding and (b) flat-position welding.

position welding and flat-position welding. The simulation is able to predict the bead shape and weld penetration accordingly.

The weld thermofluid dynamic model was used to solve a problem of resolution for weld defects of a plug weld process for a fabricated component in a mining machine. Figure 7.27 demonstrates the model's capability to predict a weld defect, namely lack of fusion, in a plug weld. The 3D weld thermofluid flow model is effective in optimization of welding parameters to obtain weld bead shapes with large toe radii and deep penetration. A better weld bead shape can be achieved by optimizing welding parameters based on the results from simulation of the weld thermofluid flow.

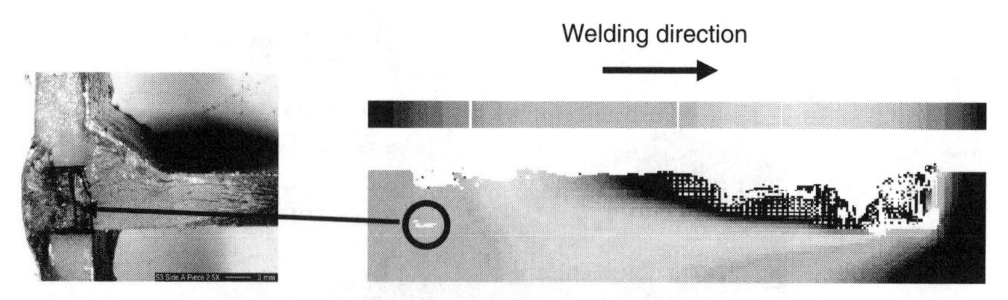

7.27 Simulation of a weld void in a plug weld.

7.7 Assessing residual stress in pipe welds

Pipes and piping systems represent a large and significant application area of welded construction. There are thousands and thousands of kilometers of buried pipelines in North America alone and essentially all industrialized countries have significant buried pipelines. According to the US Office of Pipeline Safety, there were more than 257 000 km of liquid transportation lines and 480 000 km of natural gas lines in the USA alone in 2003. The transportation pipelines typically carry natural gas or liquid petroleum products from place to place throughout the countries. Integrity of the transportation pipeline welds is often the main concern in these systems. Degradation of buried pipelines is often due to corrosion and, at times, unexpected service loads such as heavy-truck traffic, landslides and earthquakes. In addition, welded pipelines and corresponding pressure vessels represent a large part of power plants, both conventional and nuclear, throughout the world. There are extensive piping systems in nuclear-powered aircraft carriers and surface ships.

The main concern with the welds in piping is the weld-induced residual stresses since stress corrosion cracking must be controlled. Weld residual stresses can also contribute to fatigue life degradation. Distortion control of vessels piping systems is often of less concern than the residual stress control although there are instances where distortions must be managed (for instance, the shrouds in boiling-water nuclear power plants). This section will provide some examples concerning control and management of weld residual stresses in pipe.

7.7.1 Typical residual stress patterns in pipe welds

Weld residual stresses were calculated for a series of typical pipe geometries for type 316 stainless steel material. It will be seen that the results can be used to estimate the residual stresses for many materials in order to perform fatigue or corrosion assessments. The welding residual stresses were calculated using a half-symmetry axisymmetric model. Experience and comparisons with experimental data clearly suggest that, for pipe weld analysis, an axisymmetric model is adequate. The finite-element model was subjected to a numerical thermal analysis, which simulated the weld process functions of laying down the molten bead of weld rod, introducing heat energy into the weld bead and cooling the weld to an interpass temperature. The thermal analysis calculated the temperatures throughout the finite-element model through the welding process. A subsequent stress analysis was performed which used the previously defined temperatures to calculate the elastic–plastic residual strains in the welded pipe segment due to the thermal effects of welding. As discussed in Chapters 5 and 8, a proper constitutive law is required (see, for example, the work by Brust *et al.*[11]) in order to account properly for the many unique aspects of the welding process that are not accounted properly for in the constitutive law libraries of many commercial finite-element codes.

A test matrix of pipe radius, thickness and crack size was used. Pipe wall thicknesses t of 7.5 mm, 15 mm, 22.5 mm and 30 mm were studied in pipes with mean radius-to-thickness ratios of 5, 10 and 20. Cracks with half-lengths in radians of $\pi/16$, $\pi/8$, $\pi/4$, and $\pi/2$ were introduced in these virtual pipes to assess the effect of weld residual stresses on crack opening areas. The matrix of results was used to produce correction factors for crack-opening displacement equations applicable to a broad range of pipe sizes. Leak before break and the effect of weld residual stresses are briefly discussed later.

Figure 7.28 shows a typical weld model used for the 15 mm thick pipe. Standard weld groove and weld pass geometries commonly used for stainless steel welding were adapted from the report by Barber *et al.*[12]. They also showed that precise weld groove geometry has a second-order effect on the weld-induced residual stress state. In addition, measurements of stress by Barber *et al.*[12], Brust *et al.*[13] and Fredette and Brust[14,15] (and the references sited therein) clearly illustrate the accuracy of the residual stress predictions.

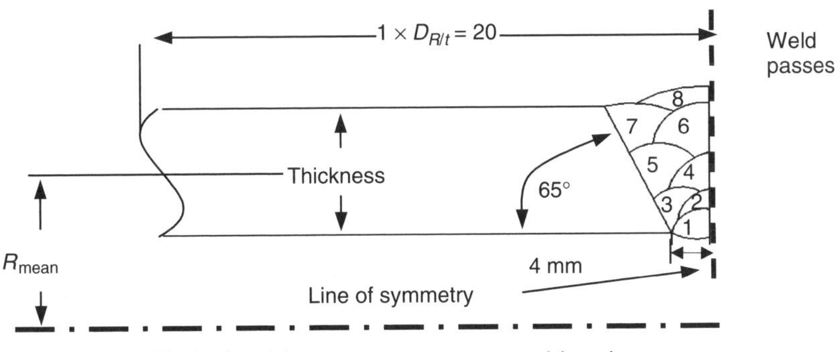

7.28 Typical weld groove geometry considered.

Table 7.1 Pipe thicknesses and welding energy inputs used in the simulations

Pipe thickness (mm (in))	Energy (kJ/mm (kJ/in))
7.5 (0.295)	0.526 (13.354)
15.0 (0.590)	0.660 (16.760)
22.5 (0.886)	0.794 (20.177)
30.0 (1.181)	1.01 (23.583)

The actual weld parameters of voltage and current were measured for multiple weld passes on several pipe thicknesses[12,13] and were used here. These values are shown in Table 7.1. These weld parameters are quite typical of those used in nuclear piping for austenitic steels.

A typical set of both axial- and hoop-direction residual stresses are shown in Fig. 7.29 for the 15 mm pipe with different R/t ratios. It is seen that axial residual stresses produce a 'bending'-type residual stress pattern for pipes of this thickness. This is quite typical. In fact, the axial residual stress pattern in pipe, for any material, typically produces a 'bending'-type distribution for pipes with thickness less than about 20 mm, and a 'tension–bending–tension'-type distribution near the weld for thicker pipe. Dong and Brust[16] summarized typical residual stress distributions in pipe, vessels and other weld joint types. Note also that, in all cases, the residual stress pattern reverses itself as one proceeds away from the weld zone. This occurs for both axial and hoop residual stresses. Note that, in this example, the weld bead was not ground out at the inner-diameter (ID) and outer-diameter (OD) regions near the weld. As such, the stress concentration near the weld bead discontinuity at both the ID and OD of the pipe should be noted.

Note also from Fig. 7.29 that axial residual stresses typically are higher near the weld zone as the R/t ratio of the pipe decreases because the pipe is stiffer and there is more axial direction constraint. Moreover, the axial residual stresses damp out more quickly as the R/t ratio decreases for this same reason. On the other hand, the hoop weld residual stresses are larger as the R/t ratio increases. One can think of the weld bead shrinkage in a pipe as applying a banded pressure to the pipe in the weld region. This 'pressure band' dies out more quickly in stiffer pipe (lower-R/t-ratio pipe). It will be seen next that these results can be used to make stress corrosion crack growth and fatigue life prediction estimates for many pipe materials at temperatures other than these. It should be noted that all stresses shown here are at room temperature.

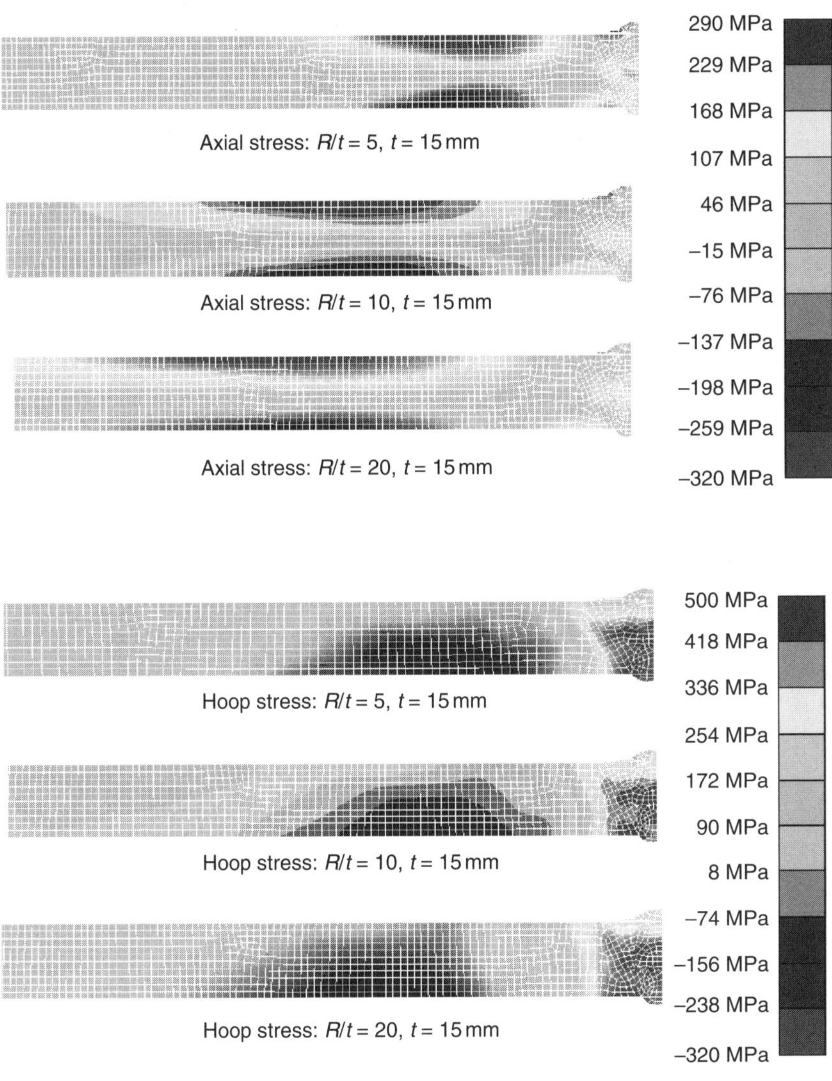

7.29 Typical axial and hoop residual stresses for 15 mm pipe.

7.7.2 Generalization of results for corrosion and fatigue predictions

The axial residual stresses for all cases considered are plotted as line plots in Fig. 7.30 along the ID surface and the OD surface. It is seen that, for the thinner pipe ($t \lesssim 20$ mm), stresses are higher for lower R/t ratios and this trend reverses for thicker pipe ($t \gtrsim 20$ mm). Note that 20 mm is roughly the thickness where the stresses change from a 'bending'-type (tension on ID)

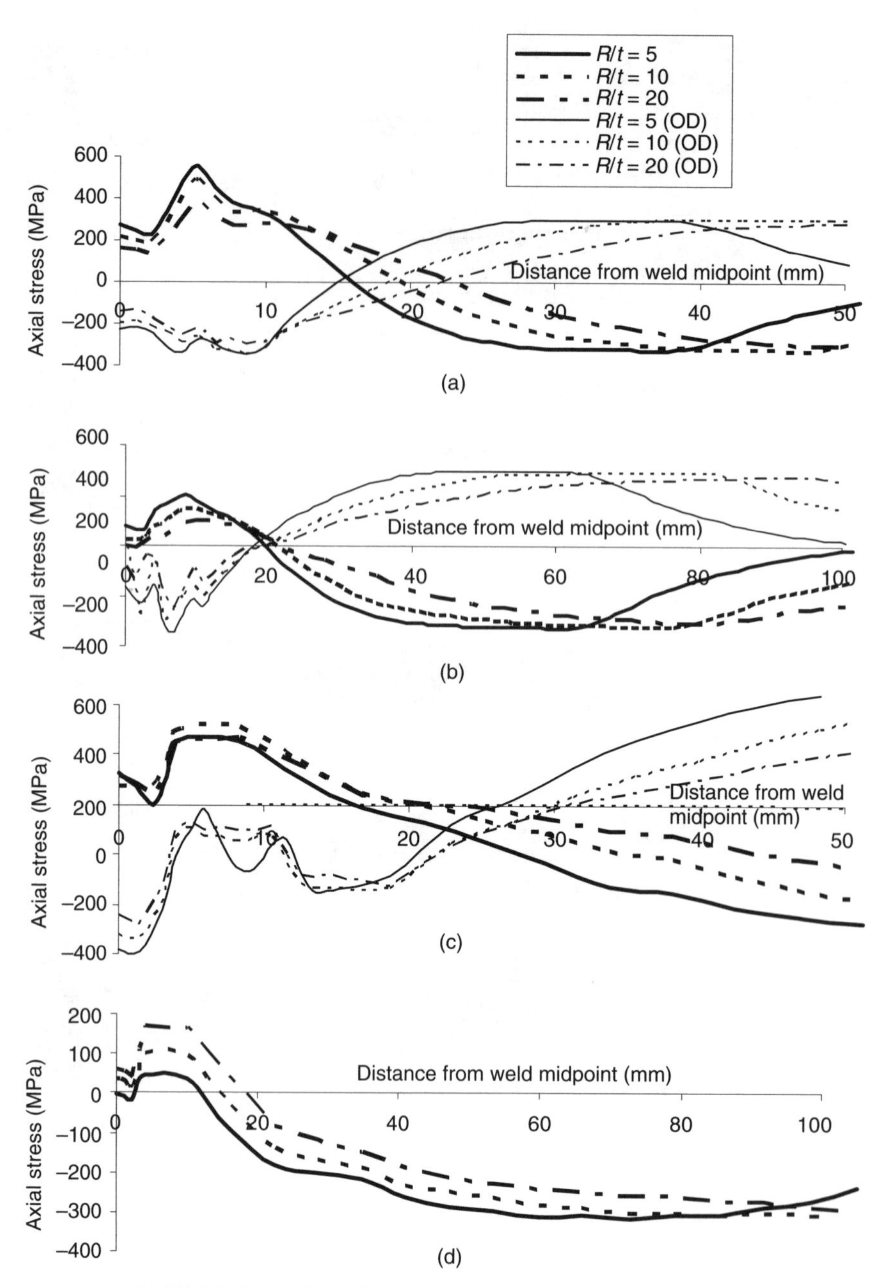

7.30 Weld simulation of pipe of various thicknesses t: (a) 7.5 mm; (b) 15 mm; (c) 22.5 mm; (d) 30 mm.

to a 'three-way bending' (tension on ID, compression and then tension on OD) type of distribution.

These results can be used to make engineering predictions of corrosion (or fatigue) for many different pipe materials and operating at different temperatures as follows. As mentioned earlier, the material used for these simulations was type 316 stainless steel, which, for many grades, is quite similar to type 304 stainless steel. The yield stress as a function of temperature for type 316 stainless steel is illustrated in Fig. 7.31. Note that, for this material, the yield stress does not vary linearly with temperature. Note also that at the melting temperature (1500 °C) the yield stress is given a negligible value (3 MPa) to facilitate numerical treatment of the weld metal not yet deposited (or melted material) (see modeling procedures discussed in Chapter 5).

Fortunately it has been observed that the residual stresses in welds vary nearly linearly with temperature for pipes. If the pipe is uniformly heated to 288 °C (typical nuclear reactor piping temperatures), the residual stresses will decrease by the yield stress ratio. Here, the yield stress at room temperature is 283 MPa, and at 288 °C is 215 MPa; the weld residual stresses scale by 215/283 or about 0.76. Hence the stresses in Fig. 7.29 or Fig. 7.30 can be reduced by 76% when performing a stress corrosion crack growth life prediction for the pipe at its operating temperature. Moreover, as an approximation, one can estimate the residual stress in a pipe not made of this material by likewise using the ratio of the yield stress of the material of interest to that of this material. While this latter approximation should be used with caution, one could obtain a rough estimate of corrosion or fatigue life using this method. However, it is always best to calculate the precise residual stress state for pipes of different materials since the

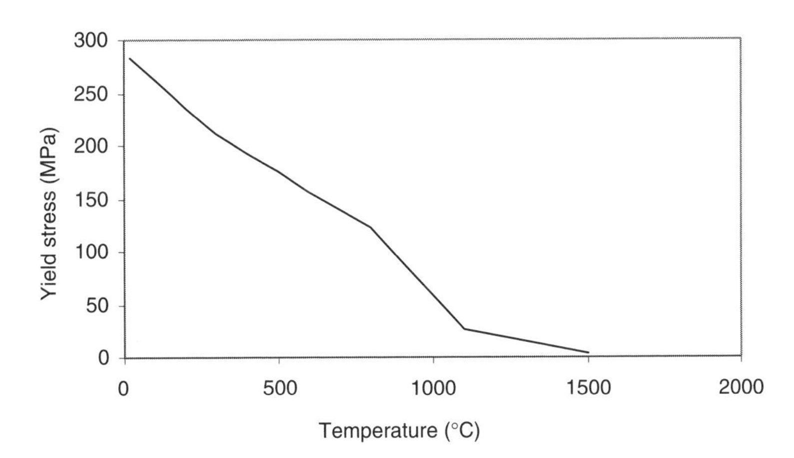

7.31 Yield stress versus temperature for type 316 stainless steel.

hardening response, phase transformation behavior, etc., can have an impact on the residual stress state in a welded pipe.

7.8 Assessing residual stress in non-steel welds

The examples presented in this chapter up to this point have all been concerned with steel. However, the methods for predicting the residual stress state in materials other than steel do not change from those summarized in Chapters 5 and 8. It is just that the material response, with regard to both the thermal properties and the constitutive response, must be considered appropriately. Moreover, modified welding methods must sometimes be used to prevent hot cracking during weld solidification of some materials. This section will first discuss general additional issues related to welding of materials other than steel and then provide an example of stress corrosion cracking analysis of a bimetal weld.

Hot cracking may occur during the weld solidification process owing to a combination of metallurgical behavior on cooling and the surrounding thermomechanical conditions near the weld. A traditional way to prevent hot cracking includes improving the weld and heat-affected zone ductility[17]. However, it has been realized that, for some materials, improving the ductility may not offer satisfactory results, especially for eliminating liquidation cracking. Consequently, thermomechanical welding techniques have received an increasing attention in recent years. For instance, Feng[18] analyzed conditions associated with weld metal solidification cracking and methods such as introducing a local heating source during welding to overcome end-cracking problems began to emerge.

The mechanisms associated with the local heating methods in preventing hot cracking can be attributed to the fact that the transverse and longitudinal strain fields were altered by the use of local heating. However, the resulting large heat-affected zone and coarser grain microstructure as a result of local heating have been of concern over its applicability in practice. It is also worth noting that local rolling can also be used to improve the thermomechanical conditions in welding. Computational models to investigate both hot-cracking mitigation methods, along with the use of a novel heat sink approach, in high-strength aluminum alloys have been recently summarized by Yang *et al.*[19,20].

The necessary conditions for hot cracking are the presence of tensile strains in the region that undergoes the brittle temperature range (BTR) near the weld as the material cools. Yang *et al.*[19] suggested using mechanical strains as a measure of the driving force for hot cracking instead of stresses since, at the BTR region, the transient stress level is usually low owing to significantly reduced material yield strength at high temperature. The idea is to prevent the tensile strain rate (with respect to temperature)

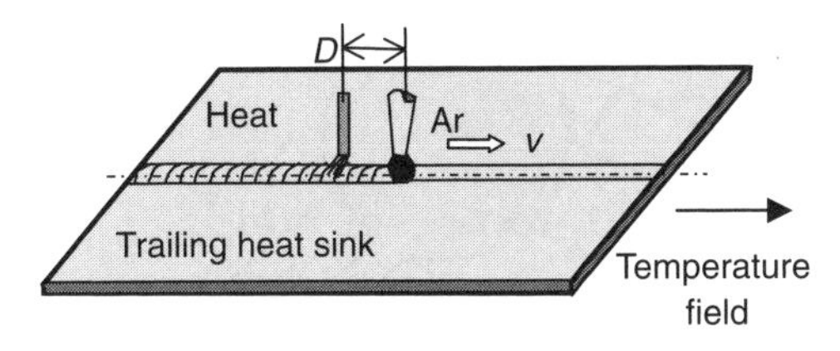

7.32 Example of a cooling source given by Yang *et al.*

exerted on the BTR region from becoming larger than the critical strain rate of the material using a local heating–cooling method. An example of using a cooling source is illustrated in Fig. 7.32.

The computational model of Yang *et al.*[19,20] is well suited to designing the parameters necessary to prevent hot cracking such as weld parameters, heating–cooling source location and intensity. By adding a heating sink (as shown in Fig. 7.32), an auxiliary compression zone can be generated between the heating source and the trailing heat sink. The effectiveness of the additional compressive strains on the strain rate (with respect to temperature) should be obvious. As a result, hot cracking can be prevented if the reduction in the strain rate with respect to temperature is sufficient within the BTR region. Note that the reduction in the strain rate (with respect to temperature) results from a complex interaction between the instantaneous strain rate and cooling rate during welding; thus a computational model is useful to define these parameters.

7.8.1 Example analysis of a bimetal weld

There have been incidents recently where cracking has been observed in the bimetallic welds that join the hot leg to the nuclear reactor pressure vessel nozzle (Fig. 7.33). The cracking has been identified as primary water stress corrosion cracking (PWSCC). The hot-leg pipes are typically thick-wall pipes of large diameter. Typically, an Inconel weld metal is used to join the ferritic pressure vessel steel to the stainless steel pipe. The cracking, mainly confined to the Inconel weld metal, is caused by corrosion mechanisms. Tensile weld residual stresses, in addition to service loads, contribute to PWSCC crack growth.

7.8.2 Weld and fracture analysis procedure

The analysis was quite involved since the actual field welds sequence was complicated and grinding and repair were considered. This analysis predicted the residual stresses for use in a PWSCC fracture assessment

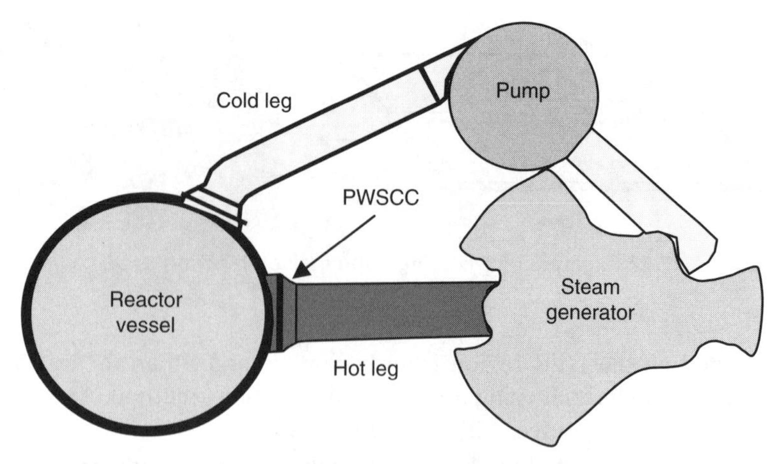

7.33 Example of PWSCC crack location in a primary water reactor nuclear plant.

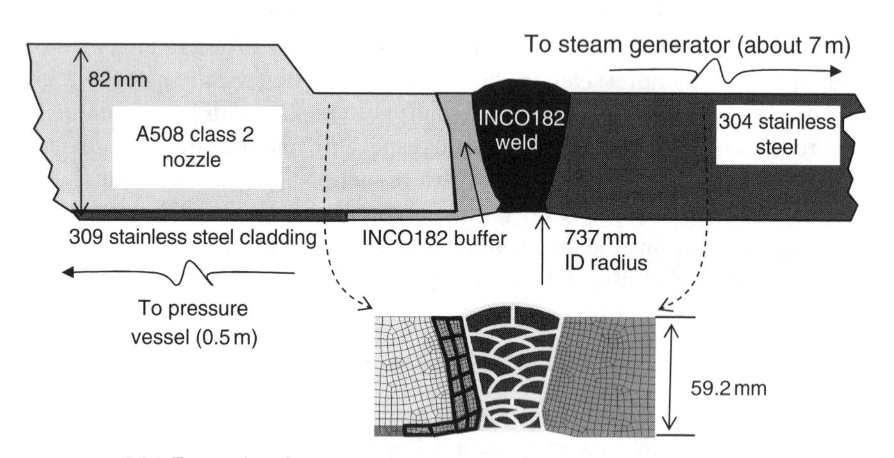

7.34 Example of a bimetallic weld model.

discussed later. Weld analyses of the design hot-leg bimetallic welds in the V. C. Summer plant were conducted. Figure 7.34 illustrates the weld geometry. The axisymmetric finite-element mesh used near the vicinity of the weld is shown in the lower portion. The distance from the center of the weld to the pressure vessel is about 0.5 m and the distance from the weld center to the steam generator is about 7 m. The finite-element mesh included the entire 7.5 m of pipe system. The boundary conditions used in the finite-element model included fixed displacements at the pressure vessel and steam generator. The material properties used, the weld sequence and many other aspects of the analysis have been summarized by Brust *et al.*[21]. The service stresses used for the PWSCC analysis are illustrated in Fig. 7.35.

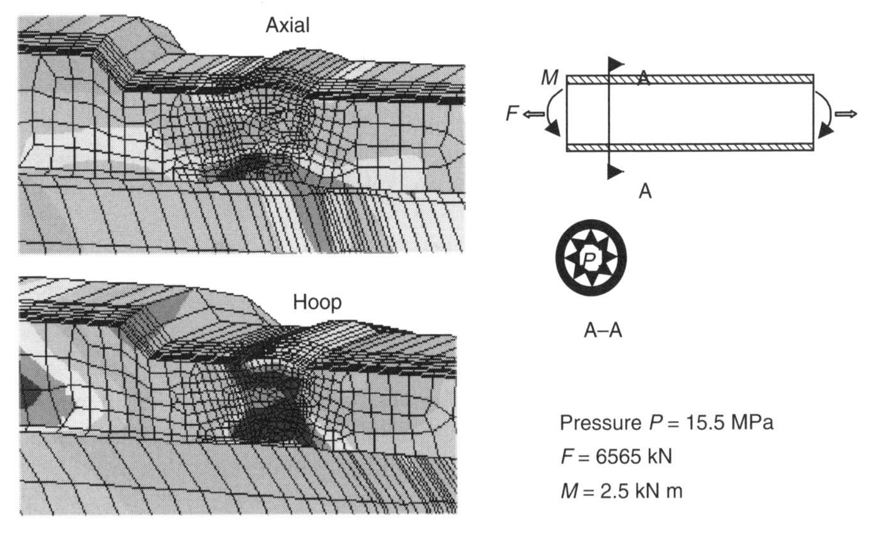

7.35 Final stress state: weld residual stress and operating loads.

Stress intensity factors were then calculated for the stress state illustrated in Fig. 7.35. Once stress intensity factors were available for many c/a values, for various crack depths, for axial and circumferential cracks, etc., for about 200 different cases, PWSCC predictions were made. The crack growth rate equation, taken from the work of Brust *et al.*[21], is

$$\frac{da}{dt} = 1.4 \times 10^{-11}(K_I - 9)^{1.16} \text{ m/s}$$

Here K_I is in MPa $m^{1/2}$ and the range for the data is for K between 20 and 45.

Some important conclusions from this study are listed below. Based on the PWSCC crack growth law[21] and the analysis results here, axial cracking should be confined to the weld region. Starting from a circular crack 5 mm in depth, the crack should break through the pipe wall within 2 years.

1. PWSCC growth is attributed to aging in INCO182 weld metal and is confined to the weld metal.
2. Circumferential cracks should take about twice as long to become a through-wall crack compared with axial cracks. Circumferential cracks will tend to grow longer than axial cracks. However, since service loads dominate circumferential cracks, they will slow their circumferential growth as they grow toward the bottom of the pipe. Here, by the bottom of the pipe, is meant the compressive bending stress region of the pipe. The service loads consist of thermal expansion mismatch, tension

caused by 'end-cap' pressure, and bending. The bending stresses caused by a bending moment are compressive 180° from the tension zone. Part-through circumferential cracks that initiate in the tension zone and grow beyond the bending neutral axis may slow down as they approach the compressive bending stress zone. However, for nonfixed bending axes, where the tension zone changes, this may not be significant.

3. Weld repairs alter pipe residual stress fields near the start–stop regions of the repairs. This may help to slow down a growing stress corrosion crack.

From this example it should be clear that weld-induced residual stresses play a very important role in stress corrosion crack growth. This example shows that, if material properties are available, there are no complications in considering weld residual stresses in multiple-material welds.

7.9 References

1. Yang, Z., Cao, Z., Chen, X.L., and Ludewig, H.W., 'Virtual welding – applying science to welding practices', *Proc. 8th Int. Conf. on Numerical Methods in Industrial Forming Processes*, American Institute of Physics, Columbus, Ohio, 2004, pp. 1308–1313.
2. Chen, X.L., Yang, Z., Nanjundan, A., and Chen, N., 'Application of computer simulation in industry – achieving manufacturing quality and reliability using thru-process simulation', *J. Physique, Coll.*, **JP.120**, 793–801, 2004.
3. Zhang, J., Cao, Z., Dong, D., and Brust, F.W., 'Weld process modeling and its importance in a manufacturing environment', SAE Technical Paper 981510, Society of Automotive Engineers, New York, 1998.
4. Cao, Z., Brust, F.W., Nanjundan, A., Dong, Y., and Jutla, T., 'A comprehensive thermal solution procedure for multiple pass and curved welds', *Proc. 2000 ASME Pressure Vessels and Piping Conf.*, Seattle, Washington, USA, 23–27 July 2000, American Society of Mechanical Engineers, New York, 2000.
5. Cao, Z. Brust, F.W., Nanjundan, A., Dong, Y., and Jutla, T., 'A comprehensive thermal solution procedure for different weld joints', *Advances in Computational Engineering and Sciences* (eds S.N. Atluri and F.W. Brust), Tech Science Press, Irvine, California, 2000, pp. 630–636.
6. *User Manual for VFT – Virtual Fabrication and Weld Modeling Software*, Battelle Memorial Institute and Caterpillar Inc., February 2002.
7. Brust, F.W., Jutla, T., Yang, Y., Cao, Z., Dong, Y., Nanjundan, A., and Chen, X.L., *Advances in Computational Engineering and Sciences* (eds S.N. Atluri and F.W. Brust), Tech Science Press, Irvine, California, 2000, pp. 630–636.
8. Yang, Z., Chen, X.L., Dong, Y., Martin, E., and Michael, D., 'An integrated FEA based procedure for weld fixture design', *Proc. 11th Int. Conf. on Computer Technology in Welding* (eds T.A. Siewert and C. Pollock), National Institute of Standards and Technology, Columbus, Ohio, 2001, pp. 203–210.
9. Yang, Z., Chen, X.L., Chen, N., Ludewig, H.W., and Cao, Z., 'Virtual welded-joint design by coupling thermal–metallurgical–mechanical modeling', *Proc. 6th*

Int. Conf. on Trends in Welding Research, ASM International, Materials Park, Ohio, 2002, pp. 861–866.
10. Cao, Z., Yang, Z., and Chen, X.L., 'Three-dimensional simulation of transient GMA weld pool with free surface', *Weld. J.*, **83**, 169-s–176-s, 2004.
11. Brust, F.W., Dong, P., and Zhang, J., 'A constitutive model for welding process simulation using finite element methods', *Advances in Computational Engineering Science* (eds S.N. Atluri and G. Yagawa), Tech Science Press, Irvine, California, 1997, pp. 51–56.
12. Barber, T.E., Brust, F.W., and Mishler, H.W., 'Controlling residual stresses by heat sink welding', EPRI Report NP-2159-LD, Electric Power Research Institute, Palo Alto, California, 1981.
13. Brust, F.W., and Stonesifer, R.B., 'Effect of weld parameters on residual stresses in BWR piping systems', EPRI Report NP-1743, Electric Power Research Institute, Palo Alto, California, 1981.
14. Fredette, L., and Brust, F.W., 'Effect of weld induced residual stresses on pipe crack opening areas and implications on leak-before-break considerations', *Proc. ASME Pressure Vessels and Piping Conf.*, PVP-Vol. 434, *Computational Weld Mechanics, Constraint, and Weld Fracture* (ed. F.W. Brust), Vancouver, British Columbia, Canada, 4–8 August 2002, American Society of Engineers, New York, 2002.
15. Fredette, L., and Brust, F.W., *Trans. ASME, J. Pressure Vessels and Piping*, 2005 (submitted).
16. Dong, P., and Brust, F.W., 'Welding residual stresses and effects on fracture in pressure vessel and piping components: a millennium review and beyond', *Trans. ASME, J. Pressure Vessel Technol.*, **122**(3), 329–339, 2000.
17. David, S.A., and Vitek, J.M., *Int. Mater. Rev.*, **34**, 213–245, 1989.
18. Feng, Z., *Weld. World*, **33**, 340–347, 1994.
19. Yang, Y., Dong, P., Tian, X., and Zhang, Z., 'Prevention of welding hot cracking of high strength aluminum alloys by mechanical rolling', *Proc. 5th Int. Conf. on Trends in Welding Research* (eds J.M. Vitek, S.A. David, J.A. Johnson, H.B. Smart and T. DebRoy), Pine Mountain, Georgia, USA, 1–5 June 1998, American Welding Society, Miami, Florida, ASM International, Materials Park, Ohio, 1998, pp. 700–705.
20. Yang, Y.P., Dong, P., Zhang, J., and Tian, X., 'A hot-cracking mitigation technique for welding high strength aluminum alloy sheets', *Proc. 6th Int. Conf. on Trends in Welding Research* (eds S.A. David, T. DebRoy, J.C. Lippold, H.B. Smartt and J.M. Vitek), ASM International, Materials Park, Ohio, 2002, pp. 844–849.
21. Brust, F.W., Scott, P.M., and Yang, Y., 'Weld residual stresses and crack growth in bimetallic pipe welds', *Proc. 17th Int. Conf. on Structural Mechanics in Reactor Technology* (ed. S. Vejroda), Prague, Czech Republic, August 2003, Brno University of Technology, Brno, 2003, Section G, CD-ROM.

8

Mitigating welding residual stress and distortion

F. W. BRUST, Battelle Memorial Institute, USA and
D. S. KIM, Shell Global Solutions (US), USA

8.1 Introduction

Residual stress is defined as any stress that would exist in a continuum body if all external loads were removed. Welding residual stresses occur in a structure because of the thermal stress developed by a nonuniform temperature distribution during welding, followed by plastic deformations. The corresponding distortions are in turn caused by these residual stresses. During arc welding, the welded piece undergoes complex temperature changes. The nonuniform temperature distribution along a continuum body causes thermal stresses and nonelastic strains. Many researchers and engineers have tried various techniques to quantify residual stresses analytically, numerically and experimentally. The root cause and methods for calculating weld-induced residual stresses and distortions were detailed earlier in this book.

The purpose of this chapter is to discuss methods to control, or even to mitigate, weld-induced residual stresses and distortions. Chapter 9 discusses methods for controlling buckling distortions in 'thin' welded components which can lead to unsightly fabrications among other problems. Chapter 7 discusses a number of practical industrial applications of management of weld residual stresses and distortions in very large 'real-world' fabrications. This chapter will discuss general techniques for controlling weld residuals in a general sense. As such, the examples will generally focus on 'smaller-scale' structures. Because of the power of numerical weld modeling today most examples presented here are based on finite-element-based solutions. All numerical solutions here were developed using the virtual fabrication technology (VFT™) software[1] (see Chapter 7 for more details).

8.2 History dependence of residual stress

Usually distortion and residual stress control assessments are made by neglecting prior history in the components that make up fabricated structures. Consider the cycle represented by Fig. 8.1 for a fabricated steel structure. Start from the upper left and move clockwise. The different components may be cast, rolled, forged, etc., from the material supplier, and often combinations are used in the service structure. An initial residual stress and distortion pattern develops depending on which process is used. For instance, in rolled plate, tensile stresses usually develop near the plate edges and compressive stresses develop in the interior.

The rough material stock is then transported from the steel mill to the fabrication shop or job site. The transportation, handling and storage may alter the original stress and deformation history inherent in the rough stock.

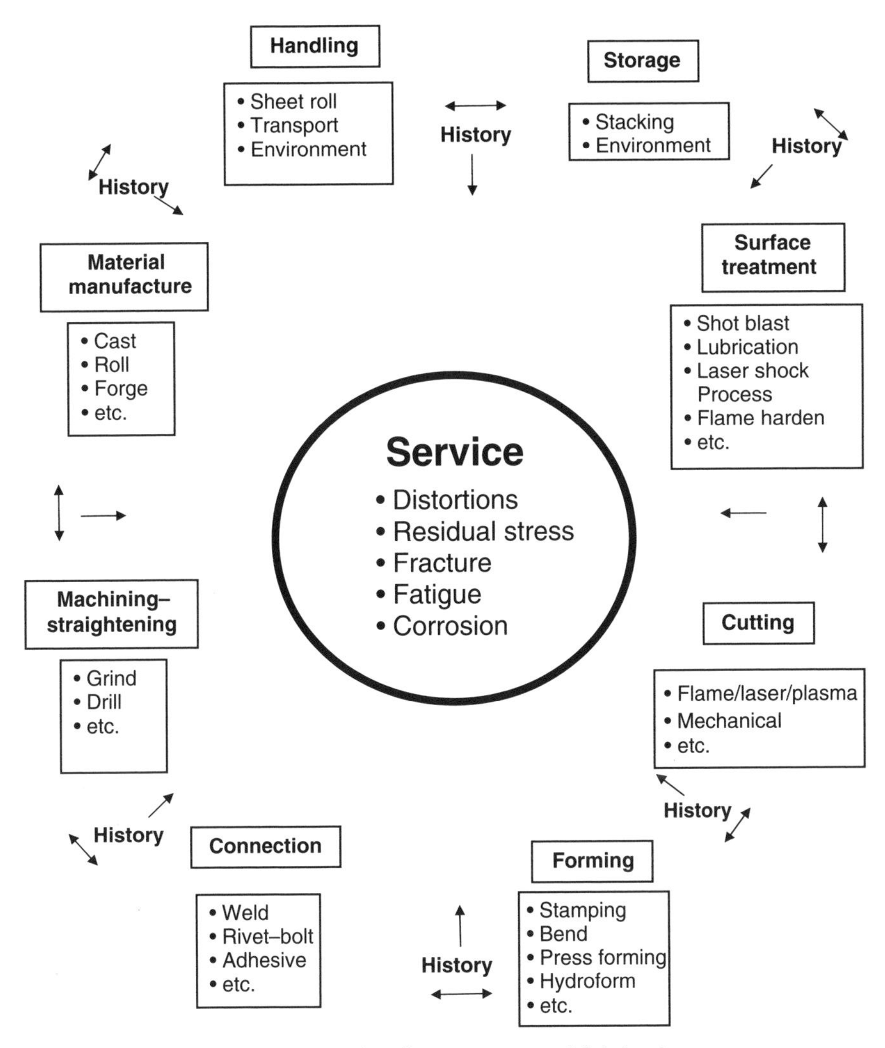

8.1 Operations history leading to structural fabrication.

The material may then be surface treated, depending on the application, further altering the original history in the rough stock. The material is then cut into the desired shape required for the fabrication (for instance, consider a truck frame consisting of a number of components (forged, cast and rolled) and welded together). This cutting process alters the original residual stress history (and thus distortion history). Moreover, the cutting process may induce additional residual stresses near the cut edge (particularly for thermal cutting since the thermomechanical flame cutting process induces nonlinear strains near the cut edges).

Again, referring to Fig. 8.1, the component might then also be bent and formed into the required shape for the fabrication. Forming again alters the original history and can add significant stresses and strains in the bent regions of the component. Finally, the welded fabrication is made on top of all the residual stresses already induced into the parts from all the prior processes. The connection process further alters the residual history in the component parts.

In particular, the welding process induces significant residual stresses and distortions into the assembled structure. The assembled component or structure might then be straightened to achieve certain desired tolerance requirements. This process further alters the history in the structure. As seen, by the time that the material from the material supplier (steel plant, aluminum foundry, etc.) makes its way into the service structure, each component has already seen a history of stresses, nonlinear strains and corresponding displacements.

As an example, consider Fig. 8.2. This is an example of the effect that bend forming and welding has on the service residual stress state for mild steel. As illustrated on the left of Fig. 8.2, a two-dimensional nonlinear finite-element analysis was performed of a plate being bent in a die. Next, a weld bead was deposited on the exterior of the bend curvature. Both the bending and the welding processes were modeled via the finite-element method (the weld models used are described next). The right-hand side shows contour plots of the X component of stress (this component is responsible for crack growth at the weld toe). The material was a standard structural steel. The stress magnitudes vary between +300 MPa and –300 MPa (the magnitudes are not important for our purposes). It is clearly seen that stresses at the weld toe are markedly different for the two cases.

This fabrication history can have an important effect on the service life of the structure. Fatigue, stress corrosion cracking, fracture, etc., can all be affected by prior history. However, in most cases this prior history is neglected in making a damage assessment of the structure which must include weld-induced residual stresses. Distortion, residual stress, fatigue and fracture assessments are often made by pretending that the service material is pristine and free of history. The corrosion or fatigue life

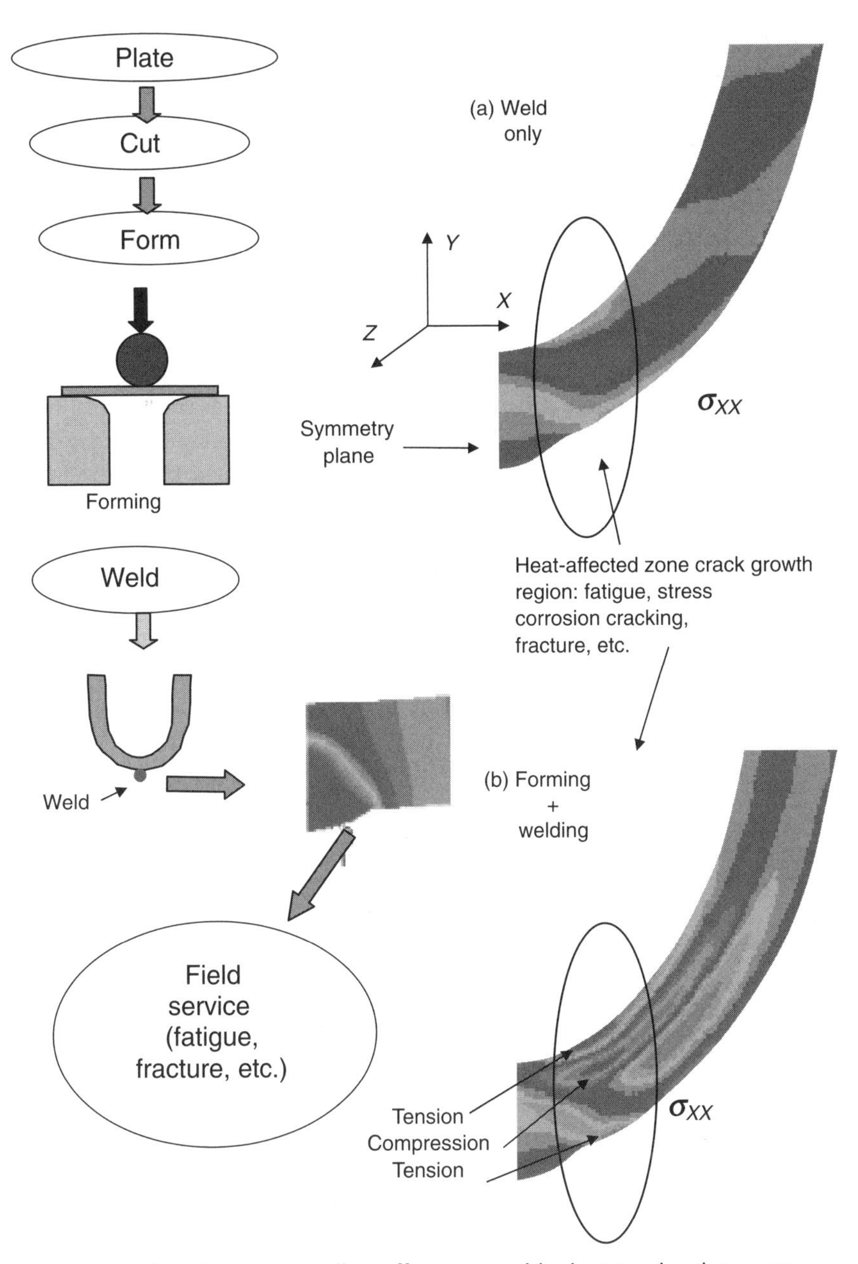

8.2 Bending and welding effects on residual stress development.

predicted assuming a pristine structure is different from that which would be predicted by including the prior history (welding alone (Fig. 8.2(a) or bending and welding (Fig. 8.2(b))). The residual stresses after welding are almost twice those when including the bending effects, as reported previously[1–5]. Examples will be shown later that clearly illustrate this effect.

Of course, for some critical structures, stress relief of many of the components is performed before the structure goes into service. This is true for instance in some nuclear components and engine components. However, the costs associated with stress relief are not practical for the vast majority of fabricated structures. In addition, stress relief, annealing and other processes do not eliminate all residual stresses and strains leading to a pristine structural component. Stress relief via post-weld heat treatment is discussed in Section 8.8.

Fortunately, the residual stresses and distortions caused by the welding process are often dominant compared with other fabrication-induced stresses. The prior history-induced residual stresses are often eliminated in the vicinity of the weld as the base material remelts or reaches such a high temperature that 'local stress annealing' occurs. As such, for many components it is valid to neglect prior residual stresses before welding. However, neglecting the prior history must always be kept in mind when using the mitigation methods discussed in the next few sections. One example of an application, which illustrates the importance of fabrication history on damage development and fracture, is illustrated in Section 8.2.1.

8.2.1 Application example illustrating history dependence effects: fracture of welded beams

Following the 1994 Northridge earthquake in California, widespread damage was discovered in the prequalified welded steel moment frames. Detailed inspections indicated that most of the structural damage occurred at the weld connections between the beam and column flanges. In particular, the weld joints between the bottom beam flange and the column face suffered the most severe damage. Cracks mostly initiated at the weld root and propagated with very little indication of plastic deformation. The desired plastic hinges assumed in the structural building codes plastic design were not formed in the weaker beam away from the weld joint. Instead, brittle weld fracture was identified as the dominant failure mechanism.

Welding-induced residual stresses are believed to be one of the factors contributing to the brittle fracture. Indeed, there exists ample evidence that residual stresses can play a dominant role in the fracture process of highly restrained welded joints[6, 7]. The design of welded moment-resistant frame connections presents perhaps the most severe mechanical restraint conditions both during welding and in service. Consequently, the presence of a high weld residual stress is expected. In addition, the high triaxial nature of the residual stress state, which can significantly reduce fracture toughness, in these joints can be significant. As such, the anticipated plastic deformation cannot develop before the fracture driving force reaches its critical value, resulting in brittle fracture.

Figure 8.3 illustrates an analysis of a typical beam–column connection (A36 beam, A572 Gr. 50 column and E70T-4 weld material). The insert illustrates the three-dimensional cross-section of the finite-element mesh that was used to model the welds in the beam–column. In addition, a two-dimensional model was used as well. The residual stresses (not shown here; see the paper by Yang *et al.*[8]) showed a very high degree of triaxial stress state and were directly caused by constraints induced by the welding process and the geometry. The two-dimensional mesh illustrated below the beam–column (Fig. 8.3) was used to study the fracture response of a typical lower flange in a beam–column connection. Nine passes were deposited (see the inset mesh of Fig. 8.3). The stress intensity factor is plotted as a function of the ratio of applied tensile stress to weld yield stress on the right of Fig. 8.3. It is seen that including the history of residual stresses in the analysis procedure has a marked effect on the stress intensity factor. Indeed, for fracture in the brittle or elastic-plastic

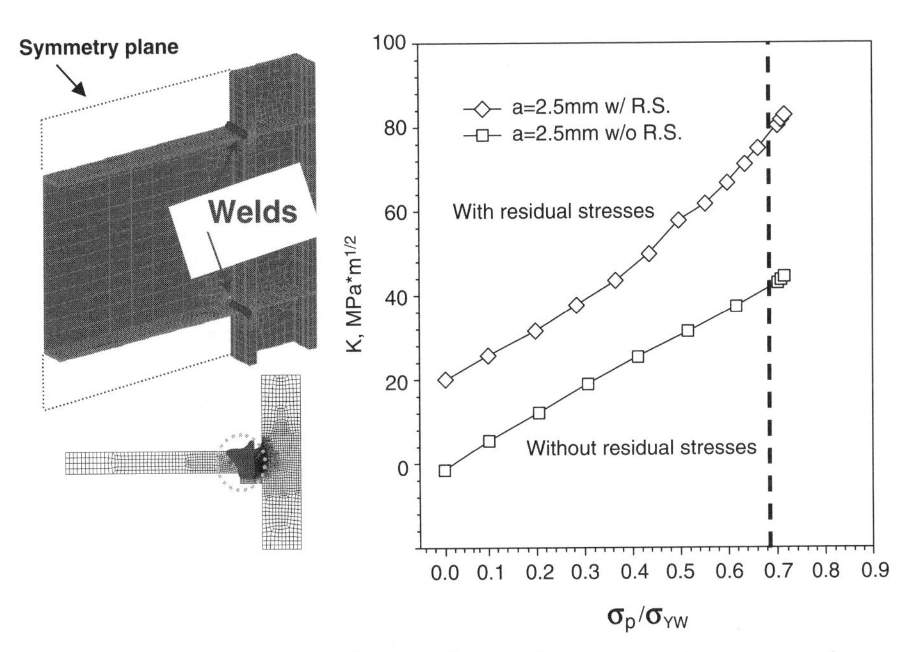

8.3 Residual stress effects on fracture for a beam–column connection.

regime, including prior history in the analysis is important to obtain correct results.

8.3 Methods for controlling residual stresses and distortions

During heating and cooling in the welding cycle, thermal strains occur in the weld and base-metal regions near the weld. The strains produced during heating are accompanied by plastic upsetting. The nonuniform plastic straining that occurs in the welded piece is what leads to residual stresses. These stresses react to produce internal forces, which must be equilibrated, that cause bending, shrinkage and rotation. It is these displacements that are called distortion.

Three fundamental dimensional changes that occur during the welding process cause distortion in fabricated structures[8,9].

1. Transverse shrinkage perpendicular to the weld line.
2. Longitudinal shrinkage parallel to the weld line.
3. Angular distortion (rotation around the weld line).

Many factors in the welding procedure contribute to the distortion. These factors are quite difficult to manage in a large complex structure. These include welding sequence, degree of restraint, welding conditions and joint details. At present, however, there is no process that completely eliminates distortion. Many distortion control techniques and weld design optimization procedures such as prestraining, weld sequencing and precambering were developed by experiment and experience. Today, however, with the advent of computational weld models which are quite accurate, new and creative distortion control methods are emerging because the scenarios can be evaluated on the computer.

With distortion control it is possible to design weld processes such that the final distortion is close to zero, with some probabilistic variation around the undistorted shape. It is difficult to weld components such that residual stresses are zero throughout the fabrication unless post-weld heat treatment is used. Post-weld heat treatment is expensive and is mainly used for high-performance structures where design requirements are stringent. For many fabrications where cost control is important, the goal with residual stress control is to reduce the stresses or, in some cases, to reverse these stresses from tension to compression in some critical areas of the structure. Such residual stress control can improve fatigue and corrosion life and even fracture performance. Methods that have been developed to control distortions also influence residual stresses and vice versa. With this in mind it is important to design carefully the control processes so as to optimize the fabrication.

8.3.1 Advantages of weld distortion and residual stress control

Two tremendous advantages are obtained by developing fabrication solutions via the computer. First, designing the fabrication to minimize or control distortions can significantly reduce fabrication costs. Second, controlling the fabrication-induced residual stress state can significantly enhance the structure's service life. For distortion control, fabrication design via modeling can achieve the following.

1. It can eliminate the need for expensive distortion corrections.
2. It can reduce machining requirements.
3. It can minimize capital equipment costs.
4. It can improve quality.
5. It can permit premachining concepts to be used.

Residual stress control via modeling has the following results.

1. It can reduce weight.
2. It can maximize fatigue performance.
3. It can lead to quality enhancements.
4. It can minimize costly service problems.
5. It can improve damage resistance during attack (e.g. naval structures).

Fabrication modeling tools were specifically developed for this purpose. Weld process models are sophisticated physics- or mathematics-based computer modeling software tools for use in optimizing and designing metal fabrication processes for industry. Savings from using a weld process model can be significant, often exceeding millions of US dollars per product.

There are a number of methods that can be used to control weld-induced distortions or residual stresses, or both. The method that is practical for a given product depends on many factors including cost. Some of the more popular methods for controlling welds include weld sequencing, weld parameter definition, precambering, prebending, thermal tensioning, heat sink welding and fixture design, among many others. It is important to note that fabrication modeling tools such as the VFT™ software[1] can be used to develop new control methods since the new methods can be first attempted on the computer. Some of these methods will be considered in the examples discussed later.

8.3.2 Residual stress and distortion control methods

The control methods listed below are some of the more popular methods of weld fabrication to control residual stresses and distortions. The procedures can be developed empirically, which is quite expensive and time consuming, or via weld modeling, which is efficient and is becoming more and more popular. Examples of many of these methods are presented in later sections.

Weld sequencing

Weld sequence simply means the order in which the welds are deposited. Sequencing is more important for distortion control although it can affect weld residual stresses as well. For some fabrications, weld sequencing is not sufficient for distortion control and it is used in conjunction with some of the other methods discussed below. Examples of the use of weld sequencing to control distortions are shown later.

Weld parameter definition

Weld parameters such as travel speed, weld groove geometry, weld size, single-pass versus multiple-pass welds, joint type and heat input can influence distortions and residual stresses, as illustrated in Section 8.4.1.

Weld procedure

The weld procedure chosen can have an important effect on distortions and residual stresses. Fusion welds often lead to the largest distortions while laser, electron beam welding and friction stir welding result in lower distortions. However, friction stir welding can impart large plastic strains to the structure even though the residual stresses may be low. These large strains, which locally strain harden the material, can influence the fracture response of the structure.

Fixture design

Fixtures control residual stresses and displacements by forcing the displacements and rotations of some portions of the welded component to be zero. The 'zero points' should be carefully designed to achieve the distortion control goals as seen in several examples shown in Section 8.5.

Precambering

Precambering consists of elastically (or plastically) bending some of the components (usually in a specially designed fixture) in a predefined manner and then welding. After welding, the precamber is released and the fabricated structure 'springs back' to a minimally distorted shape. The precamber pattern must be carefully designed. Precamber also affects weld residual stresses and as shown in an example later in this section.

Prebending

Prebending consists of plastically bending some of the components before welding and possibly before placing them in a fixture. After welding the desired 'nondistorted' shape results. The welding is performed with or without a fixture.

Thermal tensioning

Thermal tensioning consists of strategically moving a heat source ahead of, beside, behind (or combinations of these), the moving weld torch. It is most convenient to design the heat source locations and power through the use of a computational model. This method can control distortions and residual stresses during welding by controlling the heating and cooling rates of the weld. Several examples that illustrate this distortion and residual stress control method are summarized in Section 8.5.1.

Heat sink welding

Heat sink welding is similar to thermal tensioning except a cooling source is strategically moved (or kept stationary) with the weld torch. An example of use of this procedure is in the nuclear piping industry where the method produces compressive weld residual stresses in weld heat-affected zones along the pipe inner surface to mitigate stress corrosion cracking concerns.

Post-weld heat treatment

Post-weld heat treatment consists of heating parts (or all) of the welded fabrication to high temperatures (depending on the material) and holding for a period of time, while the stresses are relieved. Often the stresses cannot be fully relieved, i.e. some level of residual stress remains. A rule of thumb is that the hold time is about 1 h for each 25 mm of thickness. Again, this is expensive and is often used to prevent a service fracture problem such as corrosion, fatigue, creep or combinations.

Control of weld consumables

Welding consumables have recently been developed which result in a particular weld bead shape and can be used to control residual stresses.

Moreover, consumable materials are emerging which result in a 'low-temperature phase transformation'. This transformation can be used to control residual stresses and thus distortions. This method is especially useful in polymer welds.

Post-weld corrective methods

Finally, post-weld corrective methods are used to correct a distorted component or reduce residual stresses in a component. Corrections made 'after the weld' are often the most expensive and time consuming to implement. Section 8.8 discusses some of these methods.

This list is not exhaustive as other methods exist for controlling distortions and stresses in welded fabrications. Moreover, with the advent of computational weld models, where new control methods can be evaluated via the computer first before shop floor implementation, new methods continue to emerge. Examples of some of these procedures are presented in the following pages. Chapter 7 also shows some industrial examples using some or combinations of these methods.

8.4 Distortion control with weld parameter control and weld sequencing

There are some situations where careful design of the weld sequence is the only practical way to minimize the weld-induced residual stresses. In some cases the optimized weld sequence may not eliminate all the desired distortions but may reduce them to a level where mechanical or thermal straightening becomes much easier. Often weld sequencing is used in combination with one or more control procedures.

8.4.1 Weld parameters and sequencing on residual stress

Consider the case of a T-fillet weld in Fig. 8.4. This type of weld case was modeled extensively during the development phase of the VFT™ fabrication modeling software[1]. Numerous test mock-up welds were performed and temperatures, distortions and residual stresses were measured under varying conditions for model validation. This is a T-fillet weld of two plates of dimensions 305 mm by 610 mm by 12.5 mm, as seen in Fig. 8.4. A number of full three-dimensional analyses of this were performed and here we show results of some two-dimensional (generalized plane-strain) solutions which were validated with the three-dimensional solutions (the two-dimensional stress solution is quite accurate). A single-pass weld (Fig. 8.4, top right) and a three-pass weld (Fig. 8.4, bottom right) were considered in separate analyses.

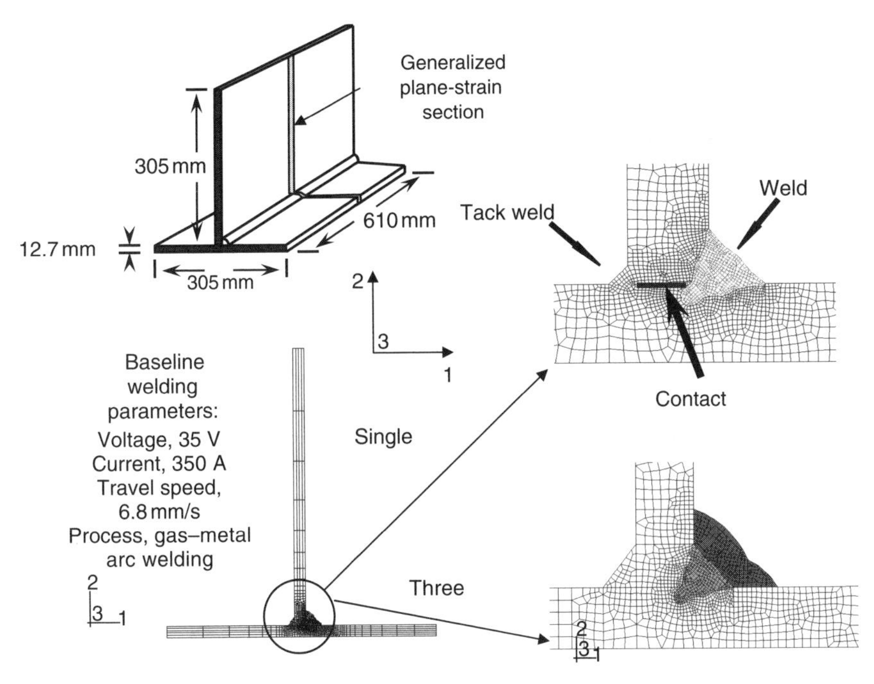

8.4 T-fillet weld example.

Weld torch speed effects

Figure 8.5 illustrates the effect of weld torch speed on the weld-induced residual stresses in the T-fillet weld shown in Fig. 8.4. The longitudinal stresses σ_L (longitudinal stresses are in the direction of the weld, in the '3 direction' in Fig. 8.4) have approximately the same magnitude (near yield). These stresses are plotted along the bottom (horizontal) plate and are measured from the toe of the weld. However, for the slower travel speeds, the longitudinal residual stresses extend a further distance from the weld toe. This is because the slower speeds permit a larger area to obtain a higher heat; so, when cool-down occurs, a larger distance from the weld toe experiences high stresses. On the other hand, the axial stresses σ_A (in the '1 direction' in the bottom plate) are higher for the faster travel speeds. Since the axial residual stresses can affect fatigue life of the component, this is very important.

Material yield stress

Section 8.2 discussed the important effects of the history of material work hardening (from material creation, forming, cutting, etc.) and how this can have an important effect on residual stresses. Figure 8.6 illustrates that the

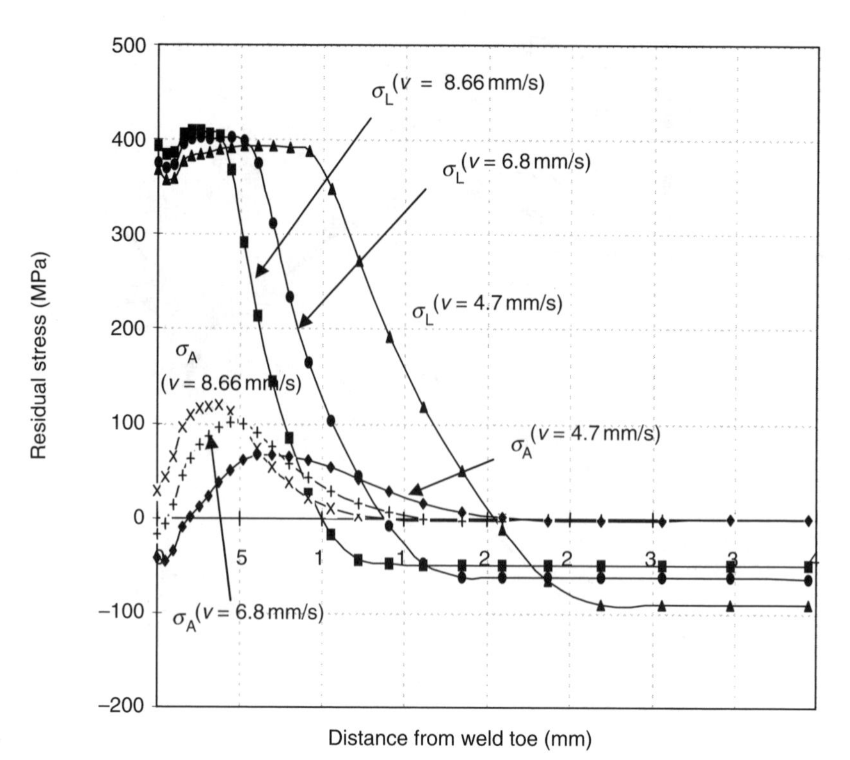

8.5 T-fillet weld example: torch speed.

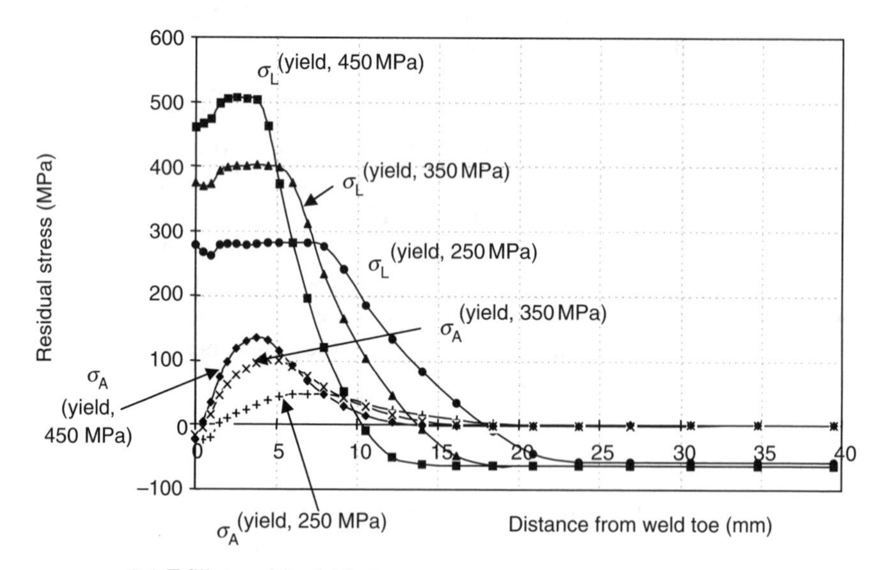

8.6 T-fillet weld: yield stress.

material yield stress has an important effect on both axial stress and longitudinal stress. In addition, as a general rule, the higher the yield stress of the material, the greater is the distortion because the magnitude of the residual stresses is higher. This is important because, in order to control variability in the fabrication process, it may be necessary to put upper limits on the yield stress maximums that are specified from the material supplier.

Multiple-pass sequence effect on stress

A natural question to ask is whether a single-pass or multiple-pass weld results in higher stresses. Figure 8.4, in the lower right region, shows the mesh for a three-pass fillet weld. Note that two additional passes were deposited over the top of the single-pass weld (Fig. 8.4, upper right). The three-pass weld is clearly larger, with more weld material deposited. The same welding parameters as in Fig. 8.4 were used for the second and third passes. Comparing Fig. 8.7 with the baseline results in Fig. 8.5 ($v = 6.8$ mm/s) or Fig. 8.6 (yield stress, 350 MPa), one sees that the magnitude of both longitudinal and axial stresses are nearly identical for the 1–3–2 sequence. The longitudinal stress is a little higher in the 1–2–3 sequence. More importantly, the peak axial stress in the 1–3–2 sequence is pushed away from the discontinuity at the weld toe (point A in Fig. 8.7). This is important because fatigue

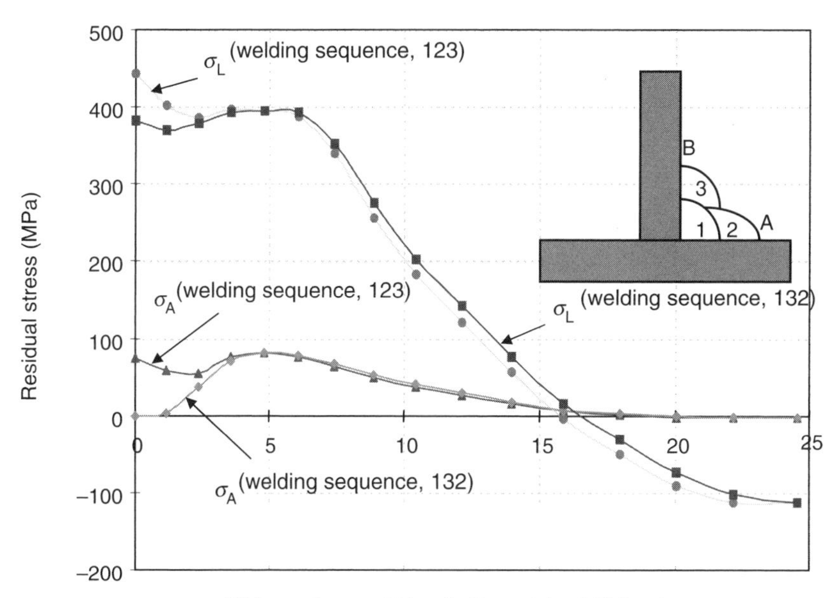

8.7 T-fillet weld: three-pass sequence.

resistance is decreased near the discontinuity of the weld. Pushing the peak axial stress away from this region by depositing the final pass near the critical fatigue location (point A here) can improve fatigue life. It is also important to note that, for multiple-pass welds, the order of each individual pass can affect not only the magnitude and location of the peak residual stress but also the constraint. Constraint can be defined as the amount of hydrostatic stress $\sigma_{kk}/3$ at a point. The higher the constraint, the more prone is that location to crack growth. Chapter 7, Section 7.5, illustrates an industrial case where the design of the weld pass sequence was used to reduce the residual stress and constraint, thereby improving the fatigue life of a product.

The effect of weld parameters on weld residual stresses is important and there are many variables involved. During the design of weld sequences it is best to look at each situation as it arises with a numerical model. Some rules of thumb were presented here and also may be found in the work by Brust and Stonesifer[10], Barber *et al.*[11], Brust *et al.*[12, 13] (see also numerous papers by Brust *et al.* on weld modeling in this volume) and Dong and Brust[14] and references cited therein.

8.4.2 Local weld sequencing effects on distortion

Now let us consider the effects of weld sequencing on distortions. Consider the same T-fillet weld of Fig. 8.4 and the single-pass weld case. Figure 8.8 illustrates a three-dimensional analysis of this using the VFT™ software[1]. The mesh, the temperature profile when the weld is half-complete, the constraints imposed and the deformed shape are shown in Fig. 8.8(a). Note that the vertical plate and left half of the bottom plate were constrained so that the distortion of the right half of the plate is vertical. The weld started at A and ended at B, and note that the angular distortions increase as we approach the weld stop location at B. Figure 8.8(b) shows the results of nine experimental measurements and three analysis results: two shell solutions and a solid model solution. It is seen that a shell solution is quite accurate here and the solution time is greatly reduced compared with a solid model. The accuracy of computational weld modeling is clearly evident. Figure 8.9 shows the distortions using different sequences. It is clear that the backstepping approach results in the most uniform distortions. However, if the goal of the final product is zero distortion, another distortion control method such as precambering, will be needed. Precambering is discussed in Section 8.5.

8.4.3 Global weld sequencing effects on distortion

The above example illustrated that distortion on a component level can be designed using weld models on a case-by-case basis. The final weld process

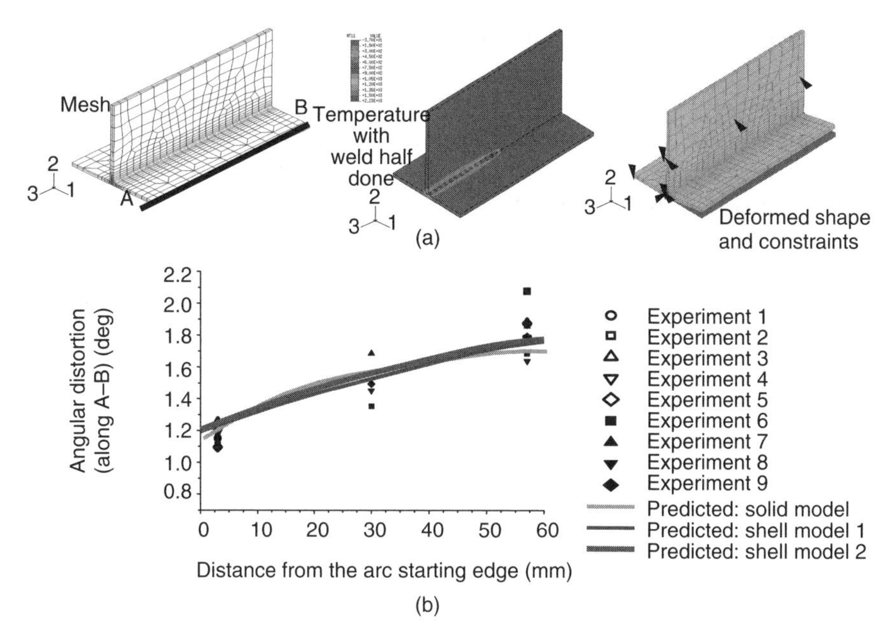

8.8 T-fillet weld: angular distortions.

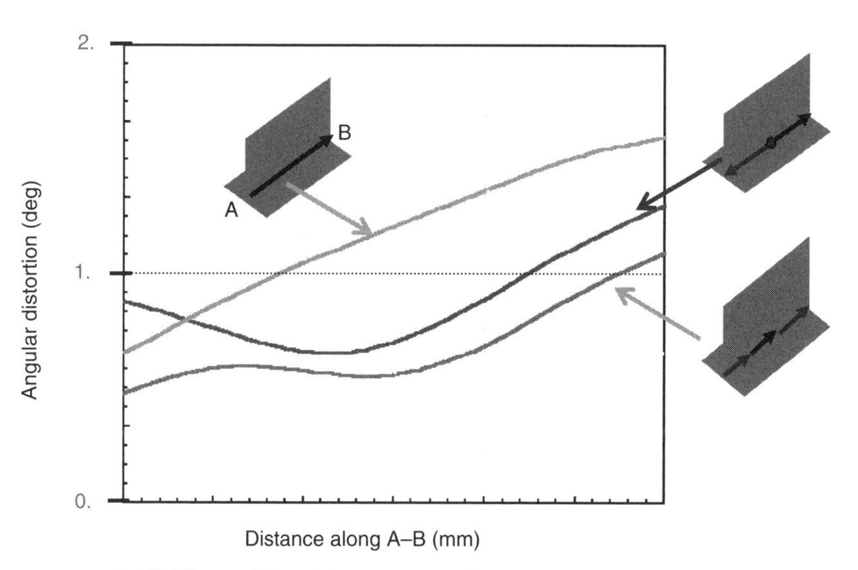

8.9 T-fillet weld: weld sequence effect.

(and distortion control strategy) must be weighed by the fabrication costs. However, it is clear that weld modeling as a tool in defining the best strategy is almost a requirement today. Welding sequencing has been widely used in controlling distortions in large fabrications. However, it is not always clear which sequence is the best. Sometimes the question cannot be answered on the basis of welding experience alone and a weld model is required. An example illustrating the effect of weld sequencing on final distortion is shown in Fig. 8.10. If the four long 'outside' welds are deposited first, followed by the numerous 'inside' welds, the distortions are much larger than in the reverse order. In practice, one would run through a number of weld sequences with the model until the desired solution is obtained. Experience suggests five to ten iterations are often required. Examples of optimum weld sequencing in an industrial setting are provided in Chapter 7.

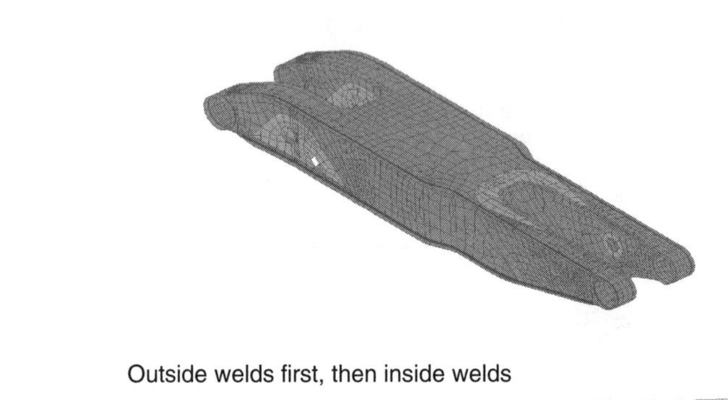

Outside welds first, then inside welds

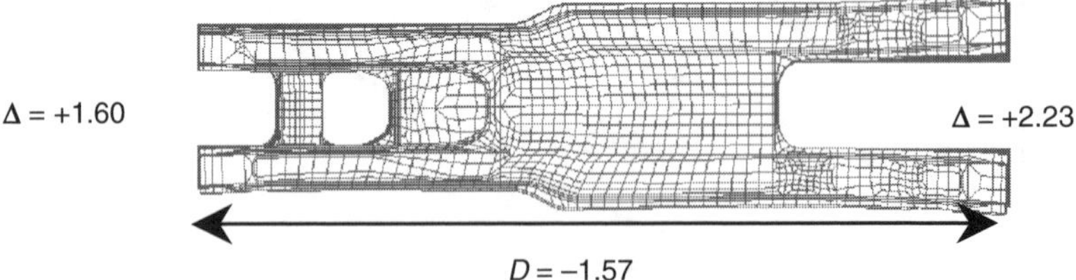

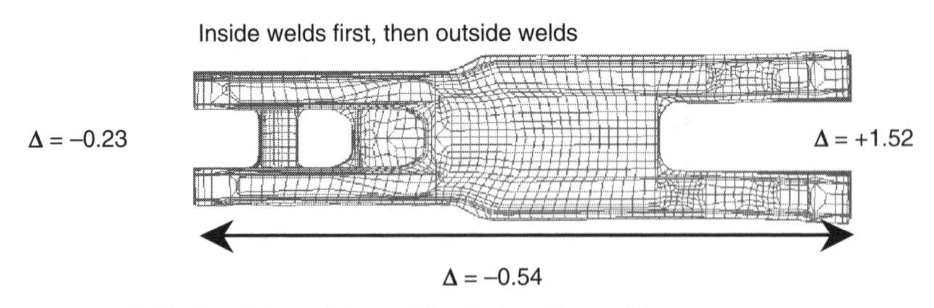

8.10 A weld model used to design the weld sequence.

8.5 Constraint effects and distortion control with precambering

The effects of constraints applied during the welding process from fixturing can be important. Designing a new fixture to control distortions and residual stresses can often be difficult. Several simple examples are shown here, which illustrate this method for controlling residual stresses and distortions.

8.5.1 Fixture constraint effects on residual stresses

In many industrial settings, fixtures are designed to help to manage distortions. However, fixture constraints also affect residual stresses. Figure 8.11 illustrates the single-pass T-fillet weld case considered throughout this section (Fig. 8.4). Figure 8.11(a) illustrates the axial residual stress state σ_1 in a welded plate which was constrained as shown. In practice, this type of constraint could be applied via clamps or within a fixture itself. The axial stresses at the toe of the weld are quite large. After release of the constraints (Fig. 8.11(b)) the stresses throughout the joint are rearranged significantly and the axial stress at the toe of the weld reduces to about 50 MPa. Comparing this stress with that which exists with no constraint applied at all (Fig. 8.5, $v = 6.8$ mm/s), where the transverse stress near the weld toe peaks at 100 MPa, one sees that the constrained welding reduces stress by about 50% in this case. This is not always the case, even where precambering is applied, and must be studied on a case-by-case basis. As such, fixture design can be used to control residual stresses as well as distortions.

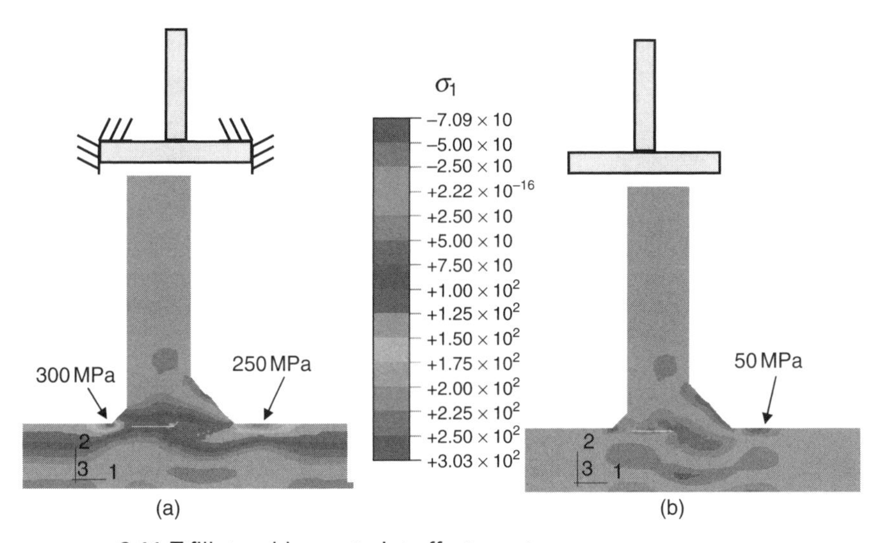

8.11 T-fillet weld: constraint effect on stress.

8.5.2 Precamber for distortion control

Precambering consists of welding a component while some of the parts being welded are bent in a fixture. The parts being welded are usually bent elastically but some may be bent by the fixture such that plastic deformation occurs before welding. After welding, the fixture controls are released and the welded component 'springs back' to the desired deformed shape (usually zero displacements are the goal). When an engineer is designing a weld process to control distortions in a critical component, usually the weld sequence is first examined. In some components, especially lap-welded components where multiple welds are on one side of a major piece, sequencing can only allow the designer to achieve a certain amount. Consider again the T-fillet example of Fig. 8.4. Figure 8.12 (upper left) shows the predicted displacements that result after standard welding. The boundary conditions are again those of Fig. 8.8. The displacements range from 5.84 mm at the weld

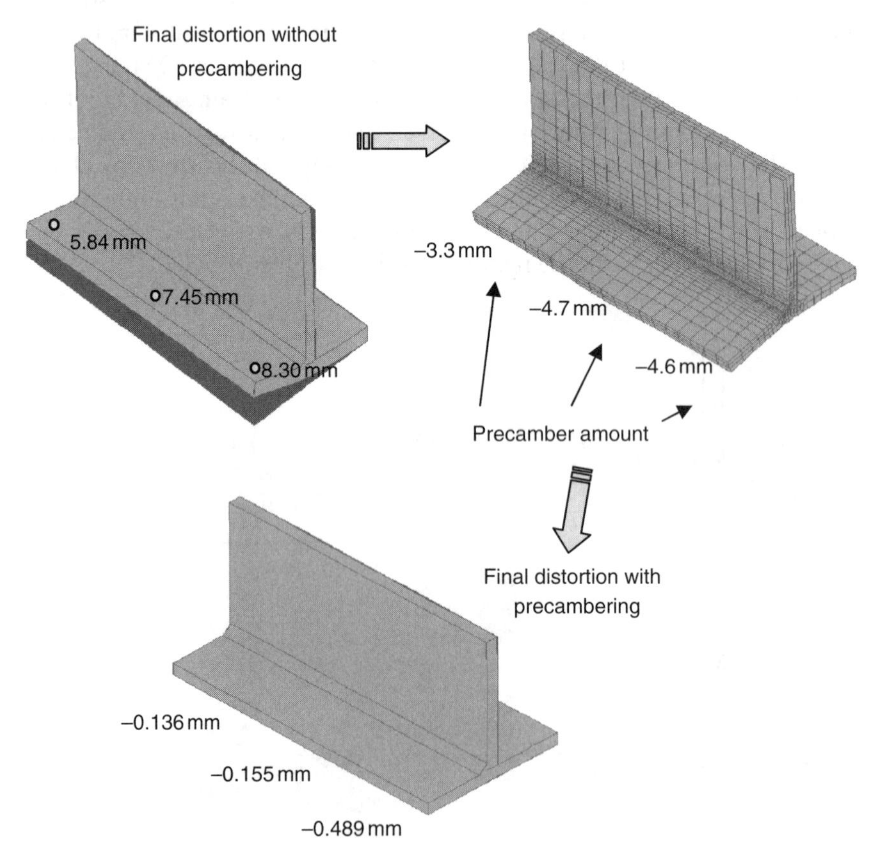

8.12 T-fillet weld: pre-camber for distortion control.

start location to 8.3 mm near the stop point. If one designs a fixture and pre-cambers the bottom plate as shown in the upper right of Fig. 8.12, reduced distortions as seen in the lower illustration result. Typically, one has distortion design tolerances to achieve (for instance, they might be zero distortion ±0.5 mm). The precamber amounts are varied from analysis to analysis until the goal is achieved, which might take five to ten analysis iterations for a complicated structure.

At this point it is useful to discuss control of process variability. For mass-produced welded components, if one proposes a precamber design scheme, each component that is welded must fall within some accepted tolerance goals. Because of the process variability (see Section 8.2) of all the processes (from the steel plant, handling, through forming, cutting and welding), control must be maintained to achieve the goals. If, for instance, the yield stress varies between 350 and 500 MPa, there may be so much variability that the process is not robust and too many welded components must be rejected because the distortion control goals are not met. This can happen if an industry specifies a minimum yield strength requirement of, say, 350 MPa from the material supplier, and the material supplier only tries to meet this minimum requirement. It may be necessary to specify both a minimum and a maximum yield strength requirement to make the process robust. There are many other sources of variability that must be managed as discussed in Section 8.2 to ensure a robust process.

8.5.3 Combined weld sequence and precamber design

Welding sequences are not always effective in mitigating welding-induced distortion. It also depends on the geometry of welded structures. Figure 8.13 shows a component that consists of three layers of plate welded together with surrounding lap joint. Figure 8.14 shows its finite-element model with five welding sequences. Weld A separated into A1 and A2 is used to weld the base layer to the second layer plate. Weld B and Weld C are used to weld the second layer to the top layer plate. During welding, two holes were fixed by bolts to establish a baseline from which distortions can be measured.

Figure 8.15 shows the severe distortion irrespective of which welding sequence was used. Welding sequences are not effective because all welds are located in one corner of the structure, on the same side. After welding, the welds contract to produce bending stresses. This makes the far ends of the structure rise up. If the far ends of the structure are clamped during welding, the distortion is reduced, as might occur in a rigid fixture. The distortions are still significant as indicated in parentheses in Fig. 8.15. Now let us consider the effect of including precamber as part of the solution. Figure 8.16(a) shows the precambering shape and uniform magnitudes of –9 mm

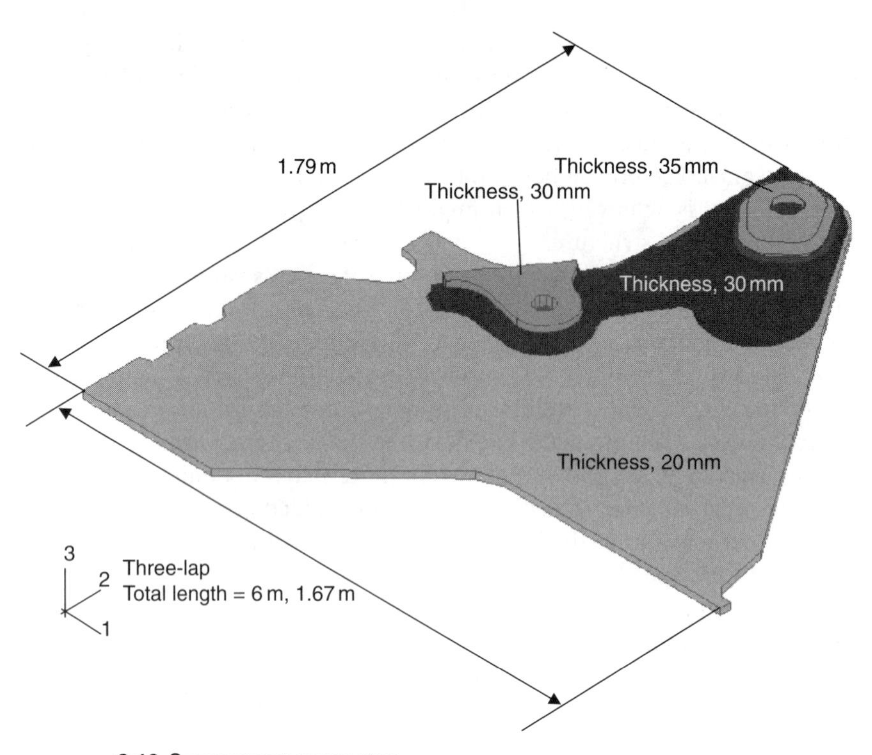

8.13 Component geometry.

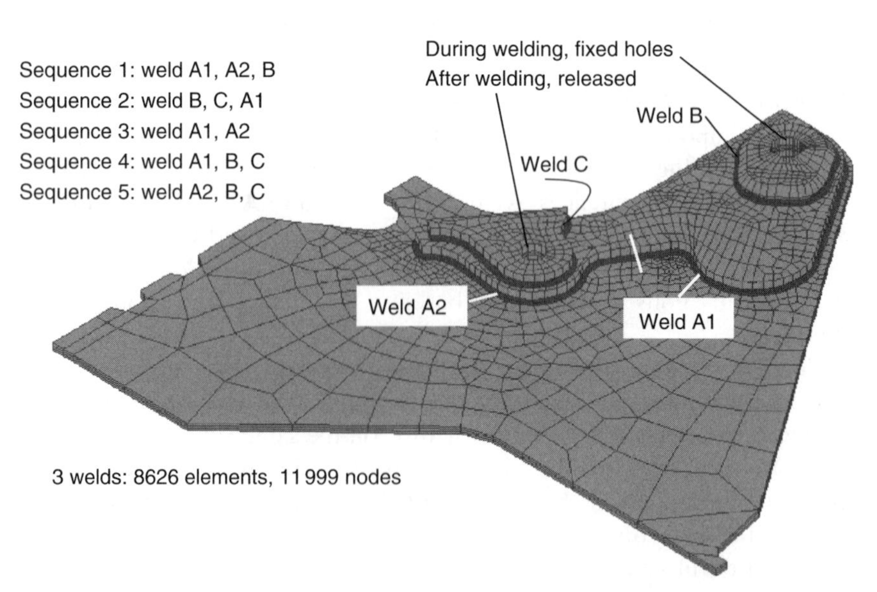

8.14 Three-dimensional solid finite-element model and welding sequence.

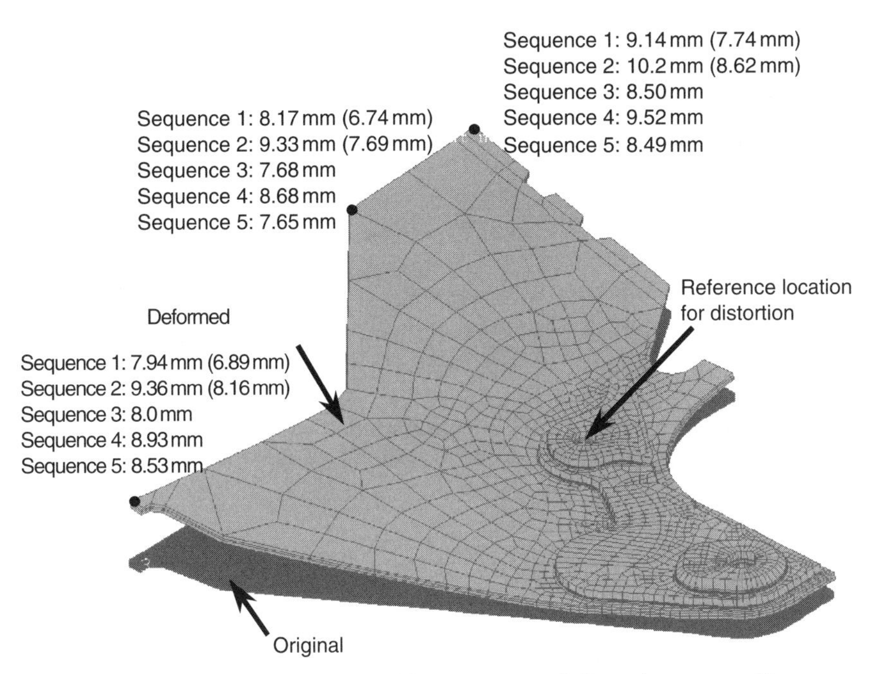

8.15 Predicted distortions. The amounts of distortion at specific locations are indicated after the sequence numbers.

before welding. Figure 8.16(b) shows the final distorted shape and magnitudes after welding in a precambered condition and after releasing the clamping and spring-back. All marked values are below 2 mm, which is a requirement for this product. Coordinate measurement shows that the flatness of the plate after welding is between 1 and 2 mm in a test component case. Note that one corner has the maximum distortion of 1.884 mm. If nonuniform precambering is applied as shown in Fig. 8.17(a), the final distortion is smaller than 1 mm, as shown in Fig. 8.17(b). Therefore, nonuniform precambering should be used in the production for optimum distortion control. The nonuniform scheme can be designed as follows.

1. Model the welding process with no precamber. Find the distortions after welding at specific critical locations, as in Fig. 8.15 (for the chosen sequence).
2. Use a negative of these displacements at the selected locations as the initial precamber amount. The locations for the precamber should be chosen to be practical in the field or with a fixture. Obtain the predicted displacements after the spring-back.
3. Add the displacements predicted from steps 1 and step 2. Apply the negative of these in the next precamber trial. Again, calculate the

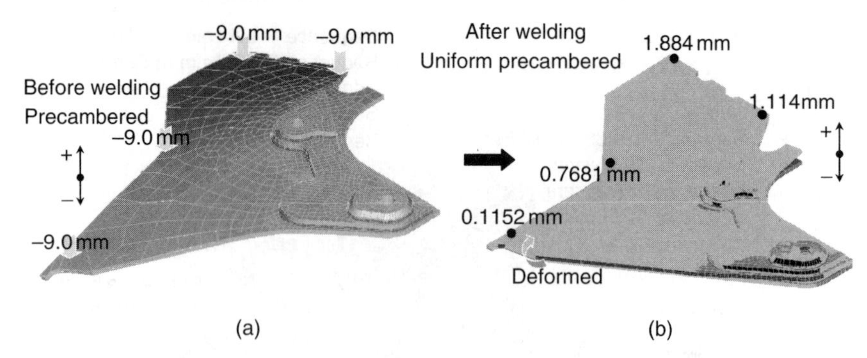

8.16 Component displacements after different precambers: (a) uniform precamber 9.0 mm before welding (the precambers at specific locations are indicated); (b) final distortion with uniform precamber (the final distortions at specific locations are indicated).

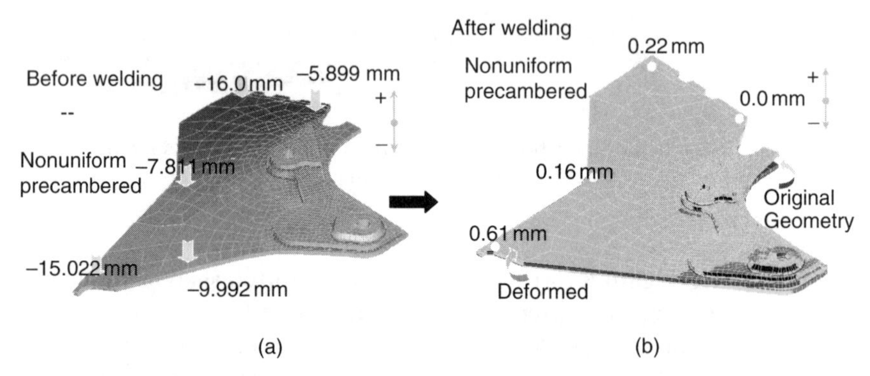

8.17 Component displacements after a variable precamber: (a) nonuniform precamber before welding (the precambers at specific locations are indicated); (b) final distortion with nonuniform precamber (the final distortions at specific locations are indicated).

displacements after spring-back. Add these displacements to those calculated in steps 1–3.

4. Perform another precamber trial with the displacements from step 3. Continue until the desired goal is attained. It normally takes about five or so iterations to optimize the nonuniform precamber sequence.

The welding residual stresses with and without precambering were compared. It was found that the von Mises residual stresses critical to fatigue performance were reduced by using the precambering technique[15]. It is important to recognize that precambering has an important effect on weld-induced residual stresses. Stresses may decrease in one location but increase in another critical location. As such, stresses should be checked after a pre-camber solution to ensure that goals were met for both distortions and residual stresses.

8.6 Thermal methods for control

Heat sink welding is a method that was developed in the early 1980s to control weld-induced residual stresses in nuclear pipe so as to mitigate stress corrosion cracking in weld-sensitized stainless steel. The early weld models were used to design heat sink welding and similar processes[10, 11]. Thermal tensioning consists of making the welds by using a heating source strategically placed ahead of, beside or behind the weld torch. If a cooling source is used, it is termed heat sink welding. Both distortions and residual stresses can be controlled using this family of methods. Two examples of both methods are shown next.

8.6.1 Heat sink welding

Heat sink welding was developed around 1980 to control intergranular stress corrosion cracking (IGSCC) in nuclear piping. IGSCC develops and grows in the heat-affected zones of welds where tensile residual stresses caused by the welding process can develop. In a pipe, if one cools the inner surface during the welding process, compressive residual stresses can develop in the heat-affected zone along the pipe inner surface where the corrosive water flows. This can be achieved in the field by running water through the pipe during the application of the latter passes in a girth butt weld. Figure 8.18 illustrates the procedure. Both axial and hoop stresses are reduced to compression. The weld heat input, water flow rate, passes sequence, etc., were optimized using the weld process models to produce the desired compressive stresses. This process is now well established and routinely used in field welds to help to prevent and delay IGSCC in nuclear piping systems.

8.6.2 Thermal welding

Figure 8.19 illustrates the use of a heating source during welding. The T-fillet weld of Fig. 8.4 is considered again here. Welding with a trailing thermal source moves the stresses away from the toe of the weld. A cooling source does not reduce stresses significantly here. Thermal methods for welding, particularly thermal tensioning, are emerging as effective methods for controlling residual stresses and distortions, especially in thin-plate welded components (see Chapter 9).

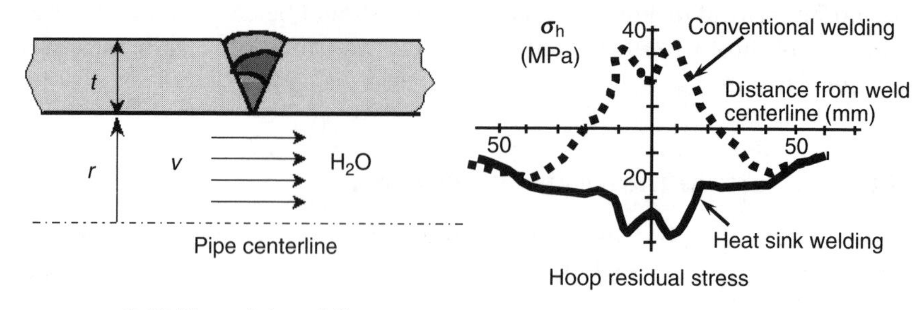

8.18 Heat sink welding.

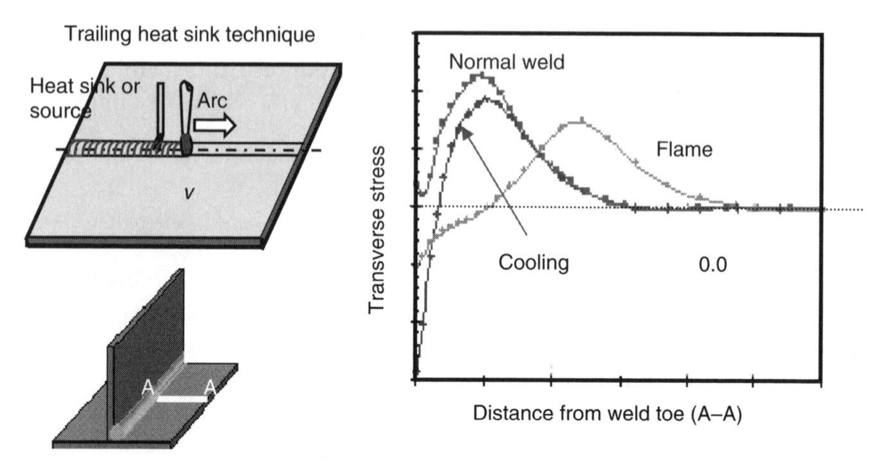

8.19 Stresses after thermal welding.

8.7 Other methods for in-process weld distortion and residual stress control

Methods for controlling *hot cracking* during welding include mechanical methods[16] and thermal methods[17]. Hot cracking of welds occurs most often during the welding of high-strength steels and special purpose alloys. The thermal methods discussed in Section 8.6.2, or variations of these, are used.

New consumables have been developed in recent years which can improve the residual stress state in some welds. In general, high tensile residual stresses are expected at the weld toe. This is particularly important for welding of high-strength steel, which ends with higher tensile residual stress at the weld toe than in mild steels with lower yield strength. Recent work has demonstrated that compressive residual stresses can be obtained at the weld toe by using a lower-temperature-transformation consumable.

The compressive residual stress is induced by the expansion of volume due to martensite transformation at the lower temperature. Chapter 7, Section 7.5.2, provides an example where this concept is discussed in some detail.

8.8 Post-weld distortion and residual stress control

After the component has been welded, and the distortions and residual stresses already exist in the structure, it is expensive and time consuming to make corrections. It is clearly best to attempt to control the process during welding to achieve the design goals. However, there are a number of methods to achieve these post-weld corrections, none of which is entirely satisfactory.

Press straightening, whereby the distorted component is placed within a press and mechanically straightened, is time consuming. Moreover, the press can be an expensive shop floor component, especially for large welded components, where such a press might cost upwards of seven figures. The time lost in the straightening process is an expense as well. Moreover, the redistribution of the residual stresses after straightening is not usually considered.

Post-weld heat treatment is often used to reduce weld residual stresses after the component is welded. This process can be performed in a furnace or locally on site using heater blankets or other heating method. A rule of thumb is to heat the component (for steels often to 1100 °F or higher)[21] for about 1 hour per 25 mm of thickness to alleviate weld residual stresses. Alleviating the residual stresses also alters the distortion field. It is important to note that post-weld heat treatment usually does not eliminate all the residual stresses. It is clear that computational weld models can be used to design the post-weld heat treat process. The mechanism of stress relief during post-weld heat treatment is often a combination of creep relaxation along with phase transformation effects. It is also important to note that post-weld heat treatment of bimetallic welds can be difficult to achieve, especially if the materials are somewhat variable, such as the pressure vessel steel welded to stainless steel pipe via Inconel in nuclear plants[18] if the materials have different coefficients of thermal expansion.

Laser shock processing (LSP), sometimes called *laser shock peening*, is emerging as another post-weld residual stress control method. LSP methods are more effective than shot peening and produced a deeper zone of compressive residual stresses. LSP has demonstrated the ability to increase significantly the damage tolerance of turbine engine airfoils. This process utilizes a laser-driven impact event to generate a compressive residual stress field similar to that of shot peening. In contrast with the shot-peening process, the LSP process uses a high-intensity short-pulse-duration laser to deform plastically the surface and subsurface of the target. The resulting compressive residual stress field is typically five to ten times deeper than that achieved during shot peening and the surface remains relatively unscathed[19].

Hammer peening at the weld toe is another commonly used post-weld treatment not only to produce a large toe radius but also to introduce compressive residual stresses at the weld toe. Examples of the use of this method, as well as a computational model to permit proper design of this process in an Al–Li weld alloy, have been discussed by Dong *et al.*[20].

8.8.1 Flame straightening example

A method popular in many industries for correcting weld distortions is called flame straightening. This method can be effective but often requires an experienced welder to perform the straightening procedure. It is very time consuming and costly to perform flame straightening. Consider a deck plate where distortions caused by welding can lead to distortions that must be controlled with flame straightening. Flame straightening can be modeled using weld modeling procedures as well. This may become more important in coming years as experienced welders retire. Figure 8.20 illustrates a situation where the deck plate is modeled with a T-beam with appropriate boundary conditions to permit the edges of the top plate to be modeled approximately (periodic symmetry). As can be seen, the welds are placed underneath the top plate. The distortions predicted are 'negative', i.e. they go down. The displacement contour plots show the negative displacement. The lowest figure illustrates the modeling of flame straightening where a heating torch was applied to the top portion of the top plate, as seen in the upper illustration of Fig. 8.20. In this analysis case, overstraightening occurred. The flame straightening parameters (heat input, speed, etc.) must be designed.

8.9 Conclusions

Welding-induced distortion control techniques, weld design optimization, prestraining, optimum welding sequence and precambering were discussed in this chapter using simple examples. Detailed industrial applications of many of these methods are discussed in Chapter 7. It is emphasized that computational fabrication models have emerged to the point that they are an important tool to the weld designer. By using this software the optimum distortion control parameters could be obtained from the computer simulation. Therefore the welding operation cost and prototype period can be reduced and the integrity of welded structures can be improved.

The major points discussed in this chapter are as follows.

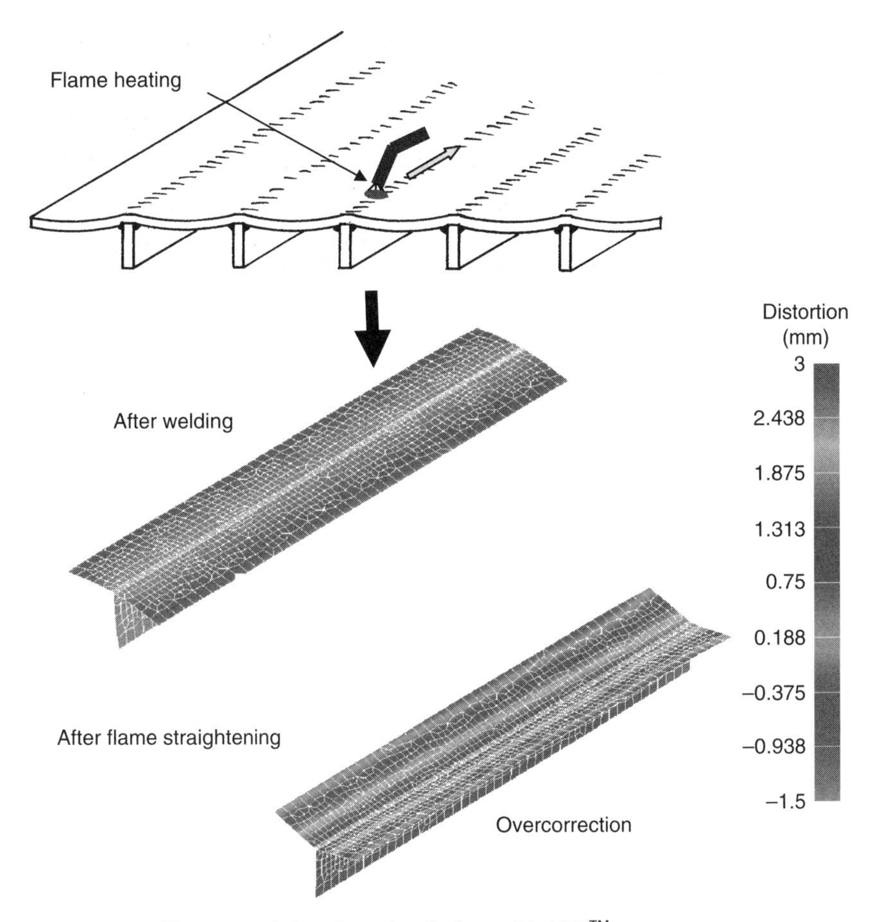

8.20 Flame straightening simulation with VFT™.

1. Weld design optimization for controlling distortion can be achieved with the help of weld modeling analysis in the product design stages.
2. Optimum welding sequences can be effective in controlling distortion, but it is not always true and depends on the welded structures being considered.
3. Prestraining and precambering are very effective for distortion control in heavy industries. The optimum prestraining or precambering shape and magnitude can be obtained from finite-element analysis results.
4. Using precambering to mitigate distortion can reduce the von Mises stress on the weld toe so that the fatigue life of the welded structures may be increased.

5. Residual stresses can be controlled by using alternative welding procedures. This can result in structures with increased corrosion and fatigue capabilities and also increased fracture resistance.

Other methods for controlling distortions and residual stresses are used today as well. Heat sink welding is routinely used in the nuclear industry to induce compressive residual stresses in weld-sensitized regions for pipe materials susceptible to stress corrosion cracking. Thermal tensioning (or thermal cooling), where a heating or cooling source that trails, parallels or precedes the weld torch is an active area of research for controlling distortions in large ship structures. Chapter 7 provides a number of industrial examples illustrating some of these residual stress and distortion control procedures.

8.10 Acknowledgments

The author would like to thank his colleagues at Caterpillar Inc., Dr Leo Chen, Y. Dong, Dr T. Jutla, Dr Z. Yang and K. Huber for the excellent interactions and joint work that we have had over the years. Some of the examples shown in this chapter were worked on as part of a Caterpillar-led program entitled Fabrication of Advanced Structures Using Intelligent and Synergistic Materials Processing. Chapter 7 shows some industrial applications from the Caterpillar, Inc., files. Dr Y. Yang, his former colleague here at Battelle Memorial Institute, is gratefully acknowledged as well.

8.11 References

1. *User Manual for VFT – Virtual Fabrication and Weld Modeling Software*, Battelle Memorial Institute and Caterpillar Inc., February 2002.
2. Hou, C., Kim, M., Brust, F.W., and Pan, J., 'Effects of residual stresses on fracture of welded pipes', *Proc. ASME Pressure Vessels and Piping Conf.*, PVP-Vol. 327. *Residual Stresses in Design, Fabrication, Assessment and Repair*, Montreal, Quebec, Canada, 21–26 July 1996, American Society of Mechanical Engineers, New York, 1996, pp. 67–75.
3. Abou-Sayed, I.S., Ahmad, J., Brust, F.W., and Kanninen, M.F., 'An elastic–plastic fracture mechanics prediction of stress corrosion cracking in a girth welded pipe', *Fracture Mechanics: 14th Symp.*, ASTM STP 791, American Society for Testing and Materials, Philadelphia, Pennsylvania, 1983, pp. 56–72.
4. Kanninen, M.F., Brust, F.W., Ahmad, J., and Papaspyropoulos, V., 'An elastic–plastic fracture mechanics prediction of fatigue crack growth in the heat affected zone of a butt welded plate', *Fracture Tolerance Evaluation* (eds P. Kanazawa, A.S. Kobya and K. Iida), Poyo Press, Tokyo, 1982, pp. 113–120.
5. Kanninen, M.F., Brust, F.W., Ahmad, J., and Abou-Sayed, I.S., 'The numerical simulation of crack growth in weld induced residual stress fields', *Residual Stress and Stress Relaxation, Proc. 28th Army Sagamore Research Conf* (eds E. Kula

and V. Weiss), Lake Placid, New York, 13–17 July 1981, Plenum New York, 1982, pp. 227–248.
6. Zhang, J., Dong, P., and Brust, F.W., 'Residual stress analysis and fracture assessment of welded joints in moment resistant frames', *Modeling and Simulation Based Engineering* (eds S.N. Atluri and P.E. O'Donoghue), Tech Science Press, Irvine, California, 1998, pp. 1894–1902.
7. Gates, W.E., and Morden, M., 'Lessons from inspection, evaluation, repair and construction of welded steel moment frames following the Northridge earthquake', Technical Report: Surveys and Assessment of Damage to Buildings Affected by the Northridge Earthquake of 17 January 1994, Report SAC-95-06: 3-1 to 3-79, 1995.
8. Yang, Z., Cao, Z., Chen, X.L., and Ludewig, H.W., 'Virtual welding – applying science to welding practices', *Proc. 8th Int. Conf. on Numerical Methods in Industrial Forming Processes*, American Institute of Physics, Columbus, Ohio, 2004, pp. 1308–1313.
9. Masubuchi, K., *Analysis of Welded Structures*, Pergamon, Oxford, 1980, pp. 235–331.
10. Brust, F.W., and Stonesifer, R.B., 'Effects of weld parameters on residual stresses in BWR piping systems', EPRI Report NP-1743, Electric Power Research Institute, Palo Alto, California, 1981.
11. Barber, T.E., Brust, F.W., Mishler, H.W., and Kanninen, M.F., 'Controlling residual stresses by heat sink welding', EPRI Report NP-2159-LD, Electric Power Research Institute, Palo Alto, California, 1981.
12. Brust, F.W., *et al.*, 'Weld process modeling and its importance in a manufacturing environment', SAE Technical Paper 981510, Society of Automotive Engineers, New York, 1998.
13. Brust, F.W., Jutla, T., Yang, Y., Cao, Z., Dong, Y., Nanjundan, A., and Chen, X.L., *Advances in Computational Engineering and Sciences* (eds S.N. Atluri and F.W. Brust), Tech Science Press, Forsyth, Georgia, 2000, pp. 630–636.
14. Dong, P., and Brust, F.W., 'Welding residual stresses and effects on fracture in pressure vessel and piping components: a millennium review and beyond', *Trans. ASME, J. Pressure Vessel Technol.*, **122**(3), 329–339, 2000.
15. Yang, Y.P., Brust, F.W., Cao, Z., Dong, Y., and Nanjundan, A. 'Welding-induced distortion control techniques in heavy industries', *Proc. 6th Int. Conf. on Trends in Welding Research* (eds S.A. David, T. DebRoy, J.C. Lippold, H.B. Smartt and J.M. Vitek), ASM International, Materials Park, Ohio, 2002, pp. 844–849.
16. Yang, Y., Dong, P., Tian, X., and Zhang, Z., 'Prevention of welding hot cracking of high strength aluminum alloys by mechanical rolling', *Proc. 5th Int. Conf. on Trends in Welding Research*, Pine Mountain, Georgia, USA, 1998, American Welding Society, Miami, Florida, ASM International, Materials Park, Ohio, 1998.
17. Yang, Y.P., Dong, P., Zhang, J., and Tian, X., 'A hot-cracking mitigation technique for welding high strength aluminum alloy sheets', *Proc. 6th Int. Conf. on Trends in Welding Research* (eds S.A. David, T. DebRoy, J.C. Lippold, H.B. Smartt and J.M. Vitek), ASM International Materials Park, Ohio, 2002, pp. 850–856.
18. Brust, F.W., Scott, P.M., and Yang, Y., 'Weld residual stresses and crack growth in bimetallic pipe welds', *Proc. the 17th Int. Conf. on Structural Mechanics in Reactor Technology* (ed. S. Vejvoda), Prague, Czech Republic, August 2003, Brno University of Technology, Brno, 2003, Section G, CR-ROM.

19. Lykins, C., and Brust, F.W., 'Finite element alternating method – crack initiation predictions in a laser shock peened zone', *Proc. Air Force High Cycle Fatigue Symp.*, San Antonio, Texas, USA, February 1998.
20. Dong, P., Hong, J.K., and Rogers, P., 'Analysis of residual stresses in Al–Li alloy repair welds and mitigation techniques', *Weld. J.*, **77**(11), 439-s–445-s, 1998.
21. Stout, R.D., 'Weldability of steels', Welding Research Council, 1987.

9

Control of buckling distortions in plates and shells

Q. GUAN, Beijing Aeronautical Manufacturing Technology Research Institute, People's Republic of China

9.1 Introduction

Manufacturing of sheet-metal-formed plates, panels and shells are always accompanied by buckling distortions where fusion welding is applied. Buckling distortions are more marked than other forms of welding distortions in thin-walled structural elements, and they are the main troublesome problems in sheet metal fabrication especially for high-speed vehicles, light superstructures of naval surface ships cruisers in which longitudinal and transverse stiffeners are fillet welded to thin plates as well as for aerospace structures such as airframe panels, fuel tanks, shells of jet engine cases, etc., where thin-section sheet materials of thickness less than 4 mm are widely applied. Buckling distortions affect the performance of structures in a great many ways; they cannot fit in with designed geometric and aesthetic requirements and make assembly of elements difficult owing to the exceeding mismatch gaps, and in some cases even impossible. So, the normal production procedures have to be broken for time-consuming and costly distortion removal. From the viewpoint of in-service reliability, buckling may lower the rigidity of welded elements and cause variable quality of products. Efforts have been made and progress has been achieved in solving buckling problems by experts in the welding science and technology worldwide for the past decades. Many effective methods for removal, mitigation and prevention of buckling distortions adopted before welding, during welding or after welding are successfully developed and widely applied in industries[1–10]. Recent progress in eliminating buckling distortions has resulted in trends from the adoption of passive technological measures to creation of active in-process control of inherent (incompatible residual plastic) strains during welding without having to undertake costly reworking operations after welding[11–14].

9.2 Buckling distortions in plates and shells

9.2.1 Typical buckling patterns in thin-walled structural elements

In sheet metal fabrication, structural elements are mostly designed and assembled using fusion welding. Welding-induced buckling differs from bending distortion by its much greater out-of-plane deflections and several stable patterns. Buckling patterns depend much more on the geometry of the elements, and types of weld joint especially depend on the thickness of sheet materials under certain conditions.

Figure 9.1 (a), Fig. 9.1 (b), Fig. 9.1 (c) and Fig. 9.1 (e) show some typical buckling patterns in plates and stiffened panels caused by longitudinal welds, and Fig. 9.1 (d) and Fig. 9.1 (f) show buckling patterns in plates caused by circular welds.

Figure 9.2 (a), Fig. 9.2 (c) and Fig. 9.2 (e) show some typical buckling patterns in shell elements caused by longitudinal welds; Fig. 9.2 (b), Fig. 9.2 (d), and Fig. 9.2 (f) those caused by circular welds.

From a comparison of Fig. 9.1 and Fig. 9.2, it should be clear that the buckling distortions caused by longitudinal welds in plates, in panels or in shells are mainly dominated by longitudinal compressive stresses produced

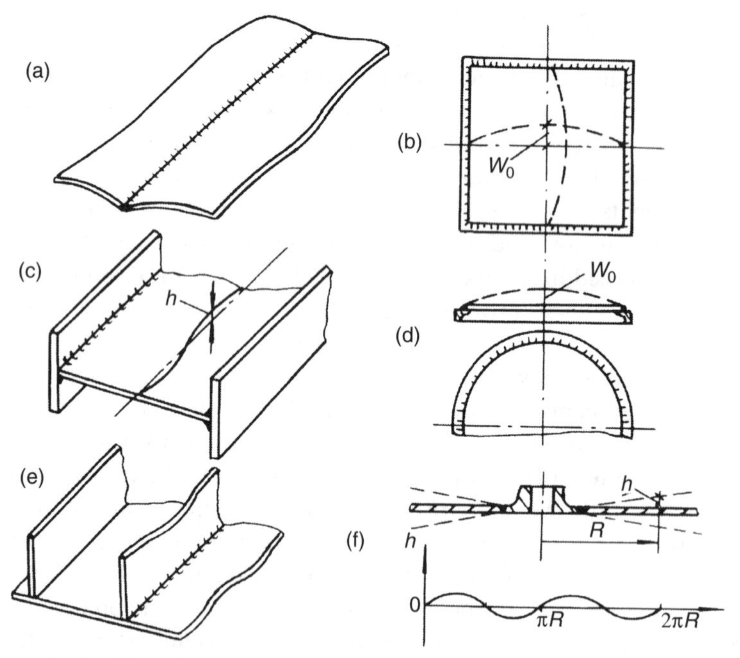

9.1 Typical buckling patterns in welded plates and panels.

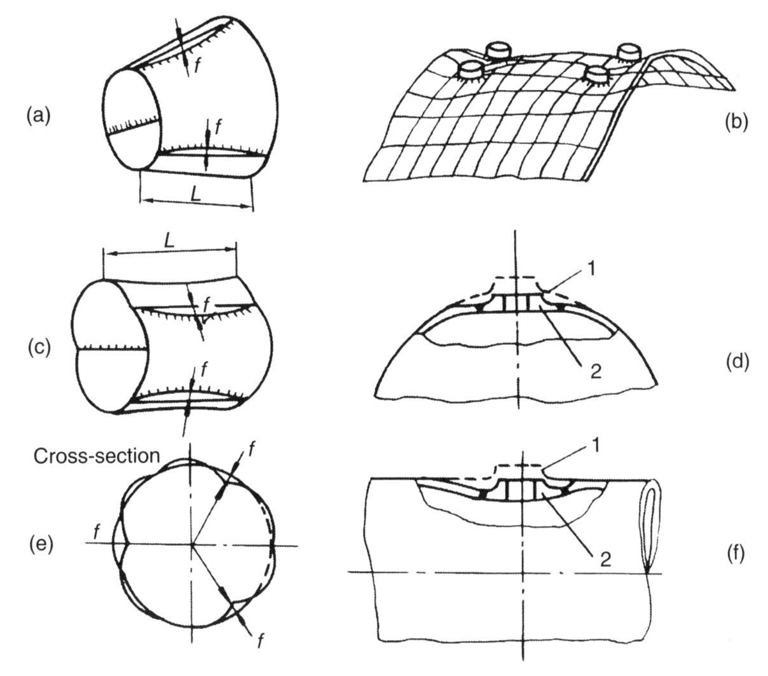

9.2 Typical buckling patterns in welded shell elements.

in areas away from the weld, but the buckling distortions caused by circular welds either in plates or in shells are mainly determined by transverse shrinkage of welds in the radial direction whereby compressive stresses are produced in the tangential direction. The buckling patterns depend much more on rigidity (thicknesses, dimensions and shapes) of elements to be welded, as well as types of weld joint and welding heat inputs.

9.2.2 Buckling caused by longitudinal welds

The nature of buckling phenomena is mostly a result of the loss of stability of the elements under compressive stresses induced in the peripheral areas away from welds. The mechanism of buckling is hidden in the action of the inherent (incompatible residual plastic) strains formed during welding. The experimentally measured longitudinal inherent strain ε_x^p distributions in the cross-section of weld joints after gas–tungsten arc welding (GTAW) of specimens of titanium and aluminum alloys are shown in Figs 9.3 (a) and 9.3 (b) respectively[3,11]. The maximum value $\varepsilon_{x\,max}^p$ and its width of distribution $2b$ (shown schematically) depend on the materials and welding heat input. If two slits are cut along the inherent strain distribution width $2b$ on both sides of the weld, the buckled plate will perform as shown

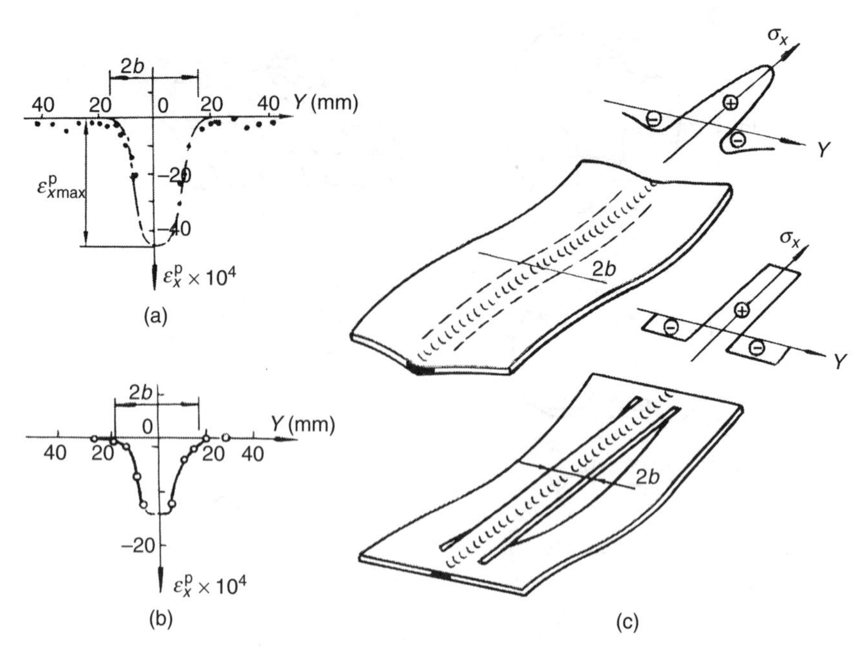

9.3 Distribution of inherent strains measured in cross-sections of weld joints after GTAW of plates 1.5 mm thick[2,3,11]: (a) titanium alloy; (b) aluminum alloy; (c) buckled patterns before and after two slits cut on both sides of weld.

in Fig. 9.3 (c)[2] and the residual stresses fields σ_x are also changed appropriately.

In the 1950s, based on a model of the minimum potential energy of a welded plate Vinokurov[5] analyzed in more detail the loss of stability of a thin plate after welding. As shown in Fig. 9.4, after welding, the plastically deformed weld zone with inherent strain distribution of ε^p_x is schematically shown as a dashed rectangle in Fig. 9.4 (a) in width $2b$, which is simpler for theoretical analysis and calculation. The plate of length L and width B (Fig. 9.4 (b)) has a fan-shaped segment with a longitudinal curvature of radius R (Fig. 9.4 (c)) and is buckled into a saddle shape of cross-section A–A (Fig. 9.4. (d)) with its center of gravity at 0.

On losing stability, the buckled plate (Fig. 9.4 (b)) is released from an unstable flat position of high potential energy with the maximum level of residual stress distribution after welding and takes a stable warped shape. The longitudinal curvature is caused by a certain bending moment formed by the shrinkage force in the weld zone which is offset from the center of gravity 0 of the cross-section A–A of the buckled plate (Fig. 9.4 (d)).

Figure 9.5 (a) shows a functional relationship between the potential energy U accumulated in the plate and the curvature $1/R$ of the buckled

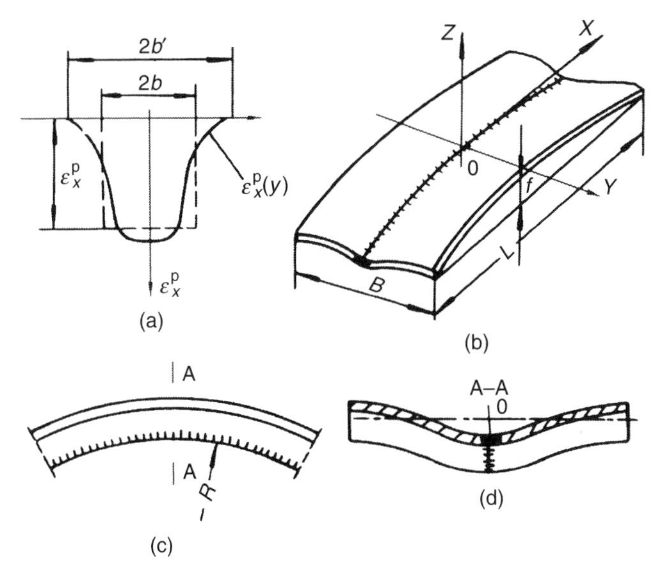

9.4 Buckling pattern of a plate caused by a longitudinal weld with an inherent strain distribution.

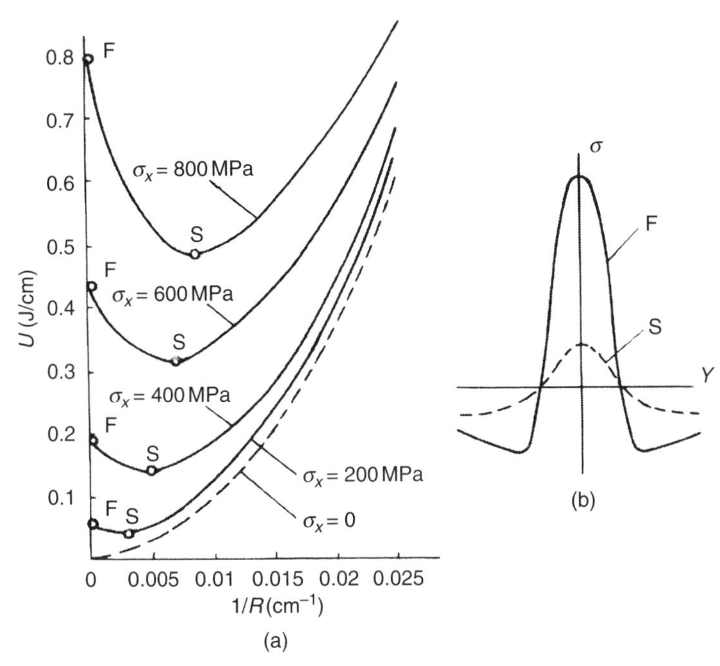

9.5 Functional relationship (a) between U and $1/R$ of buckled plate ($E = 2.05 \times 10^5$ MPa; $\mu = 0.3$; $B = 30$ cm; $2b = 4$ cm; $\delta = 0.15$ cm)[5] with different σ_x in the weld and distributions of longitudinal residual stresses and (b) corresponding to potential energy at points F (unstable, with maximum U) and points S (stable, loss of stability).

plate with a different value of the initial maximum longitudinal residual stress σ_x in the weld[5]. Figure 9.5 (b) shows the distributions of longitudinal residual stresses in the plate corresponding to the points F (unstable) and points S (stable) in Fig. 9.5 (a).

For the given geometry and material of the plate shown in Fig. 9.5, the changeable distribution of residual stresses in the weld zone is the dominating factor of the variation in potential energy. After welding, when the weld is kept in a flat position, a maximum potential energy U is accumulated in the plate (points F on the curves) with residual stress distribution shown by curve F in Fig. 9.5 (b). On losing stability, the buckled plate possesses a minimum of potential energy U. The dashed curve in Fig. 9.5 (a) shows the $U = f(1/R)$ relation in the plate without the welding residual stress ($\sigma_x = 0$). The value U increases with increase in σ_x. The points S on the curves correspond to the minimum value of U where the buckled plate takes a stable shape and the redistributed residual stress field is shown by the dashed curve S in Fig. 9.5 (b). In other words, any forced change in the stable curvature of the buckled plate will cause an increase in the potential energy and, once the force is removed, the buckled plate will be restored to its stable position, minimizing the potential energy.

In analyzing the buckling phenomena of the cruciform cross-section beam joined by longitudinal welds shown in Fig. 9.6, Vinokurov[5] suggested a simplified approximation method to determine the critical stress σ_{cr} and buckling waves.

The area of inherent strains induced by the fillet welds is shown by a shadowed cross-section in Fig. 9.6 (a). The longitudinal shrinkage force F_s can be calculated for a stainless steel beam 2000 mm long by considering the total area S_b of the beam cross-section and the material properties $\sigma_s = 400$ MPa and $E = 1.8 \times 10^5$ MPa. Consequently, the assumed residual longitudinal compressive stress σ_x in the beam as a whole is also determined approximately by $\sigma_x = F_s/S_b = 5$ MPa. The corresponding strain ε_x is determined by $\varepsilon_x = \sigma_x/E = 2.78 \times 10^{-4}$. To determine the critical compressive stress σ_{cr}, the thin-walled elements I and II with $\delta = 2$ mm (see Fig. 9.6 (a)) can be assumed to be strips, with one longitudinal edge (Fig. 9.6 (b)) rigidly clamped and restrained in the transverse direction but unrestrained in the longitudinal direction. The strip is subjected to the longitudinal compressive stress σ_x. According to the theory of stability of elastic systems[15], the critical stress is determined by the relation

$$\sigma_{cr} = K\frac{\pi^2 E\delta^2}{12(1-\mu^2)B^2} \qquad [9.1]$$

For this particular case the coefficient $K = 1.328$, the ratio of plate length L to plate width B is $L/B = 2000/200 = 10$; therefore, the value $\sigma_{cr} = 21.6$ MPa, and the corresponding $\varepsilon_{cr} = 1.2 \times 10^{-4}$.

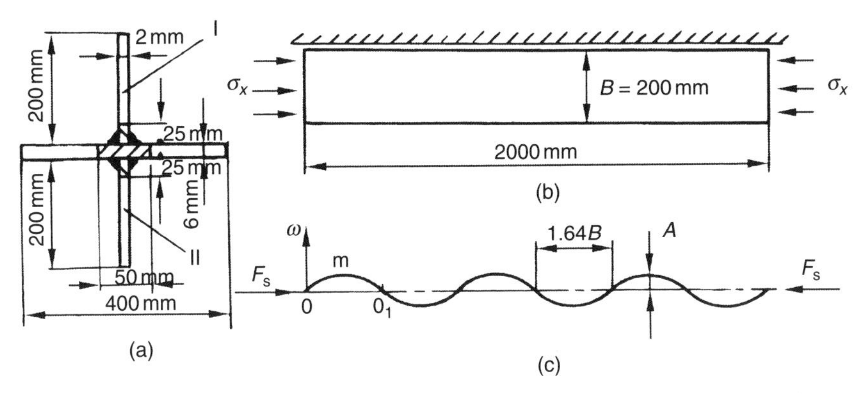

9.6 Determining buckling distortions of a cruciform beam joined by longitudinal welds[5].

Hence $\varepsilon_x > \varepsilon_{cr}$, the thin strips I and II (Fig. 9.6 (a) will lose stability and their edges become sinusoidal (Fig. 9.4 (c)) with half-wavelength $l = 1.64B$ = 32.8 cm. The sinusoidal wave can be expressed as

$$\omega = A \sin\left(\frac{\pi x}{32.8}\right) \qquad [9.2]$$

To calculate the amplitude A, it is assumed that the edge of the buckled strip is subjected to the strain ε_{cr}, and the difference in the lengths of the original section 00_1 (Fig. 9.6 (c)) and the arc $0m0_1$ on one half-wave can be calculated. For this case, the amplitude $A = 2.62$ mm.

Early in the 1950s, fundamental studies on the buckling of a thin plate due to welding were carried out by a group of Japanese experts[8, 16, 17]. Theoretical calculations were accompanied by experimental verifications as well as exploitation of methods for practical application to eliminating buckling. In the 1970s, Terai[7] studied the buckling behavior of a thin-skin plate to be welded on a rectangular rigid frame (see Fig. 9.1 (b)) which is a typical structural element for naval ships. Results indicate that buckling increases dramatically when the heat input surpasses a certain critical amount.

Later, in the 1990s, Zhong *et al.*[18] analyzed the buckling behavior of plates under an idealized inherent strain. The results of theoretical calculation for critical thickness show that it can be determined by the relationship shown in Fig. 9.7 (a). For different ratios of width B to length L of the plate, the narrower the plate the thinner is the critical thickness. For thin elements to be welded, the thickness of 4 mm (shown by a dashed line in Fig. 9.7 (a)) seems to be a threshold value; a plate thinner than 4 mm becomes more susceptible to buckling due to welding.

As shown in Fig. 9.7 (b), in principle, all efforts at either 'passive' measures or 'active' control methods of low-stress no-distortion (LSND)

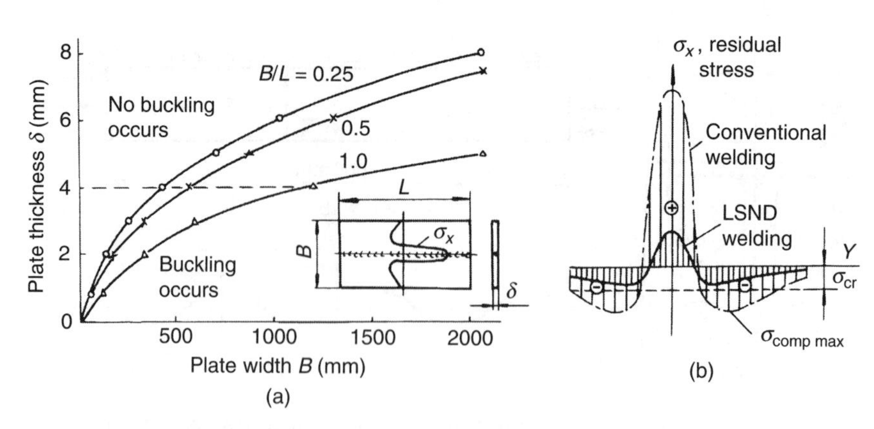

9.7 (a) Determination of the critical thickness of a plate[18]; (b) the way in which $\sigma_{comp\ max}$ can be reduced to lower than σ_{cr}; (b) for preventing buckling due to welding.

welding to eliminate buckling are aimed at reducing $\sigma_{comp\ max}$ to a level lower than the critical compressive stress σ_{cr} at which buckling occurs.

From Fig. 9.7 and equation [9.1] it can be seen clearly that, the thinner the elements, the lower will be the value σ_{cr}. For plates of thickness less than 4 mm as widely used in aerospace and modern vehicle welded structures, the value of σ_{cr} dramatically drops to a level much lower than the peak value $\sigma_{comp\ max}$ of the compressive stress after conventional GTAW. However, the actual value σ_{cr} is difficult to determine solely on the basis of either the linear stability theory of small deformations or the nonlinear theory of large deformations in the theory of plates and shells. These problems are the most complex. Sometimes, as the shrinkage force depends on the rigidity of the elements to be welded, the directions of the action of the shrinkage forces may change when the loss of stability occurs.

Obviously, any method for eliminating or controlling buckling distortions must be based on adjusting the compressive residual stresses to achieve $\sigma_{comp\ max} < \sigma_{cr}$ by means of reduction and redistribution of inherent strains. In thin-walled elements, the amount of buckling also depends on the combined effects of other types of weld shrinkage, e.g. shrinkages transverse to the weld, and angular shrinkages, where the warpage caused by loss of stability increases on account of the additional angular rotation of the sheet relative to the weld joints.

In the past decade, welding simulation and prediction by computational method have been increasingly applied in addition to classic analytical and conventional empirical procedures. The finite-element method has also been adopted by Michaleris and coworkers[19–23] for analyzing buckling distortions of stiffened rectangular plates for shipbuilding welded steel elements. Shrinkage forces were obtained from a thermal elastic–plastic cross-sectional model analysis. A simplified structural model is suitable for the parametric investigations required in practice. The critical buckling load, expressed by the cooling strain in the weld zone, is determined on the basis of the lowest positive eigenvalue of the elastic structure. The deflections of the plate were determined on the basis of a nonlinear elastic analysis with large displacements. Based on the finite-element analysis for large displacements, and using an inherent shrinkage strain method, Tsai *et al.*[24] investigated the buckling phenomena of a rectangular plate (1 m × 1 m) of aluminum alloy with two welded-on longitudinal T stiffeners positioned with different transverse spacings. With a large space between the stiffeners, the plate thickness of 1.6 mm is critical to buckling. Angular distortions enhance the buckling. The joint rigidity method suggested by these workers is effective in determining the optimum welding sequence. Using the optimum welding sequence can improve the flatness of the panel and minimize angular distortion in the skin plate.

9.2.3 Buckling caused by circular and circumferential welds

The buckling distortions caused by circular welds on flat plates shown in Fig. 9.1 (d) and Fig. 9.1 (f) have more complicated specific characters owing to the variation in residual stress distribution. Typical distributions of residual stresses in plates with circular welds to join circular elements to it, e.g. flanges, are given in Fig. 9.8. Figure 9.8 (a) shows the measured residual stress distributions on titanium plate of thickness 1 mm with a circular weld to join a flange of thickness 4 mm. Figure 9.8 (b) gives schematically the dependence of the residual stress distribution on the radius R of the circular welds, while $R = 0$ represents the case of arc spot welding; in the case when $R_2 > R_1$, the distributions of both σ_r and σ_θ are changed a great deal. In the case when $R \to \infty$, the circular weld tends to a linear longitudinal weld; the distributions of σ_r and σ_θ are altered to became the σ_x and σ_y in a plate with a butt weld. The rigidities (thicknesses) of both the inner flange and the outer plate are also dominating factors in the distributions of σ_r and σ_θ.

The loss of stability of a plate with a flange welded in it is illustrated as a whole in Fig. 9.9 (a). For a clearer presentation, the loss of stability of the inner disk flange and the outer plate with a opening of radius R cut in the centre are illustrated separately in Fig. 9.9 (b) and Fig. 9.9 (c).

For the case shown in Fig. 9.9 (b), Vinokurov[5] gave a diagram showing the variation in radial stress σ_r and the buckling deflection W_0/δ as functions of radial displacement U_r (Fig. 9.10 (a)).

As the out-of-plane displacement W increases (Fig. 9.10 (a)), the stresses σ_r and σ_θ in the circular weld ring decrease, as shown by sloping line 1. The

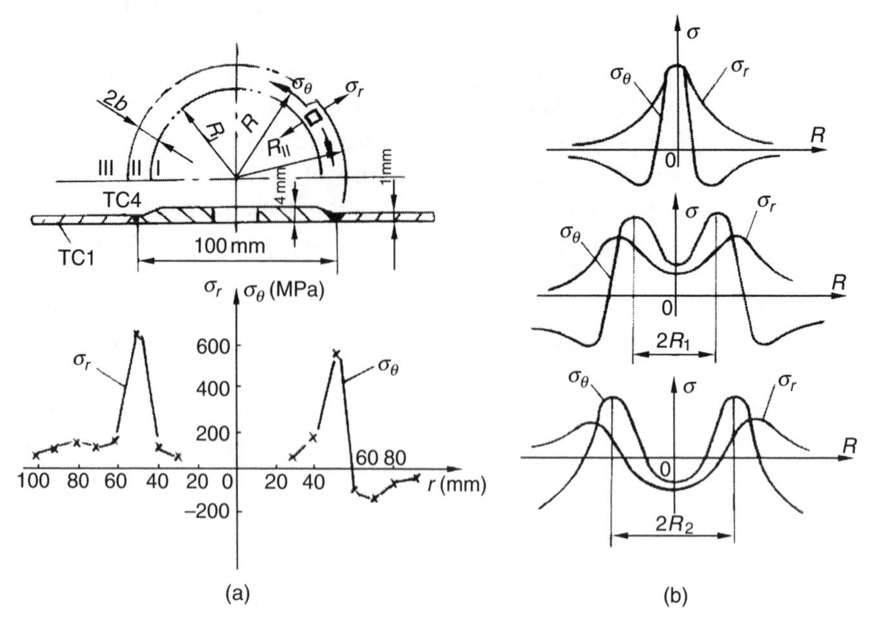

9.8 Residual stress distributions caused by circular welds (a) on a titanium plate and (b) with different radii of the welds.

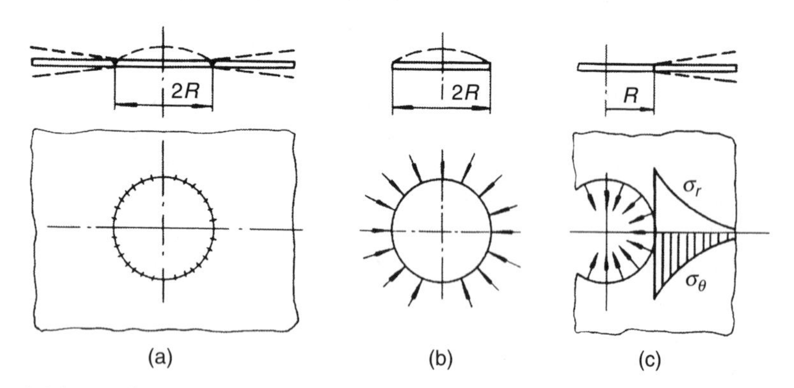

9.9 Loss of stability caused by a circular weld in a plate: (a) plate with a disk flange welded in it; (b) contour-welded circular plate disk; (c) circular weld on the inner edge of the circular opening cut in a plate.

stress σ_d in the disk plate increases as U_r increases shown by line 2 while the disk does not lose stability. In Fig. 9.10 (a), point A corresponds to equilibrium of the stresses σ_r in the ring and σ_d in the disk and the radial displacement is U_0. In fact, the disk loses stability at point B where the critical stress σ_{cr} is reached. The critical stress of the circular disk plate with hinge-supported edge is determined from the relation

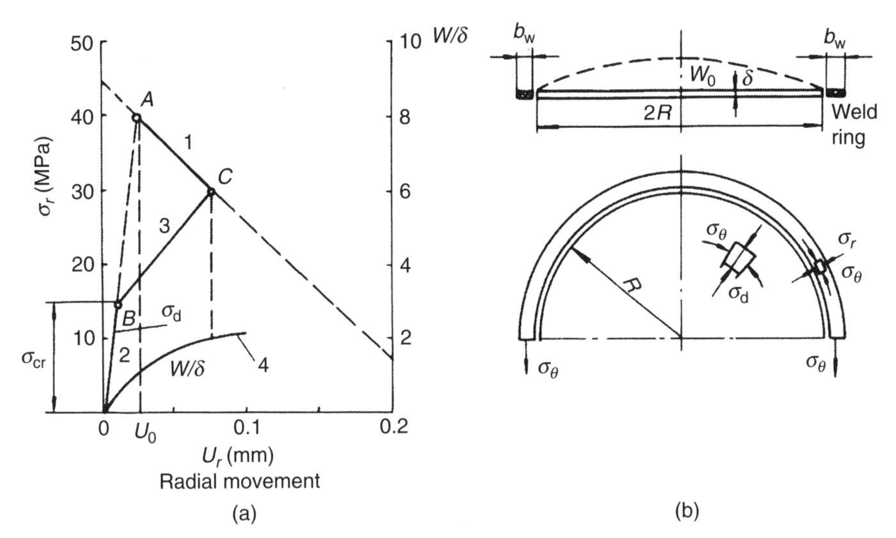

9.10 Diagram showing the buckling deflection of circular plate disk with a contour weld[5]; ($2R = 300$ mm; $b_w = 20$ mm; $\delta = 2$ mm; $E = 2.0 \times 10^5$ MPa; $\sigma_\theta = 300$ MPa; $\sigma_r = 40$ MPa; $\sigma_{cr} = 14$ MPa; σ_r, radial stress acting on the weld ring; σ_d, radial stress acting on the disk).

$$\sigma_{cr} = \frac{0.385E\delta^2}{R^2} \qquad [9.3]$$

Afterwards, the radial displacement U_r increases rapidly according to line 3. At point C, σ_r in the weld ring and σ_d in the disk plate reach equilibrium. In this state, the tangential stress σ_θ in the ring in relation to σ_r, denoted by $\sigma_\theta = \sigma_r R/b_w$, will decrease considerably. Losing stability, the buckled disk will have maximum radial displacement U and the deflection W/δ is determined by the relation $W_0/\delta = 1.96(\sigma_r/\sigma_{cr} - 1)^{1/2}$, as shown by curve 4. For the particular case shown in Fig. 9.10 (b), the deflection of the contour welded circular disk plate corresponding to point C on curve 4 will be $W_0 = 2.2\delta$.

For the case shown in Fig. 9.9 (c), the circular weld on the inner edge of the circular opening of radius R in the plate causes a radial tensile stress σ_r, which produces the tangential compressive stress σ_θ. As the thin plate is most susceptible to loss of stability, sinusoidal wave buckling always takes place round the whole circumference. Vinokurov[5] determined the relationship between the critical radial stress in terms of $\sigma_{r_{cr}}\delta R_1^2/D$, where $D = E\delta^2/12(1 - \mu^2)$ and the ratio R_2/R_1 is as shown in Fig. 9.11. Two possible clamps 1 and 2 on the periphery of the circular plates (inner opening of radius R_1 and outer radius R_2) were analyzed. The number of sinusoidal half-waves at which the plate loses stability are given on the curves. Large

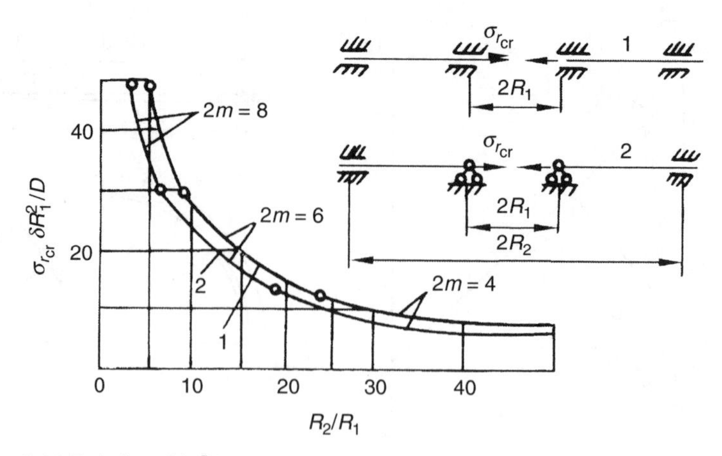

9.11 Relationship[5] between $\sigma_{r_{cr}}$ and the ratio R_2/R_1.

sheets lose stability for $2m = 4$ half-waves. The number of half-waves increases at low R_2/R_1 ratios.

The actual restraint of the plate in engineering practice roughly corresponds to the intermediate conditions between the rigid-clamping condition 1 and the hinged-support condition 2. When the radius R_1 is reduced, loss of stability occurs at higher $\sigma_{r_{cr}}$. In an infinite plate, when $R_2 \to \infty$, loss of stability caused by σ_r occurs regardless of the support conditions at the edge of the inner opening, as the $\sigma_{r_{cr}}$ is expressed by the formula

$$\sigma_{r_{cr}} = \frac{E\delta^2}{4(1-\mu^2)R_1^2} \tag{9.4}$$

From the formula it can be seen that the modulus E of elasticity and the metal thickness δ essentially cause the loss of stability of a flat plate. Titanium alloys with $E = 1.05 \times 10^5$ MPa and aluminum alloys with $E = 0.7 \times 10^5$ MPa are more susceptible to buckling than steel elements with $E = 2 \times 10^5$ MPa at identical shrinkage forces. This is the reason why buckling problems are much more troublesome in aerospace industries where titanium and aluminum alloys are adopted widely.

A typical example of the buckling of a cylindrical shell in the form of a sinusoidal wave caused by a circumferential weld is shown in Fig. 9.12. The thin shell skin of thickness less than 1 mm was fusion welded on a inner rigid supporting ring; a compressive circumferential stress σ_θ was induced in cross-section AA and a sinusoidal wave with a certain pitch formed over the whole circumference.

Based on the shrinkage force model, calculation of residual stress distribution and deflection in a cylindrical shell due to a circumferential weld was made by Guan *et al.*[25]. Figure 9.13 presents the results using

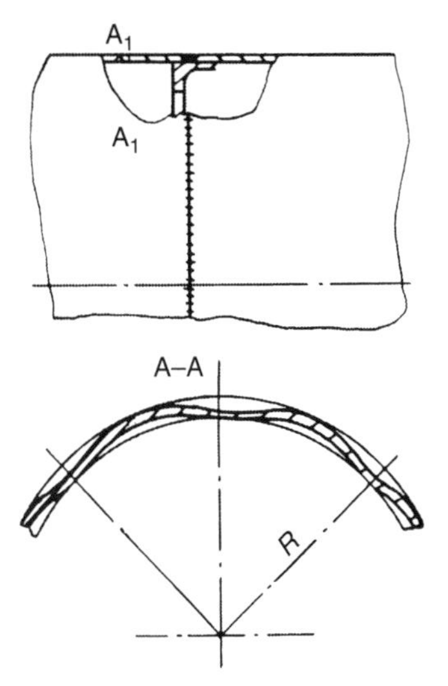

9.12 Buckling of a cylindrical shell due to a circumferential weld.

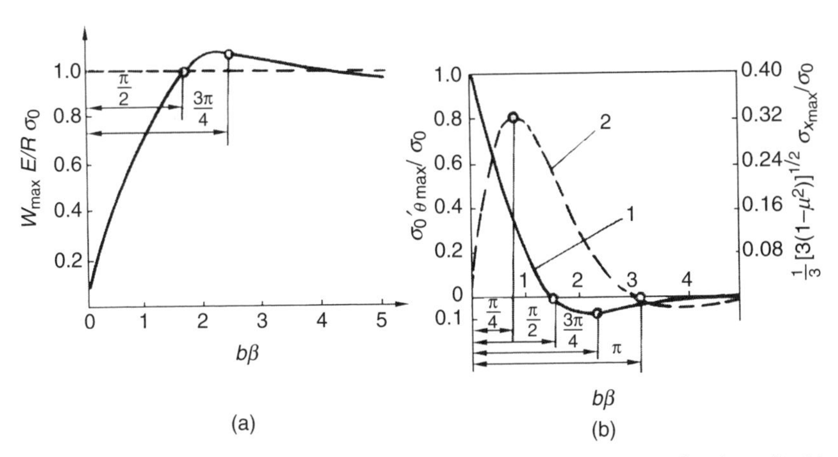

9.13 (a) Deflections *W* and (b) residual stresses σ_θ and σ_r in cylindrical shells caused by a circumferential weld[25].

dimensionless expressions. The initial residual stress σ_0 in the weld over the width of $2b$ (plastically deformed zone) can be determined either experimentally or computationally. In many cases, $\sigma_0 = \sigma_s$. Relating to σ_0 the maximum deflection W_{max} on the weld centerline is described by the following formula (see the curve in Fig. 9.13 (a)):

$$W_{max} = \frac{\sigma_0 R}{E}[1 - e^{\beta b}\cos(\beta b)]; \quad \beta = \left(3\frac{1-\mu^2}{\delta^2 R^2}\right)^{1/4} \qquad [9.5]$$

The two-dimensional stresses $\sigma_{\theta_{max}}$ and $\sigma_{x_{max}}$ along the weld centerline can be expressed in the dimensionless system as follows:

$$\frac{\sigma_{\theta max}}{\sigma_0} = e^{-\beta b}\cos(\beta b) \quad \text{(curve 1 in Fig. 9.13(b))} \qquad [9.6]$$

and

$$\frac{\sigma_{x max}}{\sigma_0} = e^{-\beta b}\sin(\beta b)\frac{3}{[3(1-\mu^2)]^{1/2}} \quad \text{(curve 2 in Fig. 9.13(b))} \qquad [9.7]$$

These relations provide a quantitative assessment of the deflections and stresses as functions of the geometrical parameters of the shell as well as the material characteristics and welding heat input. However, the buckling shown in Fig. 9.12 occurs only in the case when the deflection W_{max} is prevented by a certain rigid supporting ring underneath the circumferential weld while both the tensile and the compressive residual stresses possess their maximum values.

9.3 Methods for removal, mitigation and prevention of buckling distortions

In sheet metal fabrication, excessive buckling distortions can generally be prevented by proper arrangement of technological measures during assembly and welding, and also by selecting rational welding processes. In the case of proper and skillfully applied techniques for eliminating buckling, welded thin-walled plates, panels and shells can be fabricated economically, and their quality and reliability can be improved considerably.

9.3.1 Classification of methods and technological measures

The most important stage in eliminating buckling distortions is rational design of welded structural elements: plates, panels and shells. It is essential that designers should adhere to the concept that buckling is not inevitable and that the problem could be solved by working together with technological engineers. Buckling can be controlled by a variety of methods and technological measures for removal, mitigation or prevention.

Table 9.1 illustrates the main stages and methods for the control of buckling distortions. In the design stage, rational selection of geometry, thickness of materials and types of weld joint is essential. In the fabrication stage, technological measures and techniques to eliminating buckling can be classified as follows:

Table 9.1 Classification of methods adopted for the control of buckling distortions

Stage	Methods adopted
Design stage	Rational selection of geometry and thickness Rational selection of types of weld joint Rational selection of welding techniques
Fabrication stage, before welding	Predeformation (counterdeformation) Pretensioning (mechanical, thermal) Assembly in rigid fixture
Fabrication stage, during welding	Selection of a welding process with low heat input Application of power beam welding process Forced cooling Selection of weld sequences Thermal tensioning LSND welding technique
Fabrication stage, after welding	Removal or correction using mechanical shock (hammer, weld rolling, electromagnetic shock) Heat treatment (overall heating in rigid fixture, local heating)

1. Methods applied before welding.
2. Methods adopted during welding operation.
3. Methods applied after welding.

In the methods applied before welding, e.g. predeformation, post-weld buckling is compensated by a counterdeformation formed in the elements prior to welding by a specially designed fixture die. In the methods applied after welding, once buckling is in existence, warpage is removed by special flattening processes, using a manual hammer, applying a mechanized weld-rolling technique or utilizing an electric—magnetic pulse shock. These methods both before welding and after welding are arranged as special operations on a production line and specific installations or fixtures are needed, resulting in increasing cost and variable quality of welded elements.

Pretensioning can be classified in the category of methods applied either before or during welding. For each particular structural design of panels, a device for mechanical tensile loading is required. Owing to their complexity and reduced efficiency in practical execution, application of these methods is limited. In this sense, the thermal tensioning (see Section 9.4) is more flexible in stiffened panel fabrication.

LSND results could be achieved during the welding process based on a thermal tensioning (stretching) effect which is provided by establishing a required specific preset temperature gradient either in the overall cross section of plate to be welded or in a localized area limited in the near-arc zone; simultaneously, restraining transient out-of-plane warpage movements of the workpiece is necessary. Different from the above-mentioned 'passive' methods which have to be employed after welding, the LSND welding techniques (see Sections 9.5 and 9.6) can be classified as 'active' methods for in-process control of buckling distortions without the need to rework operations after welding.

9.3.2 Selection of rational welding processes

The more attractive achievements in the control of buckling distortions are the applications of power beam welding, e.g. electron beam and laser beam welding. Their contributions are to reduce dramatically the heat input and therefore to decrease the inherent strains, residual stresses and buckling distortions.

In Fig. 9.14, comparisons are given of the experimentally measured inherent strain ε_x^p distributions caused by different welding processes[13]. Curve 1 in Fig. 9.14(a) corresponds to a higher heat input of GTAW, and curve 2 to a lower heat input of GTAW of a titanium specimen of thickness 1.5 mm. The ε_x^p distribution width $2b$ is about 40 mm. The higher the heat input, the wider is the distribution of ε_x^p and the higher is the peak value of ε_x^p. The relationships between the dimensionless width b/δ of the ε_x^p distribution and the thickness δ of the plate obtained using different fusion welding processes are given in Fig. 9.14(b). Among the well-known fusion welding processes, electron beam welding (curve 4) and laser welding (curve 5)

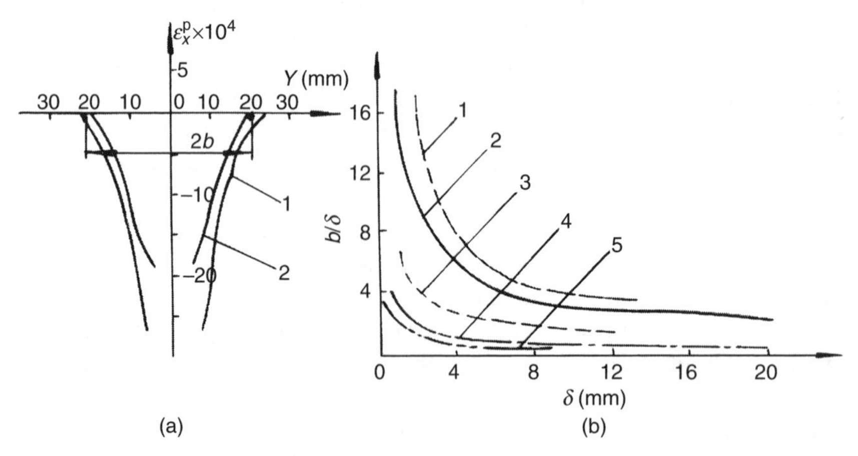

9.14 Inherent strain ε_x^p distributions caused by different welding processes[13].

impose a minimum width of the ε_x^p distribution much narrower than that imposed by arc welding (curve 2), plasma welding (curve 3) and even oxy-acetylene flame welding (curve 1). As result of the existence of inherent (incompatible) strains in weld joints even by the application of high-energy-density beam welding processes, buckling distortions thereby induced by compressive residual stresses in plates and shells cannot be avoided completely.

9.3.3 Passive technological measures for buckling removal and correction after welding

Passive technological measures have to be taken, once warpage is in existence after welding. They are adopted as additional procedures and have to be incorporated into production line. Basically, all the passive measures rely on the formation of the inverse plastic deformations to compensate the incompatible welding residual strains either in the weld joints stretched by tensile residual stresses by using mechanical expansion effects or in the compressed bowed zones of base metal far away from the weld by using thermal shortening effects. Comparisons between the thermal shortening effect by means of spot heating and the mechanical expansion effect by means of spot shocking are given in Table 9.2. A residual shortened spot zone can be obtained using spot heating whereas a residual expanded spot zone remained using the spot shocking technique.

Table 9.2 Comparison between the thermal shortening effect and the mechanical expansion effect

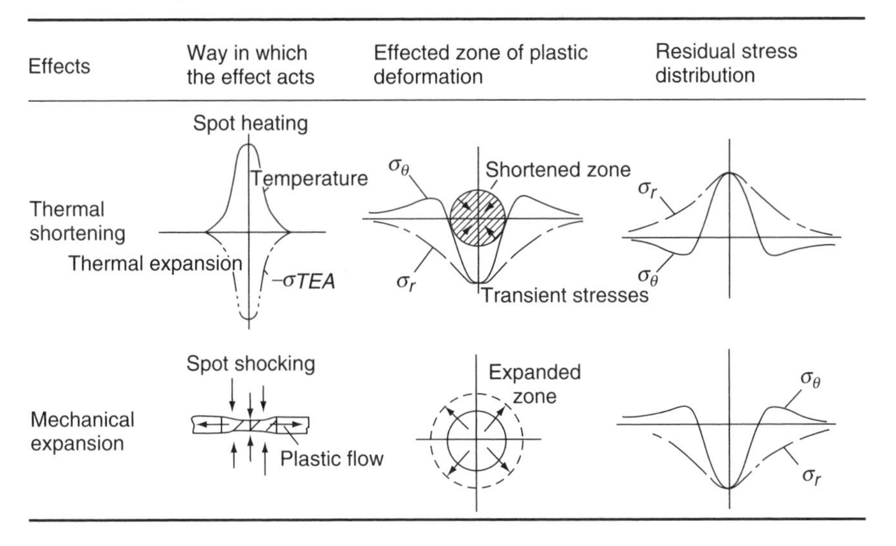

Effects	Way in which the effect acts	Effected zone of plastic deformation	Residual stress distribution
Thermal shortening	Spot heating; Temperature; Thermal expansion; $-\sigma TEA$	σ_θ; Shortened zone; σ_r; Transient stresses	σ_r; σ_θ
Mechanical expansion	Spot shocking; Plastic flow	Expanded zone	σ_θ; σ_r

To compensate for the incompatible welding strains, the inverse plastic expansion deformation in the weld zone can be achieved using a mechanized pneumatic hammer spot by spot (Fig. 9.15(b) and Fig. 9.15(c)) rather than using a manual hammer (Fig. 9.15(a)) for more stable technological results and less damage to avoid possible instability in the quality of products.

The principle of the weld rolling technique for removal of buckling is shown in Fig. 9.16. Using narrow rollers under a certain pressure P (Fig. 9.16(a)), inverse plastic deformations in the weld zone are formed (Fig. 9.16(b)) to compensate the incompatible welding stains. In some cases, besides rolling the weld, rolling both the side zones close to the weld is also feasible in practise (Fig. 9.16(c)).

The effectiveness of forming inverse plastic deformation in the weld zones by weld rolling is shown in Fig. 9.17. During rolling, the longitudinal residual stress σ_x in the weld is reduced directly under the rollers (Fig. 9.17(a)). The value of σ_x after rolling depends on the pressure applied for rolling the weld. The twisted net of broken lines in Fig. 9.17(b) shows the metal in-plane displacements near the weld joint under the rolling pressure between the narrow rollers.

Because of its stability in obtaining the positive results to remove buckling distortions, especially in the aerospace manufacturing of shell structures with longitudinal and circumferential welds, the weld rolling technique is more widely applied in sheet metal fabrication[6,10,11,14]. Figure 9.18 shows some typical examples of selecting the best results for relieving the residual stress (Fig. 9.18(a) and Fig. 9.18(b)) and therefore removal of buckling (Fig. 9.18(c)). After rolling only the weld, the residual stress

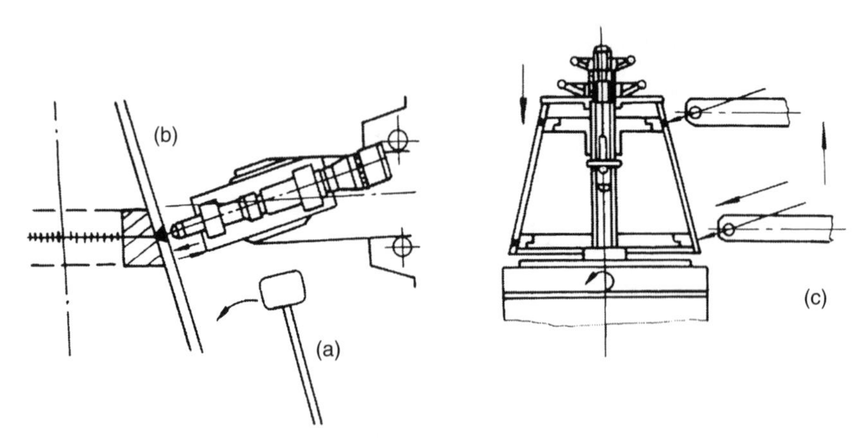

9.15 Mechanized pneumatic hammer for buckling correction after the welding of a conical shell[6].

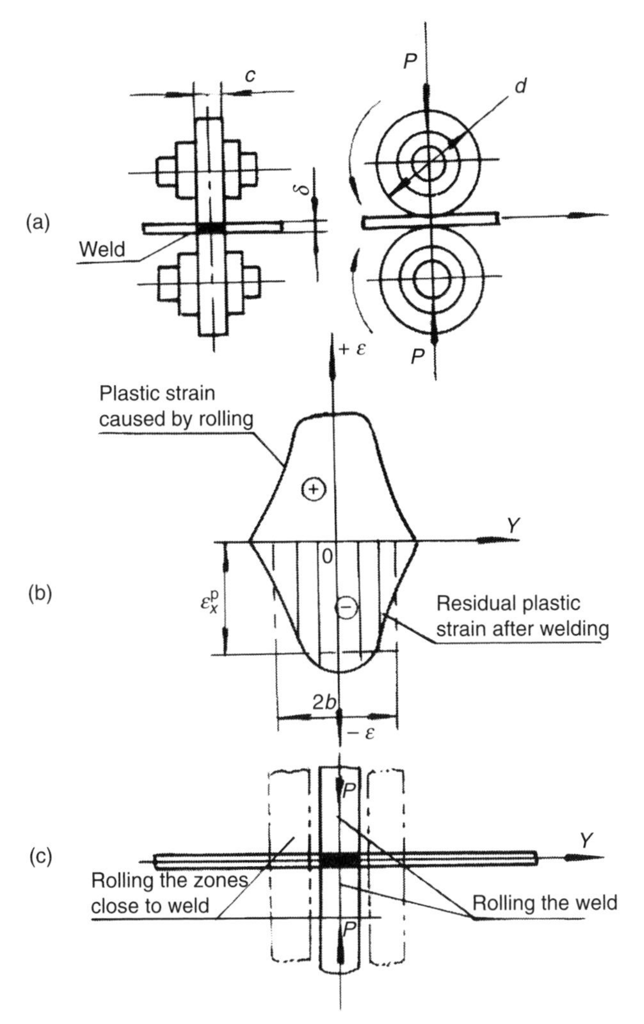

9.16 Weld rolling technique for removal of buckling after welding[2,3,5].

distribution (dashed curve, after welding; solid curve, after rolling) shows that under a higher pressure the peak value changes from tensile to compressive (see Fig. 9.18(a)). In contrast, when a lower pressure is applied and rolling is applied not only to the weld but also to the zones close to weld, the residual stress distribution is negligible (Fig. 9.18(b)). Buckling caused by a circumferential weld in a cylindrical shell (Fig. 9.18(c)) can be corrected using the weld rolling technique. Buckling and residual deformations can be removed properly after rolling the weld zone as well as rolling the near-weld zones[6,9].

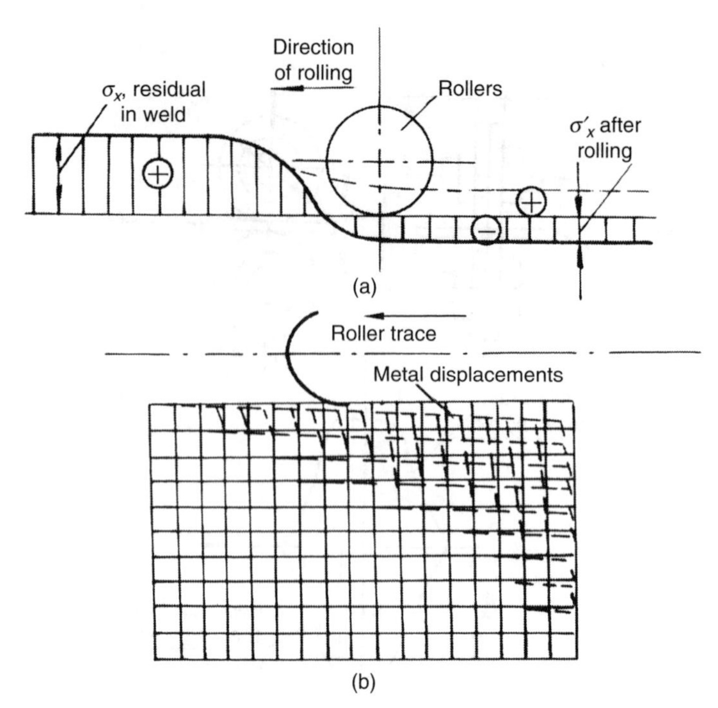

9.17 (a) Reduction in residual stress in a weld and (b) inverse plastic deformation produced by the weld rolling technique[3,5,6].

9.3.4 Active measures for in-process control, mitigation and prevention of buckling

As mentioned in the previous sections, active measures taken before or during welding are aimed at avoiding buckling without having to use reworking operations after welding. The most applicable active technological measures for in-process control of buckling are listed in Table 9.3.

The feasibility of one or another active measure is mainly dependent on the strictness of design requirements for the structural performance and also dependent on the specifics of the geometry of the elements, the materials applied, and the economic and productive results.

The predeformation (or counterdeformation) technique for mitigation and avoiding buckling in shell elements (see Fig. 9.2(b), Fig. 9.2(d) and Fig. 9.2(f)) was successfully adopted firstly in manufacturing jet engine cases. Figure 9.19 shows a scheme of the principle used for engineering execution of the predeformation technique for a shell element 4 while a flange disk 3 is to be welded in it. The circular weld 2 is performed using GTAW 1. To avoid the subsided buckling 7, the local zone on shell surrounding the

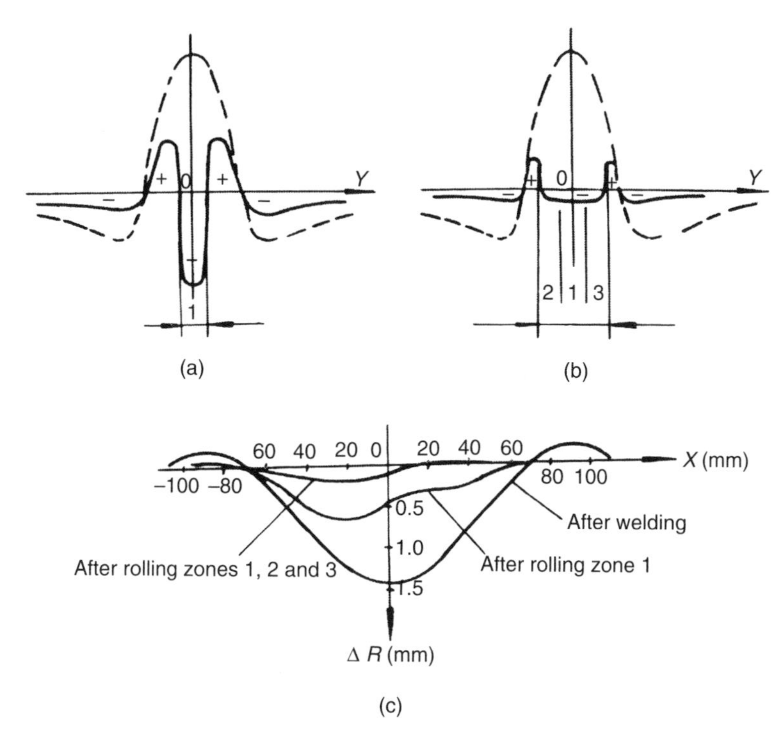

9.18 (a), (b) Residual stress relief and (c) distortion removal using the weld rolling technique[6,9].

Table 9.3 Active technological measures for the in-process control of buckling

Active measure	Main concepts	Feasible applications
Predeformation (counterdeformation)	Compensation mechanism Reduction in residual stress	Shells with circular welds and panels with longitudinal welds
Pretensioning (or tensioning during welding)	Mechanical Thermal Hybrid thermomechanical	Stiffened panels and skinned frames
LSND welding	Overall cross-section thermal tensioning Localized thermal tensioning Reduction in incompatible strains	Plates, panels and shells

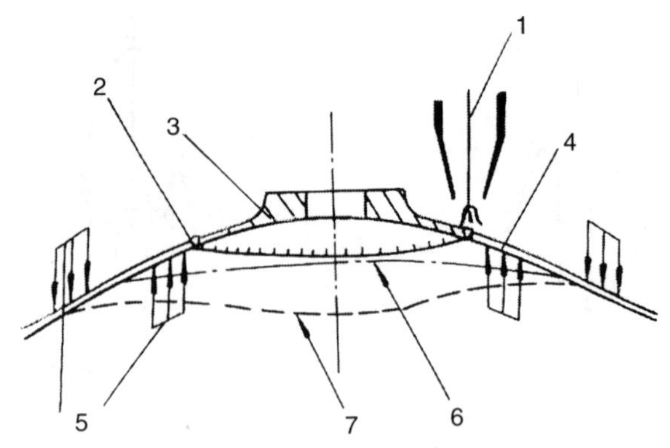

9.19 Scheme of the principle used for the execution of the predeformation technique[13].

circular weld is predeformed to position 4 by means of the fixture die under pressure 5 before welding.

After welding, while the shell element with jointed flange disk is released from the fixture die, because of the compensation displacements of both the elastic restoration movement of shell element and the shrinkage phenomena (especially in the direction transverse to the circular weld), the shell element backtracks to its required position 6 (see Fig. 9.19).

The optimized value of predeformation can be calculated and selected depending mainly on the geometry of the shell, e.g. the thickness of material applied and the diameter of the circular weld[6]. For a rationally selected and skillfully executed predeformation technique, the incompatible plastic strains of the circular weld in both the tangential and the radial directions are compensated properly by the predeformations either in the elastic state or in the elastic–plastic state. The residual stresses are also controlled to be much lower than the critical compressive stress at which undulated buckling occurs.

As an alternative option to restore the required geometry of an aluminum shell element with a flange jointed by a circular weld, a reinforcement ring on the flange could be prepared and compressed under pressure after welding to compensate the transverse shrinkage of the circular weld[6]. In addition, a reinforced circular weld bead with excess filler metal could also be preformed and compressed under a pneumatic press spot by spot after welding for the same purpose.

The pretensioning technique is applied in some cases as an active in-process control method to avoid buckling distortions in shells with longitudinal welds (Fig. 9.20(a)) or in plates with web stiffeners and stringer panels (Fig. 9.20(b) and Fig. 9.20(c)). Before welding, the edges of shells and panels to be welded are tension loaded by a specially designed installation

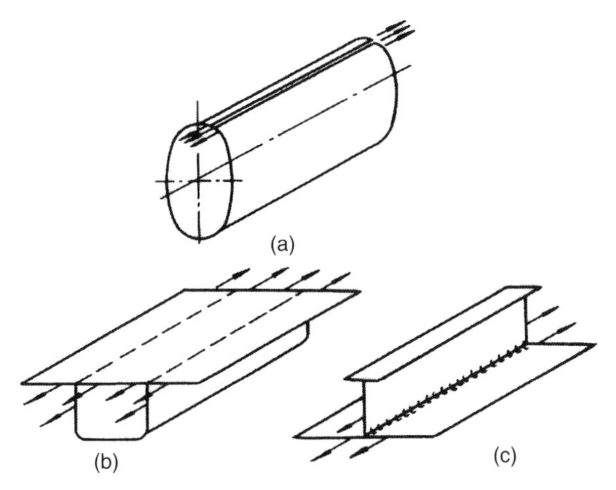

9.20 Pre-tensioning technique to prevent buckling in (a) shells and (b), (c) stiffened panels.

or stand. As result, the incompatible plastic strains in the weld zone are reduced properly[26–28].

In some engineering applications, a hybrid technique, namely combined pretensioning and prebending techniques, could also be adopted to avoid subsided buckling caused by a longitudinal weld in cylindrical shells. The bending moments result in inverse deformation in the longitudinal direction of the weld.

In general sheet metal fabrication, thin skin plates are frequently assembled and welded to rigid frames. Buckling always presents the most troublesome problems for the fabricators. To solve buckling problems in the railway carriage fabrication and shipbuilding industries, in the 1970s, the pretensioning effects were implemented successfully at Kawasaki Heavy Industries by Terai[7] and Masubuchi[8]. In Fig. 9.21 the Kawasaki methods are presented, which are based on the pretensioning effects executed either mechanically (SS method) (extension) or thermally (SH method) (expansion), as well as a hybrid mechanical–thermal method (SSH method).

In all these methods, the welding operation proceeds in a pretensioned state of the skin plate to be welded by fillet welds to the frame. The pretensioning effects are implemented either by a specialized tensioning stand or by a heating device to provide the necessary mechanical extension or thermal expansion of the skin plate prior to welding. After welding, while the skin plate is released from the tensioning stand or the heater is removed, the skin plate is nestled up to the frame as flat as before welding without buckling.

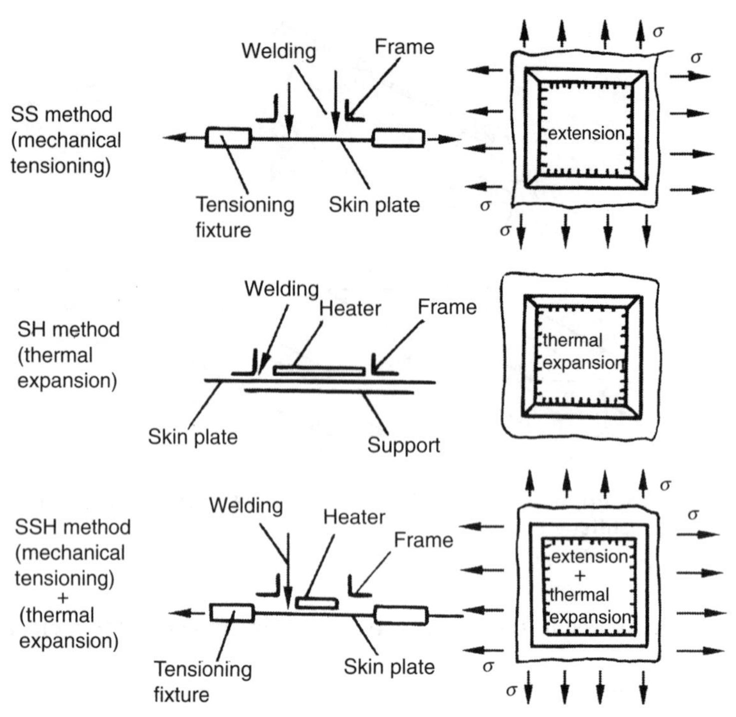

9.21 Pretensioning effects executed mechanically (the SS method), thermally (the SH method) and mechanically-thermally (the hybrid SSH method)[7,8].

More detailed descriptions of other active methods listed in Table 9.3 such as thermal tensioning and LSND welding techniques will be given in the following sections.

9.4 Thermal tensioning to prevent buckling

In shipbuilding and vessel manufacturing industries the well-known low-temperature stress-relieving technique[10,29,30] with flame heating and combined water cooling is widely adopted in welded structures of thicker plate sections for mitigation of longitudinal residual stresses after welding. It is based on a temperature gradient stretching effect induced by local linear heating and cooling parallel to the weld line on plates of thickness 20–40 mm. This technique is not applicable for stress relieving, nor for removal of buckling after welding of thin-walled elements where the metal sheets are not stiff enough to avoid the transient out-of-plane displacement during local heating and forced cooling. However, the idea of the temperature gradient stretching effect (or, as is the commonly adopted term, the thermal tensioning effect) is logically feasible for avoiding buckling of plates and shells during welding while the transient out-of-plane displacements are prevented by properly arranged rigid fixtures or clamping systems[31–33].

9.4.1 Effects of thermal tensioning

In comparison with mechanical tensioning (stretching) where special installations are needed, the thermal tensioning effect is normally implemented much more economically using localized heating and properly matched cooling to create the required temperature gradient and therefore the expected tensioning effect. Efforts in this direction have been made during the last few decades[8,13,34–37].

The basic principle of the thermal tensioning effect is shown in Fig. 9.22. Two curves σ_{x_1} and σ_{x_2} of thermal stress distributions are created by the corresponding temperature curves T_1 and T_2 on a thin plate of width $Y = 1$. In this case, the thermal tensioning effect is defined as the value of σ_x^* in the plate edge at $Y = 0$ where the weld bead will be implemented. For a given σ_x^*, the greater the temperature gradient ($\partial T_1/\partial Y > \partial T_2/\partial Y$), the higher is the induced maximum value $-\sigma_{1x_{max}}$ of the compressive stress. An optimized temperature curve can be calculated mathematically for an estimated value σ_x^* while the value $-\sigma_{x_{max}}$ does not exceed the yield stress and no plastic deformation occurs in the locally heated zone.

Based on the results of a mathematical analysis for the thermal tensioning effect, Burak *et al.*[34,35] conducted an experiment to control longitudinal

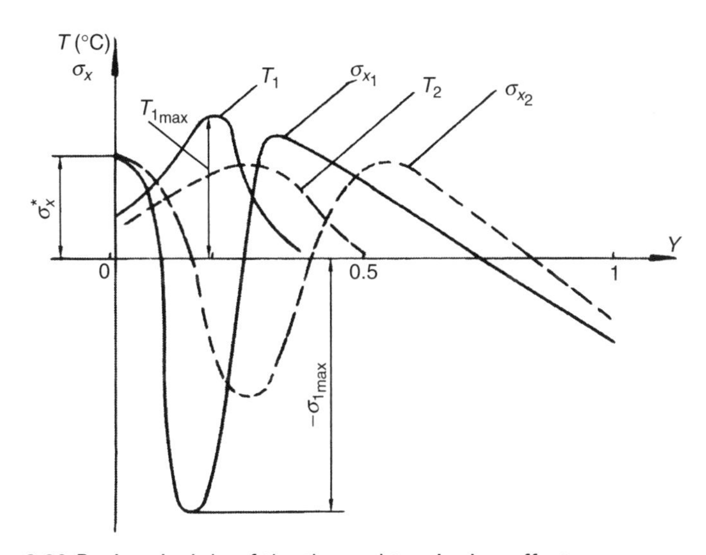

9.22 Basic principle of the thermal tensioning effect.

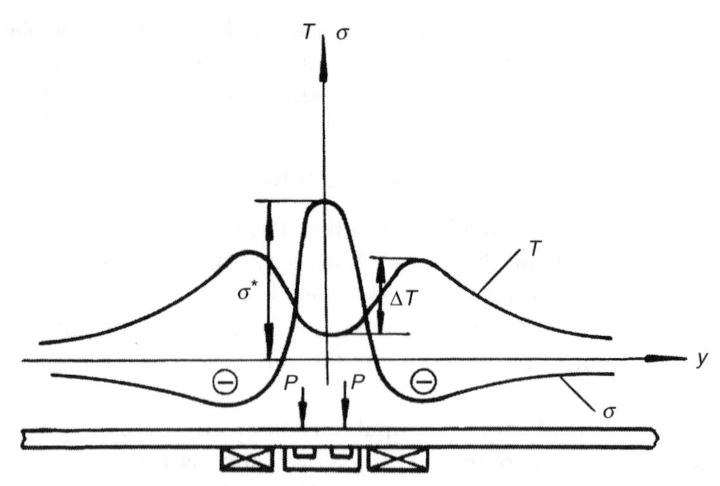

9.23 Control of longitudinal plastic strains during the GTAW of an aluminum plate 4 mm thick using the thermal tensioning effect[34,35].

plastic strains in a weld on an aluminum plate 4 mm thick. As shown in Fig. 9.23, before and during GTAW, a temperature profile (curve T) was established using a linear heater on both sides along the weld with forced-water cooling by a copper backing bar underneath the weld zone. Corresponding to the curve T, a thermal stress distribution curve σ is also shown in Fig. 9.23. The maximum value σ^* represents the expected thermal tensioning effect on both the plate edges to be jointed together.

As verified by experiment, the maximum value of the longitudinal residual stress in the weld can be reduced obviously with increase in the temperature gradient ΔT (see Fig. 9.23), and therefore the longitudinal plastic incompatible strains in the weld zone were mitigated respectively. Investigation proves that the longitudinal incompatible strains can be controlled quantitatively by an appropriately selected temperature gradient–thermal tensioning effect.

9.4.2 Progress in the control of buckling distortions: from passive measures to active in-process control

The buckling problems caused by fusion welding become more significant for thin-walled elements of thickness less than 4 mm (see Fig. 9.7(a)) which are widely applied in designing aerospace structures. In the 1980s, to apply the thermal tensioning effect to avoid buckling in aerospace structures of thickness less than 4 mm, a series of experiments were carried out by Guan *et al.*[31–33]. It has been proved by the results of repeated experiments that the scheme reported by Burak *et al.*[34,35] for a plate 4 mm thick is not applicable to eliminating buckling in elements of thicknesses less than 4 mm. The reason is that, owing to the susceptibility to loss stability of thin elements of thickness less than 4 mm, transient out-of-plane displacements occur in areas away from the weld zone. The compressive stresses in these areas are the results of superposition of both the welding temperature field and the preset temperature distribution T (see Fig. 9.22).

The transient out-of-plane displacements outside the clamping fingers (indicated as P in Fig. 9.23) release the potential energy induced by the preset temperature field with the thermal plane-stress distribution. In the lost-stability position, the expected preset thermal tensioning stress σ^* (Fig. 9.23) ceases to exist in the edges to be jointed together whereas the preset thermal stress field is redistributed to take a stable form with minimum potential energy (Fig. 9.5).

Progress was made in solving the above-mentioned problem to improve the thermal tensioning technique and to make it applicable for preventing buckling in elements of thickness less than 4 mm especially in manufacturing aerospace structures[31]. Figure 9.24 shows the improvement in the clamping systems. In conventional clamping system with a 'one-point' finger fixture (indicated by P_1 in Fig. 9.24(a)), transient out-of-plane warpage of the workpiece is inevitable whereas, using the improved 'two-point' finger

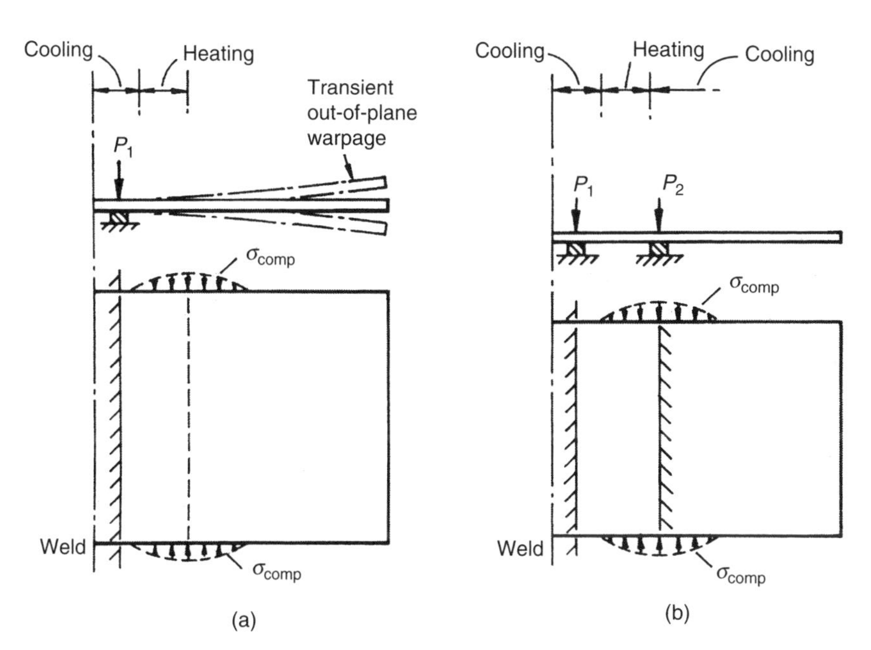

9.24 (a) Transient out-of-plane warpage displacement of workpiece in a conventional clamping system and (b) its prevention in a 'two-point' finger clamping system[31].

clamping system (indicated by P_1 and P_2 in Fig. 9.24(b)), the desirable thermal tensioning effect in terms of σ_x^* (Fig. 9.22) can be established without transient out-of-plane warpage displacements.

As an active in-process control method, this improved technique is more widely acknowledged as the LSND welding method for thin materials[13,31,32]. It is essential to note that the remarkable achievement of the LSND welding technique as an active in-process control method in most cases in aerospace engineering is ongoing in order to replace formerly adopted passive measures (see Section 9.3.3) for removal of buckling after welding. In other words, progress was achieved in avoiding buckling from passive measures to active in-process control without additional reworking operations after welding. Detailed explanation of the LSND welding technique patented in 1987 and its applications will be given later in Section 9.5.

9.4.3 Alternative options in the practical implementation of thermal tensioning

To create the thermal tensioning effect along the plate edges to be welded, the preset temperature field can be built up either statically by stationary linear heaters arranged underneath the workpiece parallel to the weld direction or by using two movable heating devices on both sides of the weld and synchronistically traveling with the welding torch. The LSND welding technique can be implemented either in the former or in the later way[32]. In 1990 a Japanese Patent was applied[37] for to prevent buckling distortions in the welding of thin metal sheets using movable heating sources parallel to the welding torch. Figure 9.25 shows the scheme for the practical execution of this method to avoid buckling in the welding of a flat plate (Fig. 9.25(a) and Fig. 9.25(c)) and to prevent buckling in the welding of webbed panels (Fig. 9.25(b) and Fig. 9.25(d)). This method is not exactly based only on the thermal tensioning effect but is mainly based on the additional heating-induced (indicated by 2 in Fig. 9.25) plastic strains equivalent to that in the weld zone. A proper shortened zone in the areas away from the weld is formed to equilibrate the incompatible plastic zone in the weld.

This method is applicable for the welding of aluminum structural elements of thickness 4–10 mm for carriages and high-speed craft. The movable heaters could be argon-shielded tungsten arc torches. Obviously, for the reason mentioned above, this technique is hardly applicable for materials of thickness less than 4 mm. Owing to the fusion effect by the additional arc heating, this method seems to be impractical for aerospace manufacturing.

For the thermal tensioning effect by means of moving heating (flame or others) parallel to welding with or without cooling (heat sink) of the weld zone, the geometrical process parameters were computationally optimized

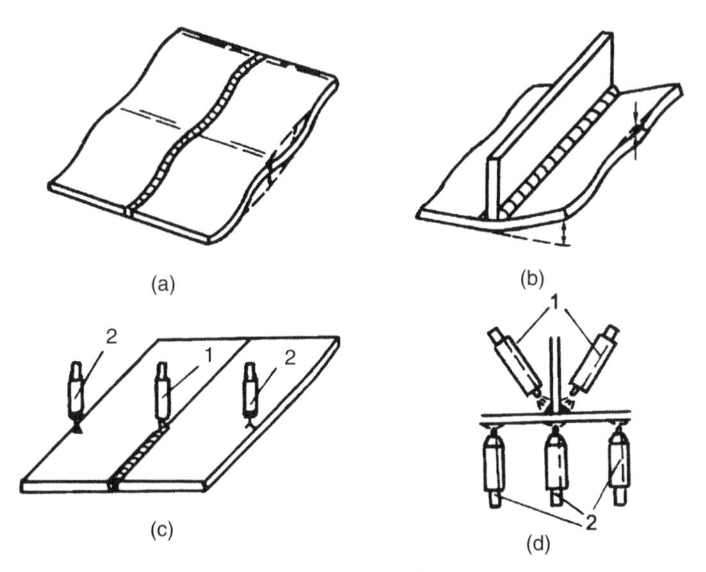

9.25 Buckling avoiding technique for the welding of (a), (c) a plate and (b), (d) webbed panels using a movable heating torch[37]: 1, welding torch; 2, additional heating torches.

by Michaleris *et al.*[38]. In this way, the mitigating residual compressive stresses in panels and therefore the smallest possible buckling distortions can be achieved.

In the broad sense of the term 'thermal tensioning', the effect can be created not only in the longitudinal direction of the weld to control the longitudinal plastic strains in weld zone, but the effect in mitigating the transverse shrinkage of the weld could also be utilized for preventing hot cracking[39]. Furthermore, manipulating the combination of heat sources and heat sinks, the thermal tensioning effect as well as the thermal compressing effect could also be established properly for the purposes required. Mitigating residual stresses in Al–Li repair welds[40] is an example in applying the alternative options of the thermal tensioning effect. The thermal tensioning effect can be created either in an overall cross-section of a plate (cross-sectional thermal tensioning) or in a localized zone limited in a near-arc high-temperature area (localized thermal tensioning). For the latter alternative the option of thermal tensioning a multiple-source system (coupled heat source and heat sink) should be adapted. Later, as described in Section 9.5, the overall cross-sectional thermal tensioning is utilized to implement LSND welding. In Section 9.6, localized thermal tensioning is investigated in more detail and utilized to execute the dynamically controlled low-stress no-distortion (DC-LSND) welding technique[41].

9.5 Low-stress no-distortion welding

To fit in with the strict geometrical integrity and to ensure dimensionally consistent fabrication of aerospace structures, the LSND welding technique for thin materials, mainly for metal sheets of thickness less than 4 mm was pioneered and developed in the 1980s at the Beijing Aeronautical Manufacturing Technology Research Institute (BAMTRI) by Guan *et al.*[31–33]. This technique aims to provide an in-process active control method to avoid buckling distortions.

9.5.1 Basic principles for implementation of low-stress no-distortion welding

The specific and essential feature of LSND welding is to provide a desired thermal tensioning effect during welding and simultaneously to prevent transient out-of-plane warpage displacements of the workpieces which occur as a result of the superposition of the temperature fields of welding and preset heating. Figure 9.26 shows schematically the basic principle for practical implementation of LSND welding[31]. The thermal tensioning effect, which is defined herein and is as shown by the maximum tensile stress σ^*_{max} in the weld zone, is formed owing to the contraction of zone 1 by the water-cooling backing bar underneath the weld and the expansion of the zone 2 on both sides adjacent to the weld by linear heaters. Both curve T and curve σ are symmetrical to the weld centerline. The higher σ^*_{max}, the better will be the results of controlling buckling distortions. If out-of-plane warpage displacements take place on the workpieces (see Fig. 9.24(a)) during preset

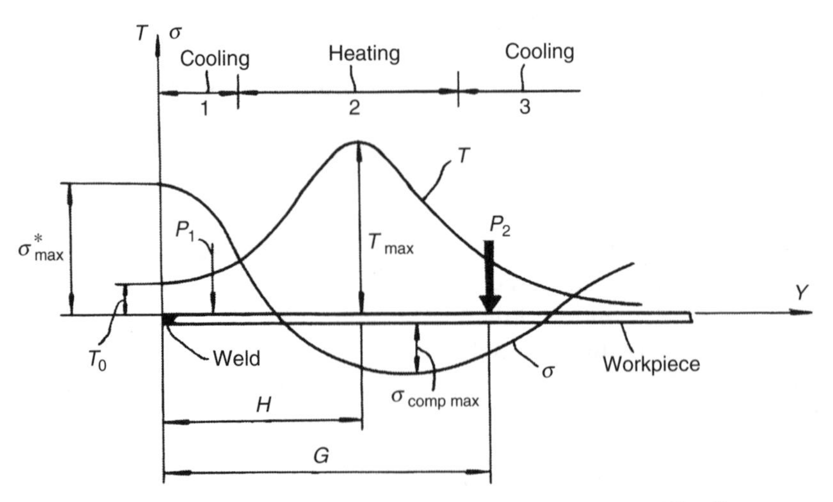

9.26 Basic principle for implementation of LSND welding[31].

heating and welding, the expected thermal tensioning effect in terms of σ^*_{max} will no longer exist. This phenomenon has an adverse influence on the control of buckling.

It is proved by experiments and engineering applications that, for LSND welding of materials of thicknesses less than 4 mm, the thermal tensioning effect is the necessary condition whereas the sufficient condition is preventing transient out-of-plane displacements by applying flattening forces in 'two-point' finger clamping systems shown by P_1 and P_2 in Fig. 9.26. The selected curve T is mainly determined by T_{max}, T_0 and H (the distance of T_{max} to the weld centerline). The tensioning effect σ^*_{max} becomes stronger as the temperature gradient $T_{max} - T_0$ increases while H decreases. The optimization of σ^*_{max} and technological parameters such as H can be implemented computationally and verified experimentally. The desired LSND welding results can be achieved appropriately.

Figure 9.27(a) and Fig. 9.27(b) show schematic views of practical implementation of the LSND welding method and apparatus for longitudinal joints in flat plates and in cylindrical shells respectively.

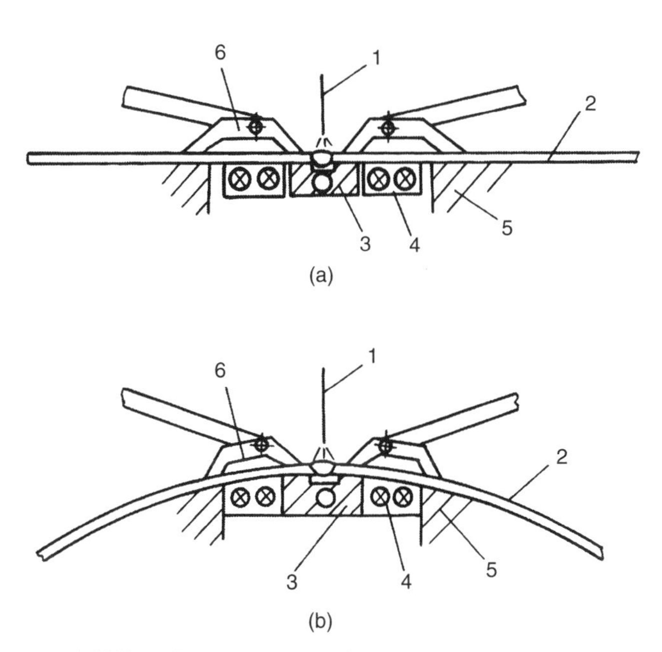

9.27 LSND welding technique for a longitudinal weld in (a) plates and (b) cylindrical shells: 1, arc; 2, workpiece; 3, water-cooling backing bar; 4, linear heaters; 5, supporting mandrel; 6, 'two-point' finger clamping system.

In the stationary seam welder apparatus for LSND welding, the technological functions of the 'two-point' finger clamping system are found to be as follows:

1. To prevent transient out-of-plane warpage displacements of workpieces.
2. To improve heat transfer in zones 1 and 3 (Fig. 9.26) for increasing the temperature gradient $T_{max} - T_0$.
3. To increase frictional resistance for avoiding in-plane rotating movement of workpieces.

In engineering practice, for a given power of heating elements, the desired temperature profile can be established on stainless steel more quickly than on aluminum alloys as the thermal conductivity of the former is almost an eighth of that of the latter.

9.5.2 Investigations on the low-stress no-distortion welding technique

The typical temperature field of two-dimensional heat transfer in GTAW of a thin plate is shown in Fig. 9.28(a). In fact, in engineering practice, the GTAW of a longitudinal weld on a thin plate is performed in a longitudinal seam welder. Workpieces are rigidly fixed in a pneumatic

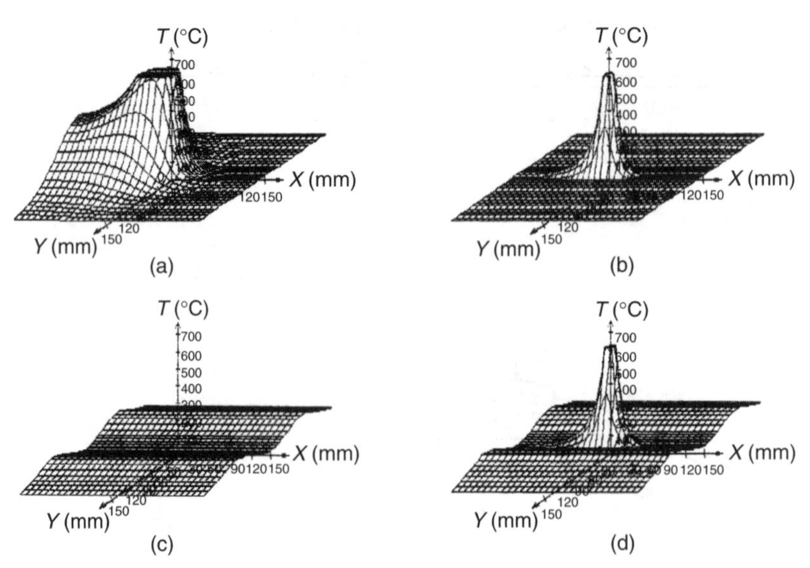

9.28 Temperature fields of (a) conventional GTAW with two-dimensional heat transfer, (b) GTAW on a copper backing bar with an intensive heat sink, (c) a preset temperature field needed for LSND welding and (d) LSND welding.

finger-clamping system with a copper backing bar on a mandrel support. Because of the intensive heat transfer from the workpiece to the copper backing bar, the temperature field is not the normal shape and has a narrowed distribution, as shown in Fig. 9.28(b). To implement LSND welding, an additional temperature field as shown in Fig. 9.28(c) is formed by preset heating and cooling (see Fig. 9.26). Therefore, the LSND welding temperature field shown in Fig. 9.28(d) results by superposition of the temperature fields of Fig. 9.28(b) and Fig. 9.28(c).

For a clearer quantitative assessment of the LSND welding technique, a systematic investigation[33] was carried out by Guan *et al.*[32] at BAMTRI and The Welding Institute.

Figure 9.29(a) and Fig. 9.29(b) show the comparisons between the experimentally measured inherent strain ε_x^p distributions and residual stress σ_x distributions respectively after conventional GTAW (curve 1) and LSND welding (curve 2) of an aluminum plate 1.5 mm thick. Reductions in either ε_x^p or σ_x are obvious (as indicated by curve 2 in comparison with curve 1).

The photographs in Fig. 9.30(a), Fig. 9.30(b) and Fig. 9.30(d) show that the specimens welded conventionally are severely buckled in all cases of aluminum alloy and stainless steel respectively. However, the specimens welded by use of LSND welding are completely buckling free and as flat as before welding.

Comparisons are also given in Fig. 9.31(a) and Fig. 9.31(b) of the results of measured deflections *f* on specimens 1.6 mm thick welded conventionally using GTAW and those welded using the LSND welding technique for stainless steel and aluminum alloy respectively. Completely buckling-free

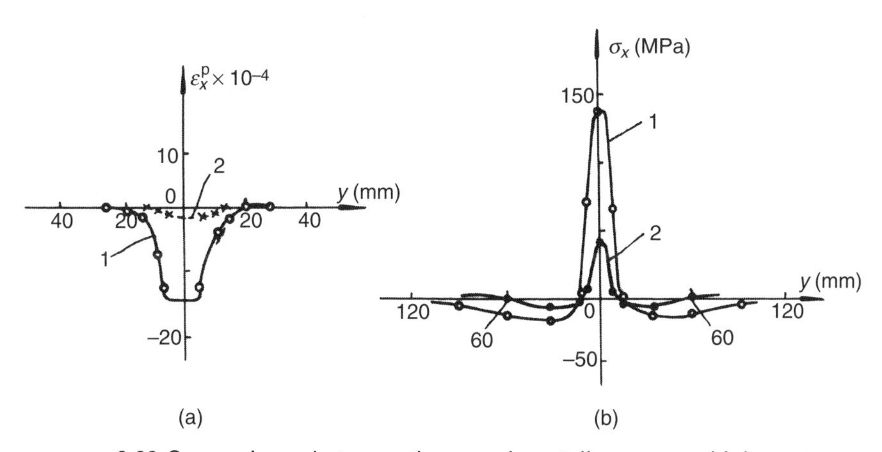

9.29 Comparisons between the experimentally measured inherent strains ε_x^p and residual stresses σ_x distributions after conventional GTAW (curves 1) and LSND welding (curves 2) of an aluminum plate 1.5 mm thick[33].

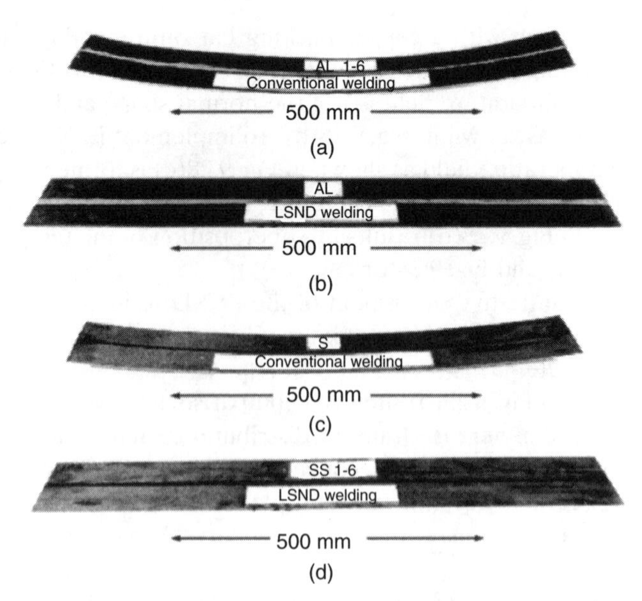

9.30 Specimens of (a), (b) aluminum alloy and (c) (d) stainless steel, both 1.6 mm thick and 1000 mm long, (a), (c) which were welded by conventional GTAW and are severely buckled and (b), (d) which were welded by LSND welding and are buckling free[32].

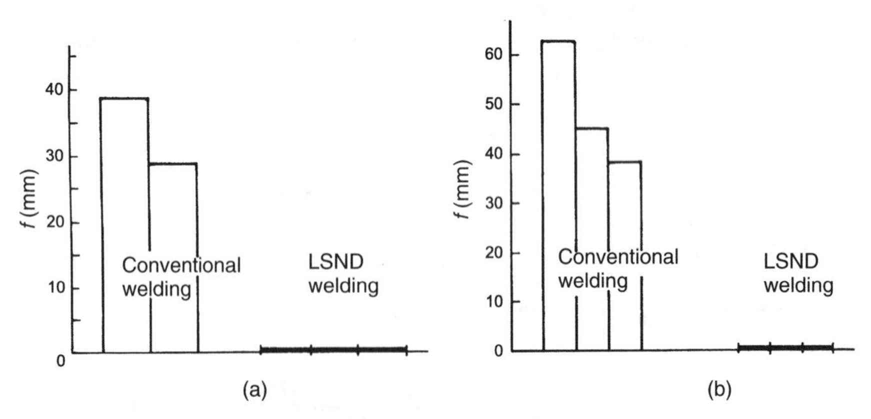

9.31 Completely buckling-free results ($f = 0$) which can be achieved using the optimized LSND welding technique on both (a) stainless steel and (b) aluminum alloy specimens 1.6 mm thick[32].

($f = 0$) results were achieved when the optimized technological parameters for LSND welding techniques were applied.

9.5.3 Application and potential development of the low-stress no-distortion welding technique

As demonstrated above, the idea of 'active' in-process control of buckling distortions leads to the invention of a new method in manufacturing thin-walled structures free from buckling. It is not necessary to have post-weld reworking for stress relief and distortion removal. In contrast with the commonly accepted concept that buckling distortions are inevitable, designers and manufacturers who suffer from problems of buckling could now adopt a new idea that buckling is no longer inevitable with the LSND welding technique. Buckling can be prevented completely and residual stresses can be reduced significantly or controlled to a level lower than σ_{cr} at which buckling occurs.

A new generation of longitudinal seam welders for executing the LSND welding technique was designed and manufactured at BAMTRI for industrial applications. These apparatus are applicable for longitudinal welds on plates and shells. Different from a conventional seam welder, the main functions of a newly designed clamping apparatus provide both the necessary condition and the sufficient condition dominating the success of LSND welding. In aerospace industries, the longitudinal seam welders existing on the production line can be modified for executing LSND welding. Successful results in preventing buckling distortions were achieved in manufacturing thin-walled jet engine cases of nickel-based alloys, stainless steels as well as rocket fuel tanks of aluminum alloys where the acceptable allowance for residual buckling deflections f at a weld length of L (see Fig. 9.2(a) and Fig. 9.2(c)) must be limited by the ratio[42] $f/L < 0.001$. The LSND welding technique could be implemented and coupled with many known fusion welding processes, e.g. arc welding, plasma welding, laser welding and electron beam welding. Because of the way in which the preset temperature profile is established, welding parameters in LSND welding remain the same as used in the conventional welding procedure. If hardenable materials are subjected to LSND welding, the required temperature gradient as shown by T_1 in Fig. 9.32 and a uniform preheating temperature field T_2 (for post-weld heating) necessary for the improvement in weldability are established by regulating the cooling and heating systems, preventing the weld joints from possible overhardening and cracking.

The LSND welding technique can also be used for straight fillet and T-type welds in panels with ribs as shown in Fig. 9.33. Cooling by the backing bar and heating by either an electrical heater or a flame torch should be arranged appropriately to keep the basic principle for LSND welding (see Fig. 9.26).

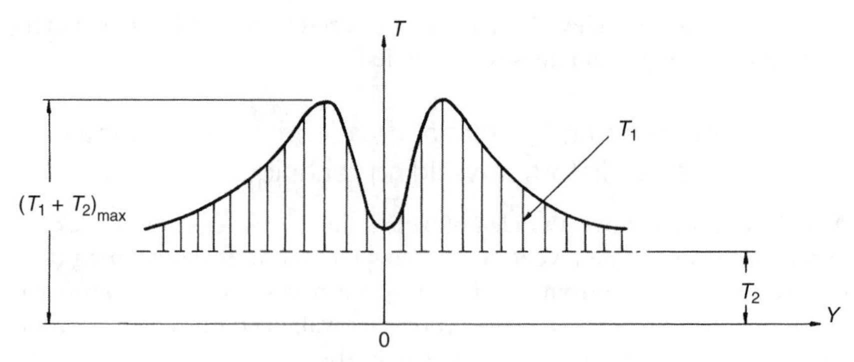

9.32 Temperature field for LSND welding of hardenable materials[32].

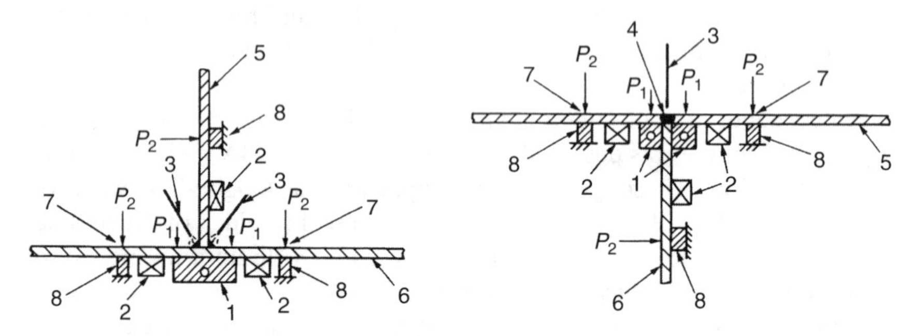

9.33 LSND welding for fillet and T-type welds in panels with ribs[32]: 1, backing bar; 2, heating; 3, welding torch; 4, weld; 5, 6, workpiece; 7, clamping; 8, support.

For LSND welding of components with sufficiently long longitudinal welds which cannot be readily placed into a stationary seam welder, an alternative option of a movable apparatus can be applied[32]. Relative movement between the workpiece and restraining means (e.g. rollers) during the welding operation is permitted. The structural elements can be fed through the LSND apparatus and simultaneously the required temperature profile will have been established and buckling-free weldments will be produced.

Figure 9.34 shows the alternative approach of LSND welding in engineering practice. Under certain circumstances, it may also be useful to create a temperature profile with a higher gradient T_{1max} (see Fig. 9.22) which is sufficient to cause an excess of local thermal compressive stress $-\sigma_{1max}$ over the value of the yield stress of the material at T_{1max}. As a result, compressive plastic strains will occur in the area of T_{1max}. The formed plastic strains will radically alter the residual stress magnitude and its distribution. After welding with the higher T_{1max}, the residual stress distribution assumes

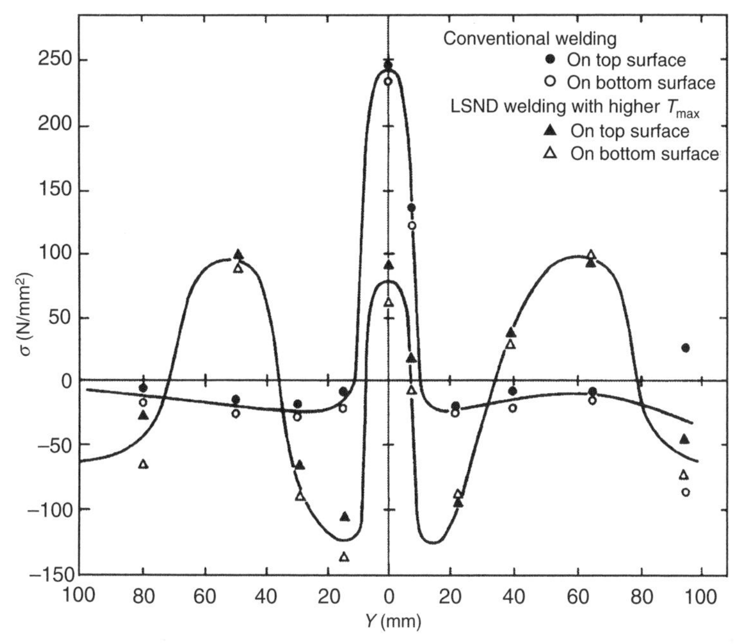

9.34 Experimentally measured residual stress distribution after conventional GTAW and LSND welding with a higher temperature gradient on stainless steel specimens 1.6 mm thick[32].

a complicated pattern as shown by the measured experimental results in Fig. 9.34[32]. The residual stresses alternately interchange from tensile to compressive symmetrically to the weld centerline. Two more zones of tensile stresses appear in areas where T_{1max} causes plastic strains. The peak value of tensile residual stress in the weld has been reduced significantly.

The specific pattern of residual stress distribution shown in Fig. 9.34 is also favorable for preventing buckling. The workpiece remains buckling free and keeps its original shape so that it is as flat as it was before welding. Obviously, the reason is that, because of the stretching effect of the newly appeared zones with tensile residual stresses located between the compressively stressed strips, the residual compressive stresses, in this case, cannot cause buckling; even the maximum value of the compressive stresses are higher than σ_{cr}. In other words, under the thermal compressive stresses, the plastically yielding strips with tensile stresses act as specific stiffening supports against possible buckling. This idea was practically implemented as a technique for avoiding the buckling of aluminum elements, as shown in Fig. 9.25.

The LSND welding technique should be considered at the early stage of design of thin-walled structures in order to achieve rational design,

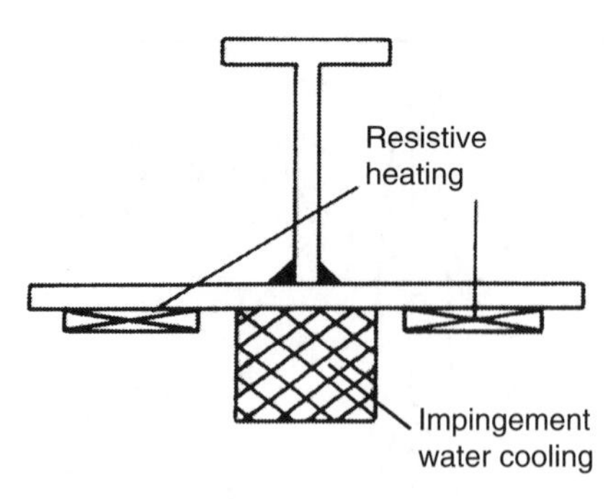

9.35 Thermal tensioning apparatus for fillet welding of a stiffened panel[20].

rational manufacturing in terms of material saving, weight reduction resulting from the use of possible thinner sheets without suffering from buckling, and time saving by eliminating post-weld correction operations. Recent progress in computational welding mechanics provides a powerful tool for prediction and optimization of technological parameters in the use of the LSND welding technique. As an example, later in the 1990s, Michaleris and Sun[20] presented a finite-element analysis of the thermal tensioning technique. Figure 9.35 illustrates the scheme of the thermal tensioning apparatus employed in their study. Prior to welding, a temperature differential is developed by cooling the weld region with a jet of tap water below the weld, and resistive heating with electric heating blankets. Both cooling and heating are applied over the entire length of the plate. However, a moving thermal tensioning apparatus with limited heating and cooling lengths, moving along with the welding torches was also suggested.

The study by Michaleris and Sun demonstrated that thermal tensioning can significantly reduce the longitudinal residual stress developed by welding and thus eliminate welding-induced buckling. It also illustrated that the finite-element method can be used to determine the optimum thermal tensioning conditions of a specified weld joint and heat input.

9.6 Developments in low-stress no-distortion welding

The previous section contributed to dealing with the overall cross-sectional thermal tensioning techniques. More recently, progress in seeking active in-process control of welding buckling shows trends toward exploiting a localized thermal tensioning technique using a trailing spot heat sink moving synchronistically with the welding torch which creates a stronger temperature gradient along the weld line within a limited area of the high-temperature zone close to the weld pool. As a successful research result achieved at BAMTRI in the early 1990s, the patented DC-LSND welding method[41–43] was developed. In this innovative technique, the preset heating to set up a temperature profile as shown in Fig. 9.26 is no longer necessary. The formation of specific inverse plastically stretched incompatible strains ε_x^p in the near-arc zone behind the welding pool is dynamically controlled by a localized moving thermal tensioning effect induced between the welding heat source and the spot heat sink.

9.6.1 Localized thermal tensioning effect with coupled heat source-heat sink

Apparatus for engineering implementation of the DC-LSND welding technique was designed and further developed at BAMTRI, as shown in Fig. 9.36[41].

With this apparatus attached to the welding torch, a trailing spot heat sink always follows the arc at a certain distance behind it. A cooling jet is directed straight onto the just-solidified weld bead. A liquid medium, such

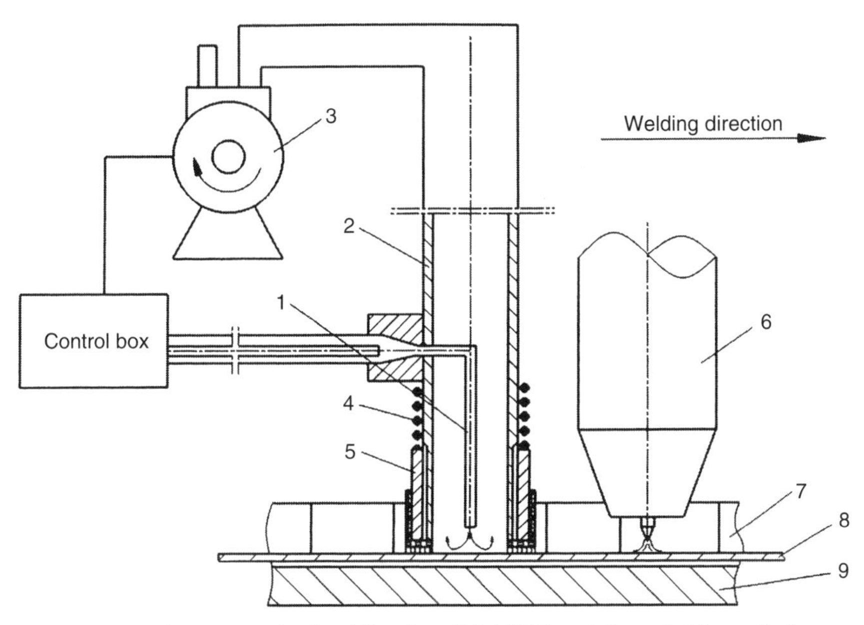

9.36 Apparatus for buckling-free DC-LSND welding of thin-walled elements[41]: 1, atomizing nozzle for a cooling jet of a liquid medium; 2, coaxial tube to draw the vaporized cooling medium; 3, vacuum pump; 4, spring; 5, axle oversleeve tube; 6, GTAW torch; 7, clamping fingers; 8, workpieces; 9, backing bar underneath weld.

as liquid carbon dioxide, liquid argon, liquid nitrogen or water, could be selected for the atomized cooling jet. To protect the arc from the possible interference of the cooling media, there is a coaxial tube to draw the vaporized media out of the zone near the arc. The technological parameters for the trailing spot heat sink and all the welding procedures are automatically synchronistically controlled with the GTAW process. The dominating factors, namely the distance between the heat source and heat sink, and the intensity of the cooling jet, can be adjusted appropriately to reach an optimum result. In systematic investigations of DC-LSND welding, finite-element analysis with a model of a cooling jet impinging on the welding bead surface has been combined with a series of experimental studies[43–48]. Comparisons between the temperature fields on a conventionally gas–tungsten arc welded titanium plate and on a plate welded using the DC-LSND technique are given in Fig. 9.37.

DC-LSND welding was carried out using the same parameters as in conventional GTAW. The flow rate of cooling media (atomized water) was selected to be 2.5 ml/s. The distance between the arc and cooling jet was regulated from 80 to 25 mm. It can be seen clearly that, in the abnormal temperature field and isotherms induced by DC-LSND welding (Fig. 9.37(b) and Fig. 9.37(d)), there is a deep valley formed by the cooling jet behind

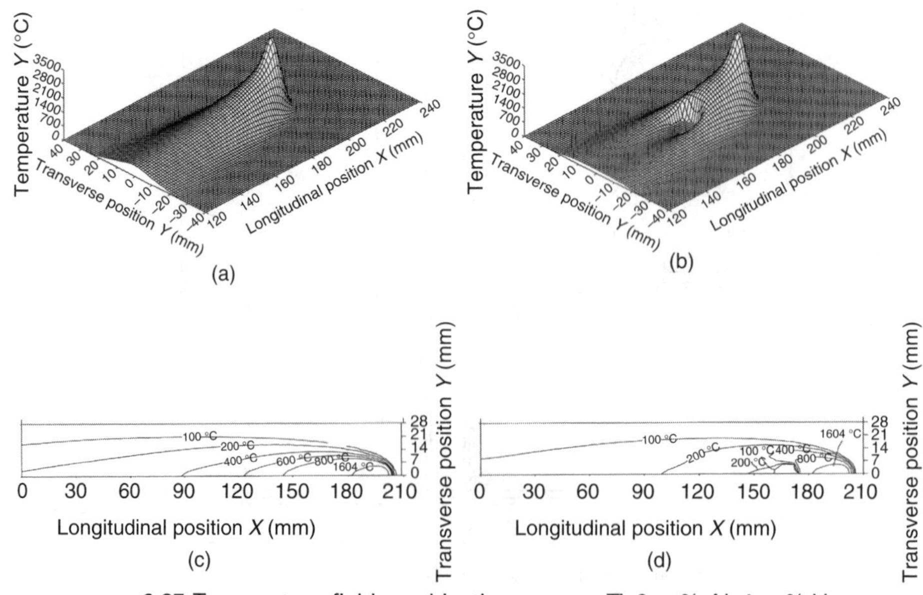

9.37 Temperature fields and isotherms on a Ti–6 wt% Al–4 wt% V plate 2.5 mm thick[44] (welding parameters: 200 A; 12 V; 12 m/h): (a), (c) conventional GTAW on a copper backing bar; (b), (d) DC-LSND welding.

the weld pool. An extremely high temperature gradient from the peak to the valley is created. The 800 °C and 400 °C isotherms in front of the heat sink are severely distorted, pushing forward closer to the weld pool (see Fig. 9.37 (d)).

The abnormal thermal cycles obtained by DC-LSND welding (Fig. 9.38 (b)) produce correspondingly abnormal thermal elastic–plastic stress and strain cycles (Fig. 9.38 (d)) in comparison with the cycles formed by conventional GTAW (Fig. 9.38(a) and Fig. 9.38(c)). It can be seen that the localized thermal tensioning effect acts within a limited zone behind the weld pool. It can be seen also from Fig. 9.38(d) that, in front of the arc, the influence of the heat sink is too weak to affect the normal thermal elastic–plastic stress and strain cycles; the incompatible compressive plastic strain in near-arc zones forms more or less as normal. However, behind the arc, in the area of the just-solidified weld bead, because of the intensive heat sink by the cooling jet, the shrinkage of metal during cooling from a high temperature causes a very sharp temperature slope (Fig. 9.38(b)) and therefore creates a strong tensioning effect in the temperature valley. In this way, the compressive plastic strains formed before in the just-solidified weld zone

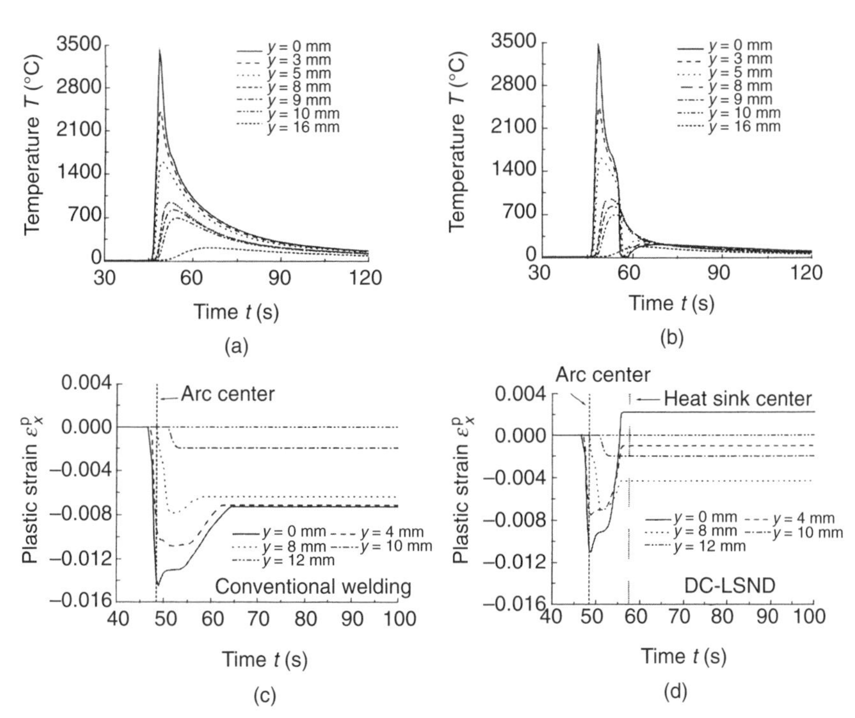

9.38 Comparisons of (a), (b) thermal cycles and (c), (d) transient plastic strain cycles for (a), (c) conventional GTAW and (b), (d) DC-LSND welding[47,48].

can be compensated properly by the incompatible tensile plastic strains in the area of the temperature valley (Fig. 9.38 (d)). Moreover, owing to the intensive heat sink, the heat transfer from the weld pool to the peripheral area is also reduced, resulting in narrowing, to some extent, of the width of the zone of incompatible plastic strain distribution.

9.6.2 Optimization of parameters dominating the dynamically controlled low-stress no-distortion welding technique

As described above, in DC-LSND welding, both the value of incompatible plastic strains and the width of its distribution can be controlled quantitatively by selecting the appropriate technological parameters: the distance D between the welding heat source and the heat sink (Fig. 9.39) and the

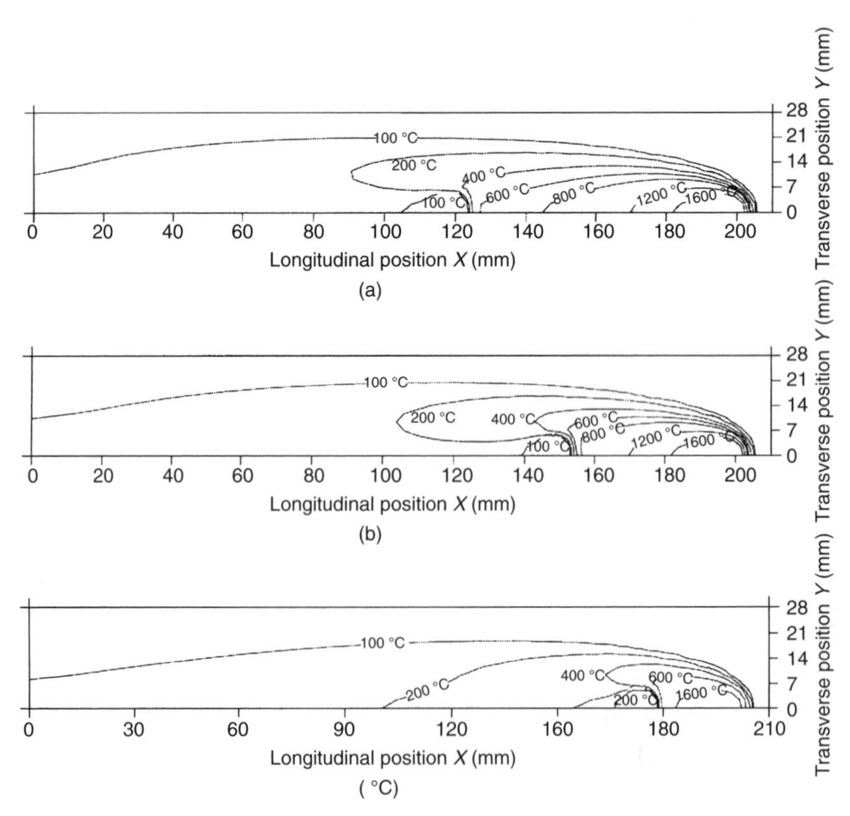

9.39 Isotherms on a titanium plate with different distances D between the arc center and the cooling-jet center[47]: (a) $D = 80$ mm; (b) $D = 50$ mm; (c) $D = 25$ mm.

intensity of the heat sink (in terms of the convection coefficient α of the heat sink in calculation).

Figure 9.40(a) and Fig. 9.40(b) show the residual strain and stress distributions respectively in the cross-section of the weld on a titanium plate. Comparisons are given between conventional welding (shown by the solid curve) and DC-LSND welding with different distances $D = 25\,\text{mm}$, 50 mm and 80 mm in Fig. 9.39. For a selected intensity of heat sink, the closer the heat sink is to the heat source (the shorter the distance D), the stronger the localized thermal tensioning effect becomes. For example, at the distance $D = 25\,\text{mm}$, the residual plastic incompatible strain ε_x^p on the weld centerline even changes sign from negative to positive (Fig. 9.40(a)), and the residual stress on the weld centerline changes from tensile to compressive (Fig. 9.40 (b)). For engineering application, complete freedom from buckling distortions can be achieved if the compressive residual stress in the peripheral area of the plate away from the weld is controlled to a level below the critical value σ_{cr} at which buckling occurs.

Results of experimental measurements of deflections and residual stress distributions on stainless steel plates, mild steel plates and aluminum plates show not only the feasibility that a wide variety of materials can be welded using the DC-LSND technique but also its flexibility for practical application. Figure 9.41 gives some typical examples from the systematic investigation program. As shown in Fig. 9.41(a), the peak tensile stress in a weld on a mild steel plate welded using conventional GTAW reaches 300 MPa (curve 1) and the maximum compressive stress in the peripheral area is about 90 MPa, which causes buckling with deflections of more than 20 mm in the centre of a specimen 500 mm long. In the case of DC-LSND welding with different technological parameters, the patterns of residual stress distribution (curves 2, 3 and 4) alter dramatically even with the compressive

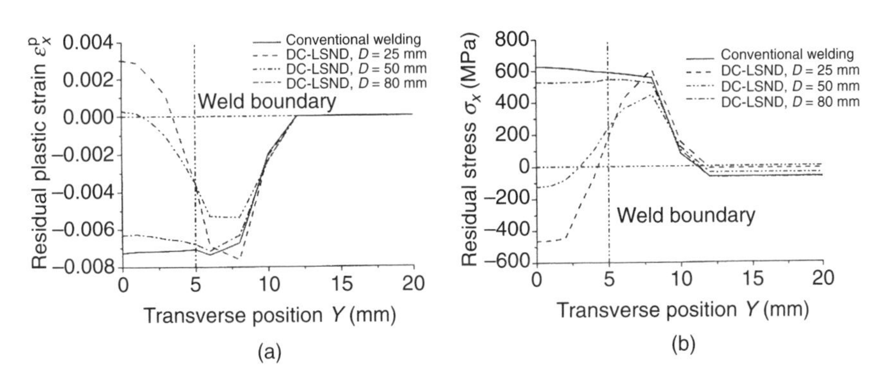

9.40 (a) Residual strain ε_x^p and (b) stress σ_x distributions in the cross-section of the weld on a titanium plate welded conventionally and using the DC-LSND welding technique[47,48].

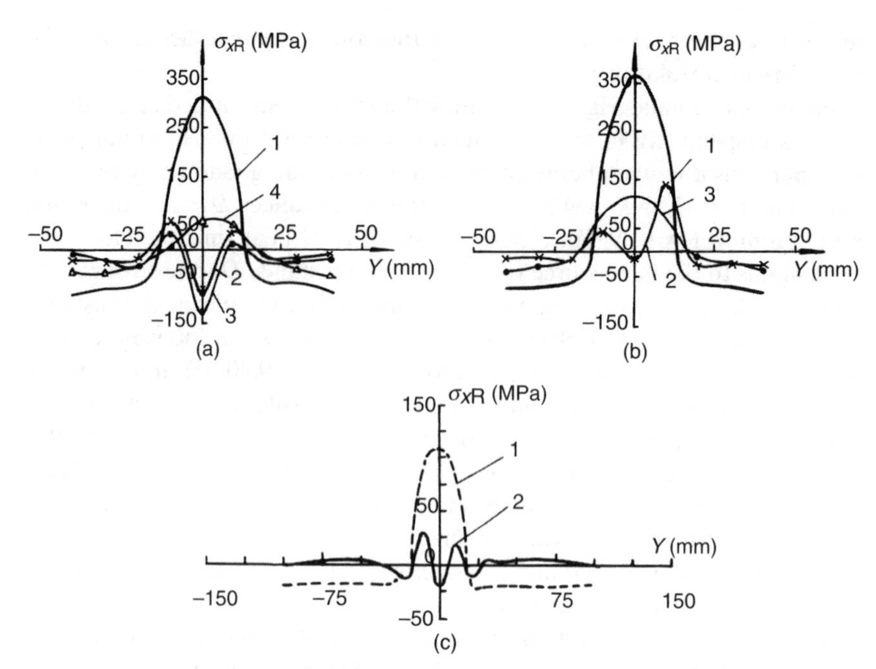

9.41 Measured residual stress distributions on plates of (a) mild steel 1 mm thick, (b), stainless steel 2 mm thick and (c) aluminum alloy 2 mm thick welded using conventional GTAW (curves 1) and by use of the DC-LSND welding technique[43,49].

residual stresses in the centerline of the weld. After DC-LSND welding, the specimens are completely buckling free and as flat as the original before welding. Similar results were obtained on stainless steel and aluminum plates as shown in Fig. 9.41(b) and Fig. 9.41(c) respectively.

The strong effect of the localized thermal tensioning either by a shorter distance D or by a very intensive convection coefficient α of heat sink causes the compressive stresses in the weld as shown by curves 2 and 3 in Fig. 9.41(a), and the curves in Fig. 9.41(b) and Fig. 9.41(c) respectively. The reason is that the shrinkage induced by the large temperature gradient between the arc and the cooling jet tends not only to compensate the welding compressive plastic strains but also to alter the sign of the residual strain.

9.6.3 Engineering application and further exploitation

As an example of the successful implementation of the DC-LSND welding technique for preventing buckling distortions, comparisons of results between the deflections on specimens welded conventionally and those welded using the DC-LSND technique are given in Fig. 9.42. The stainless steel strip with $L = 2350$ mm, $B = 150$ mm and $\delta = 0.8$ mm was severely buckled with a maximum deflection of 260 mm after conventional GTAW while the welded strip is at its longitudinal edge to avoid sinusoidal wave flexures caused by the gravity of the strip itself in the horizontal plane position (Fig. 9.42(a)). Warpage buckling was completely prevented using the DC-LSND welding technique, as shown in Fig. 9.42(b).

A specially designed welding facility with the DC-LSND apparatus (shown in Fig. 9.36) attached to the welding torch was installed at a metal sheet works. Thin bands of stainless steel are butt welded along the band width of 2 m to form an endless ribbon conveyer. Strict flatness requirements present fabricators troublesome problems that are difficult to solve even using power beam welding technology. By welding in this facility, completely deflection-free bands are produced without any post-weld flattening reworking procedure to correct the former troublesome distortions. Based on the theoretical and experimental investigations, the optimization of dominating parameters for engineering application of DC-LSND welding can be executed, e.g. using the plotted curves shown in Fig. 9.43 for the case of a titanium plate examined in previous section (see Fig. 9.37).

Results show that the distance D has a more significant influence on both ε_x^p and σ_x than the heat sink intensity α in controlling buckling on thin materials. For the recommended range for selecting the optimum distance D (Fig. 9.43(a) and Fig. 9.43(b)), the cooling intensity α of the heat sink is approximately a fixed critical value related only to the welding heat input, and it is not necessary to increase the cooling intensity over the critical value because no further improvement in ε_x^p nor in σ_x were observed (Fig. 9.43(c) and Fig. 9.43(d)).

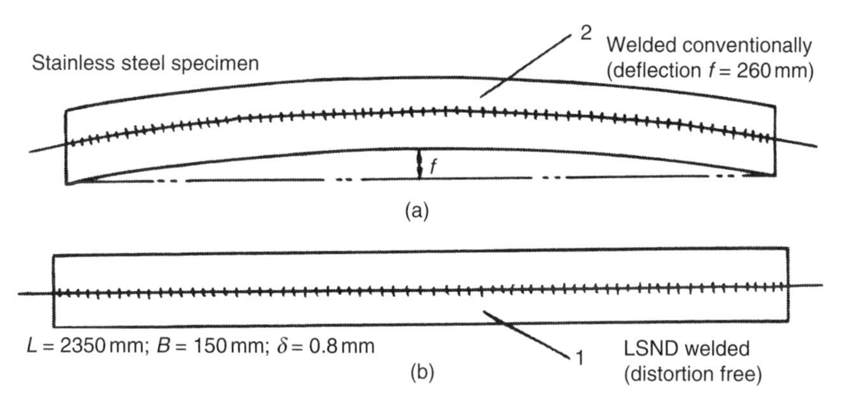

9.42 Stainless steel strip with $L = 2350$ mm, $B = 150$ mm and $\delta = 0.8$ mm (a) after GTAW, which resulted in severe buckling and (b) using the DC-LSND welding technique, which prevented buckling.

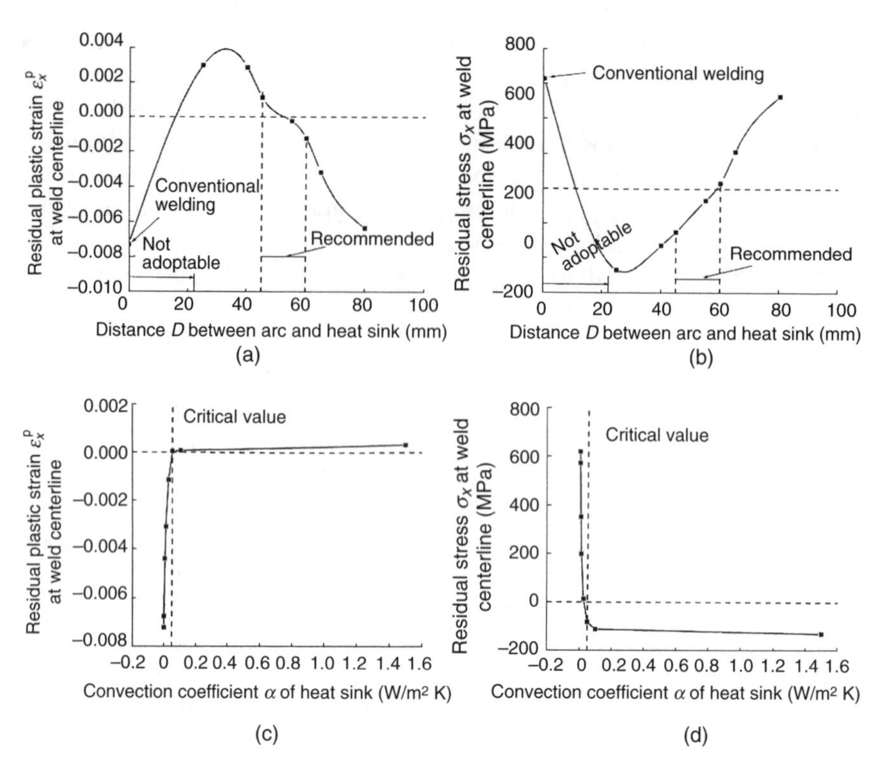

9.43 (a) The peak value of the residual plastic strain ε_x^p at the weld centerline and (b) the residual stress σ_x at the weld centerline as functions of distance D (according to Fig. 9.40), along with (c) ε_x^p and (d) σ_x as functions of the intensity α of the heat sink[47].

Metallurgical and mechanical examinations show that the cooling-jet medium has no notable influence on the titanium weld joint properties even using an atomized water jet. The reason is clearly shown by the isotherms of 400 °C in Fig. 9.39 as a distorted abnormal isotherm in front of the heat sink. In fact the cooling jet is impinging directly the solidified weld bead at a temperature below 400 °C.

As an addition to the LSND welding technique described in the previous section, the DC-LSND welding technique provides more flexibility in the control of buckling in engineering practice. Besides the longitudinal linear welds where both the LSND methods are suitable, the DC-LSND method promises a wider variety of application to nonlinear and nonregular welds in thin-walled structural elements.

Recent progress in mathematical modeling and numerical simulation of welding phenomena offers researchers powerful tools for studying in more detail three-dimensional welding thermal and mechanical behaviors. Investigations in the field of innovative creation of new methods for welding distortion especially to prevent buckling are the world's most exciting and challenging subjects, involving many disciplines and industrial practice and process experiments. These tools allow for the prediction of precise control of the abnormal temperature fields and therefore the abnormal thermal elastic–plastic cycles created by the possible combinations of the heat source–heat sink: multiple-source welding techniques. It is expected that a variety of coupled heat source–heat sink processes are feasible not only to control welding distortion but also to produce defect-free specimens.

9.7 References

1. Kurkin, S.A., and Guan, Q., 'Removal of residual welding stresses in thin-walled elements of titanium alloy' (in Russian), *Weld. Prod.*, **10**(310), 1–5, 1962.
2. Kurkin, S.A., and Guan, Q., 'Eliminating welding deformation in thin-walled elements of titanium alloy by weld rolling' (in Russian), *Selected Works on Welding of Ferrous Alloys and Some Alloyed Steels*, Moscow Bauman Technical University, Oborongiz, Moscow, 1962.
3. Guan, Q., 'Residual stresses, deformations and strength of thin-walled elements of welded structures of titanium alloys' (in Russian), PhD Dissertation, Moscow Bauman Technical University, Moscow, 1963.
4. Guan, Q., 'Welding stress and distortion of thin-walled structural elements of titanium alloys' (in Chinese), *Int. Aviation*, **198**(2), 37–41, 1979.
5. Vinokurov, V.A., *Welding Stress and Distortion: Determination and Elimination* (transt. J.E. Baker), The British Library, London, 1977 (original Russian version, Mashinostroenie, Moscow, 1968).
6. Sagalevich, V.M., *Methods for Eliminating Welding Deformations and Stresses* (in Russian), Mashinostroenie, Moscow, 1974.
7. Terai, K., 'Study on prevention of welding deformation in thin-skin plate structures', *Kawasaki Tech.* Rev., **61**, 61–66, 1978.
8. Masubuchi, K., *Analysis of Welded Structures*, Pergamon, Oxford, 1980.
9. Nikolayev, G.A., Kurkin S.A., and Vinokurov, V.A., *Welded Structures, Strength of Welded Joints and Deformation of Structures* (in Russian), High School, Moscow, 1982.
10. Tian, X.T., *Welded Structures* (in Chinese), Harbin Institute of Technology, Harbin, 1980.
11. Guan, Q., 'Efforts to eliminating welding buckling distortions – from passive measures to active in-process control', *Today and Tomorrow in Science and Technology of Welding and Joining, Proc. 7th Int. Symp. of the Japan Welding Society*, Kobe, Japan, November 2001, Japan Welding Society, Tokyo, 2001, pp. 1045–1050.
12. Guan, Q., 'Welding stress and distortion control in aerospace manufacturing engineering', *Proc. Int. Conf. on Welding and Related Technologies for the 21st Century*, Kyiv, Ukraine, November 1998, *Weld. Surfacing Rev.*, **12**, 47–65, 1998.
13. Guan, Q., 'A survey of development in welding stress and distortion control in aerospace manufacturing engineering in China', *Weld. World*, **43**(1), 14–24, 1999.

14. Guan, Q., 'Reduction of residual stress and control of welding distortion in sheet metal fabrication', *Proc. 7th Int. Aachen Welding Conference on High Productivity Joining Processes*, Vol. 1, Aachen, Germany, May 2001, Shaker, Aachen, 2001, pp. 531–549.
15. Vol'mir, A.S., *Stability of Elastic Systems* (in Russian), Fizmatgiz, Moscow, 1963.
16. Masubuchi, K., 'Studies on the mechanisms of the origin and methods of reducing the deformation of shell plating in welding ships', *Int. Shipbuilding Prog.*, **3**(19), 123–133, 1956.
17. Watanabe, M., and Satoh, K., 'Fundamental studies on buckling of thin steel plate due to bead welding' (in Japanese), *J. Japan Weld. Soci.*, **27**(6), 313–320, 1958.
18. Zhong, X.M., Murakawa, H., and Ueda, Y., 'Buckling behavior of plates under idealized inherent strain', *Trans. JWRI*, **24**(2), 87–91, 1995.
19. Michaleris, P., and DeBiccari, A., 'Prediction of welding distortion', *Weld. J.* **76**(4), 172-s–181-s, 1997.
20. Michaleris, P., and Sun, X., 'Finite element analysis of thermal tensioning techniques mitigating weld buckling distortion', *Weld. J.*, **76**(11), 451-s–457-s, 1997.
21. Michaleris, P., Dantzig, J., and Tortovelli, D., 'Minimization of welding residual stress and distortion in large structures', *Weld. J.*, **78**(11), 361-s–366-s, 1999.
22. Deo, M.V., Michaleris, P., and Sun, J., 'Prediction of buckling distortion of welded structures', *Sci. Technol. Weld. Joining*, **8**(1), 55–61, 2003.
23. Deo, M.V., and Michaleris, P., 'Mitigating of welding induced buckling distortion using transient thermal tensioning', *Sci. Technol. Weld. Joining*, **8**(1), 49–54, 2003.
24. Tsai, C.L., Park, S.C., and Cheng, W.T., 'Welding distortion of a thin-plate panel structure', *Weld. J.*, **78**(5), 156-s–165-s, 1999.
25. Guan, Q., and Liu, J.D., 'Residual stress and distortion in cylindrical shells caused by a single pass circumferential butt-weld', IIW Document X-929-79, International Institute of Welding, Bratislava, 1979.
26. Stanhope, A., Hazelhurst, R.H., and Swann, B.M., 'Welding airframe structures in titanium alloys using tensile loading as a means of overcoming distortion', *Metal Constr. Bi. Weld. J.*, **4**(10), 366–372, 1972.
27. Chertov, I.M., Kalpenko, A.C., Zublenko, G.L., and Nigovola, A.A., 'Strains and stresses in the welding of longitudinal joints in cylindrical shells of the AMg6 alloy with pre-tensioning' (in Russian), *Avtom. Svarka*, (5), 31–33, 1980.
28. Paton, B.E., Lobanov, L.M., Powlarsky, V.I., Bondaler, A.A., Naralenko, O.K., and Wutkin, V.F., 'Fabrication of thin-walled welded large panels of high strength aluminum alloys' (in Russian), *Avtom. Svarka*, **439**(10), 37–45, 1989.
29. Radaj, D., *Heat Effects of Welding: Temperature Field, Residual Stress, Distortion*, Springer, Berlin, 1992.
30. Radaj, D., *Welding Residual Stresses and Distortion*, revised edition of the original German edition, Deutscher Verlag für Schweisstechnik, Düsseldorf, 2003.
31. Guan, Q., Guo, D.L., Cao, Y., Li, C.Q., Shao, Y.C., and Liu, J.D., 'Method and apparatus for low stress no-distortion welding of thin-walled structural elements', Original Chinese Patent 87100959.5, February 1987, International Patent PCT/GB88/00136, 1988.
32. Guan, Q., Leggatt, R.H., and Brown, K.W., 'Low stress non-distortion (LSND) TIG welding of thin-walled structural elements', TWI Research Report, The Welding Institute, Abington, Cambridge, July 1988.

33. Guan, Q., Guo, D.L., and Li, C.Q., 'Low stress no-distortion (LSND) welding – a new technique for thin materials' (in Chinese), *Trans. Chin. Weld. Soc.*, **11**(4), 231–237, 1990.
34. Burak, Ya.I., Bisijina, L.P., Romanjuk, Ya.P., Kazimirov, A.A., and Morgun, V.P., 'Controlling the longitudinal plastic shrinkage of metal during welding', *Avtom. Svarka*, **288**(3), 27–29, 1977.
35. Burak, Ya.I., Romanjuk, Ya.P., Kazimirov, A.A., and Morgun, V.P., 'Selection of the optimum fields for preheating plates before welding', *Avtom. Svarka*, **314**(5), 5–9, 1979.
36. Masubuchi, K., 'Prediction and control of residual stresses and distortion in welded structures', *Weld. Res. Abroad*, **43**(6–7), 2–16, 1997.
37. Takeno, S., 'Method for preventing welding distortion of sheet metals', Japanese Patent JP4052079, Application JP 19900163121, 20 February 1992.
38. Michaleris, P., Tortovelli, D.A., and Vidal, C.A., 'Analysis and optimization of weakly coupled thermo-elasto-plastic systems with application to weldment design', *Int. J. Num. Meth. Engng*, **38**, 1259–1285, 1995.
39. Yang, Y.P., Dong, P., Zhang, J., and Tian, X., 'A hot-cracking mitigation technique for welding high-strength aluminum alloy', *Weld. J.*, **79**(1), 9-s–17-s, 2000.
40. Dong, P., Hong, J.K., and Rogers, P., 'Analysis of residual stresses in Al-Li repair welds and mitigation techniques', *Weld. J.*, **77**(11), 439-s–445-s, 1998.
41. Guan, Q., Zhang, C.X., and Guo, D.L., 'Dynamically controlled low stress no-distortion welding method and its facility', Chinese Patent 93101690.8, February 1993.
42. Guan, Q., Guo, D.L., Zhang, C.X., and Li, C.Q., 'Low stress no-distortion welding for aerospace shell structures', *China Weld.*, **5**(1), 1–9, 1996.
43. Guan, Q., Zhang, C.X., and Guo, D.L., 'Dynamic control of welding distortion by moving spot heat sink', *Weld. World*, **33**(4), 1994.
44. Li, J., Guan, Q., Shi, Y.W., Guo, D.L., Du, Y.X., and Sui, Y.C., 'Studies on characteristics of temperature field during GTAW with a trailing heat sink for titanium sheet', *J. Mater. Processing Technol.*, **147**(3), 328–335, 2004.
45. Li, J., Guan, Q., Guo, D.L., Sun, Y.C., Du, Y.X., and Shi, Y.W., 'Influence of heat sink position on welding residual stress of titanium alloy thin plate' (in Chinese), *J. Mech. Strength*, **25**(6), 637–641, 2003.
46. Li, J., Shi, Y.W., Guan, Q., Guo, D.L., Du, Y.X., and Sun, Y.C., 'Influence of heat conductivity of the backing bar on welding residual stress of titanium alloy thin plate' (in Chinese), *Trans. Chin. Mech. Engng Soc.*, **40**(3), 133–136, 2004.
47. Li, J., 'Studies on the mechanism of low stress no distortion welding for titanium alloy' (in Chinese), PhD Dissertation, Beijing University of Technology, Beijing, February 2004.
48. Li, J., Guan, Q., Shi, Y.W., and Guo, D.L., 'A stress and distortion mitigation technique for welding titanium alloy thin sheet', *Sci. Technol. Weld. Joining*, **9**(5), 451–458, 2004.
49. Zhang, C.X., Guan, Q., and Guo, D.L., 'Study on application of dynamic control of welding stress and distortion in thin aluminum elements', *Proc. 6th Int.'l Welding Symp.* of the Japan Welding Society, Nagoya, Japan November 1996 Japan Welding Society, Tokyo, 1996, pp. 539–544.

Index